Doctor's Son

By

David E. Mann, Jr.

ISBN: 1-4033-2012-8 (e-book)
ISBN: 1-4033-2013-6 (Paperback)

Library of Congress Control Number: 2002092128

This book is printed on acid free paper.

Printed in the United States of America
Bloomington, IN

1stBooks - rev. 06/18/02

ACKNOWLEDGMENTS

To Mary, my wife of 51 years. Thank you, sweetheart, for more than a half century of happiness!

To our children—David, Caryn, and Scott—who suggested this project be undertaken, and a special nod to Scott, whose invaluable assistance and guidance made the literary execution of this endeavor an enjoyable experience.

CHAPTER ONE

Earliest Recollections

1922 was a memorable year: in the air, biplanes of wood, wire, and canvas maneuvered by daredevils, permitted paying customers to experience the thrills of World War I dogfights at carnivals and flying circuses; on the ground, Henry Ford's mass-produced miracle, the "tin lizzie", gave the average American a ride over gravel and mud-filled roads in fragile vehicles as though being conveyed in royal limousines; in the home, Marconi's wireless emitted scratchy music and speech from a Bakelite box with the help of speakers the size of opened umbrellas. The year 1922 also marked a truly momentous event: the discovery of Tutankhamun's tomb by the Englishman Howard Carter in Egypt's Valley of the Kings. Remarkably, his mummy, encased in solid gold and surrounded by equally impressive archeological delights, had escaped the greedy hands of grave plunderers. That was also the year I first saw daylight in an Appalachian hospital crib at the National Sanatorium in Johnson City, Tennessee, where I unknowingly held the distinction of being the first baby born in a medical facility whose residents were veterans of the Civil War, Spanish-American War, and WWI. Furthermore, they suffered from syphilis, TB, alcoholism, cancer, and mental and physical wounds of battle, not pregnancy.

Of course, I have no recollection of my first year of existence, but one incident, as related to me years later by my mother, occurred that might have seriously altered my life. Because both parents were occupied in earning a living in the health professions (father as physician, mother as nurse), I required daycare. There were many baby-sitters, but one, Addie Johnson, a tall, pretty, African American teenager, carried me daily in the summer to a nearby field so I could crawl in the tall grass. Unbeknownst to her, the field was infested with Eastern Diamondback rattlesnakes searching for mice. The warm, chubby face of a baby made a perfect target for the fangs of a disturbed rattler! Although Addie was reprimanded for her lapse of judgment, her photograph appears prominently in my baby book as a

1

dignified and confident young lady, unquestionably the best and most loving baby-sitter of the lot.

It is appropriate that I introduce my ancestral background, which plays a role not only in setting goals, but also in defining behavioral traits responsible for their attainment. My father, David Edwin Mann, the youngest of nine children, was the offspring of Jesse Mann and Marion Logan, native to Norfolk, England, and Kirkcaldy, Scotland, respectively. Jesse, a twin, studied art in his youth with the landscape artist, George Frederic Watts (1817-1904), before deciding to seek his fortune in America. He arrived in Boston before the turn of the century as a handsome young man whose distinguished bearing and command of the English language enabled him to obtain work in the homes of New England's wealthy, including the governor's mansion in Vermont, as a butler. It was while working in that state that he met Marion Logan and was married a few months later in St. Albans, Vt.

Foregoing working for others, Jesse bought a 100-acre farm in Norfolk, Massachusetts, where, years later, I spent the happiest days of my childhood. In fact, it was on this farm that my earliest recollection of being alive occurred. At the age of 3, I gazed from my crib in a breezeway of the 17th century farmhouse and stared at a herd of deer crossing at the rear of a field known as the "mowing lot." To this day, I can see thirty or so whitetails prancing over a stone wall as they head from the road on the left to the woods on the far right.

It seems odd that the Manns came from Norfolk, England, to settle in Norfolk, Massachusetts. Franklin Weston Mann, who became a famous scientist, was the son of Levi and Lydia Mann. He was born in Norfolk on the 24th day of July, 1859, and represented the eighth generation from William Mann, who had left England, settled in Cambridge, Mass., and eventually sold his land to Harvard College. One of William's offspring was a Congregational minister and, down the line, another was the famous educator, Horace Mann. Dr. Franklin W. Mann, the scientist, was a small arms ballistics expert who authored the widely-read book, *"The Bullet's Flight from Powder to Target."* I have yet to determine on which branch Jesse's family springs forth from the family tree, but many of these Manns lie in a cemetery in Norfolk, with the exception of my parents who are buried in the delightful town of Needham, 12 miles away (I once walked the distance to earn credit toward a first class boy scout badge).

My mother, Alma Beatrice Reed, the second of four children, was born in Saco, Maine. Her parents, Napoleon Reed and Rose Lord, were of Scottish, French, German, and English descent. Both were born on the the day Lincoln was shot, and this was the mutual attraction that sparked a relationship leading to marriage. It's ironic that a bullet fired from John Wilkes Booth's *derringer played a minor role in my creation. Napoleon eventually settled in Quincy, Mass., where he became the co-owner of Reed & Vendret, a quarry that provided gravestones, statuary, and pavement rocks, and the superb black granite of the Bunker Hill Monument. Tragically, when mother was a teenager, her father was summoned to the quarry on a New Year's Day to repair a piece of equipment. While attempting to fix a stubborn air compressor, it suddenly exploded, killing him instantly. Grandma Rose, left with four young children, opened a millinery shop in Quincy and was able to eke out a living. Bertha, the oldest daughter, and Alma studied nursing at Quincy City Hospital. Bertha became a registered nurse and Alma didn't. One day I asked mother why and she replied, "I didn't want to spend the five dollars to become registered." The youngest daughter, Lillian, ushered at the Alhambra theatre where she met another usher, Harold Deacon, who became her husband after each had acquired advanced educations. A third usher was a young man who adopted the name of the theatre owner and attained fame as the film and radio star, Billy De Wolfe. Baby Bert, the youngest child, did odd jobs as he approached manhood, helping Howard Johnson's mother make her soon-to-be world-famous ice cream. He also became a volunteer fireman and, for relaxation, attended dancing school where he met Ruth Gordon. Years later, encountering the famous actress in the lobby of the Waldorf-Astoria in N.Y., he asked Miss Gordon (Mrs. Garson Kanin), "Do you remember dancing with me?" "Was it Paree?" she asked. "No," answered Bert, "It was Quincy." "Omigosh, it's Bert Reed!" she gasped.

I never had the pleasure of meeting grandpa Napoleon for he was a young man in his forties when he died. I knew he had a brother and asked mother one day what his first name was. "Adolf," she replied curtly. Napoleon and Adolf in one family was too much!

She told a story about her grandfather, grandpa Lord, the only lawyer in the family. When her mother was a little girl, her father blamed his daughters for prematurely munching on cookies that had been stored in the attic for the holidays. Each child was informed that she should tell who the culprit was or confess to the crime. When no one confessed, they were beaten severely. Weeks later, it was discovered that mice in the attic were responsible. Unfortunately, grandpa Lord was what Dickens would have called a bastard.

Henry Deringer, Philadelphia gunmaker, lent his name with a double "r" to all small pistols.

As the ninth child of a farmer, Dad was the spoiled baby in a family with a preponderance of females, three of whom lived long enough for me to meet. All of the children abhorred farm work, but David and his brother, Jesse, Jr., loved the outdoors and tolerated working in the fields and milking cows for the opportunity to hunt game in the surrounding forest. As a result of these Nimrod activities, Dad became authoritative on the wildlife of Norfolk, and Jesse Jr., similarly endowed with knowledge of the local fauna and flora, held the additional distinction of killing the last rattlesnake in the vicinity. Once, while pulling watercress from a brook that flowed through the farm, he was startled to see the snake when he was about to kneel to get a better grip on the plant.

Although the farm prospered under the guardianship of parents Jesse and Marion and their young helpers, it was evident that none of the children desired to make farming a lifelong career. Papa Jesse struck a deal when he realized that not all wanted to attend college. "You have a choice," he said one day. "Either go to college or take a trip abroad." The girls chose advanced education and Dad almost chose the European route when his mother became ill. When the doctor arrived in a horse-drawn coach, stepped out in a top hat and Sunday clothes, and performed the miracle of healing, Dad quickly resolved to enter the medical profession. Years later, he was frequently reprimanded by Mom for not charging for his medical services because some of his patients were in poverty.

After graduation from English High, Dad entered Tufts Medical School and, at the age of 23, became a physician in 1915. His internship was spent at Cambridge Hospital where he served as house officer and met a young nurse named Alma who was soon to become his wife. There was a religious problem, however. Dad was a Protestant and a Congregationalist, whereas Alma was brought up in the doctrine of Roman Catholicism. Upon her father's tragic quarry death, Alma discovered that because her father had been negligent in attending church, the Catholic diocese refused to inter him in their cemetery in Quincy where, ironically, his granite monuments stood crowned with angels, saints, and effigies of cupids, many of which he had laboriously carved himself. Furious at this rejection, Rose and her family became Episcopalians, and Lillian, after receiving a baccalaureate degree in English from the University of Pittsburgh, married the Reverend Mr. Harold Deacon, who by now had become an Episcopalian minister. It was Uncle Harold's original intent to become a lawyer. However, a day spent listening to boring lectures on mortgages and torts was interpreted by him as God's sign that he should become a theologian. When Uncle Harold was in his early sixties and about to be inducted as the Bishop of Massachusetts, he succumbed to emphysema. Coincidentally, Harold's family was also in the monument business, which prompted me, years later, to remark to Lillian: "Wow! Two great rock groups got together!" She loved it!

Neither Marion nor Jesse disapproved of the impending marriage of David and Alma, yet the daughters demanded a role in the selection of a suitable mate for their beloved baby brother. The consequence of this family quandary was the elopement of the parties involved with a brief honeymoon spent at Lake Winnipesaukee near Wolfeboro, N.H. Then, an opportunity arose for Dad to practice medicine in a TB sanatorium in Rutland, Vt., where he could delve more deeply into the etiology and treatment of pulmonary diseases, which later became his speciality, in addition to his general practice (now called family medicine). The newlyweds, with another recently married couple, the husband of which was Dad's medical school classmate, moved to the Green Mountain State to launch comparable medical careers. In those days, before the discovery of anti-TB drugs such as streptomycin and iproniazid, the concept of treatment was the colder the air inhaled, the better for the lungs, and the less muscular activity the body performed, the more rapid the healing process. Consequently, patients were exposed to fresh, cold air in the fall and winter while lying in screened porches lined with hospital beds draped with white sheets. At the end of the year, it was announced that a TB sanatorium in

Massachusetts was looking for a director. Dad assumed he was going to be selected because, during the year, his Tufts classmate had frequently asked him for advice concerning managerial problems involved in operating such a facility. When his friend was chosen for the plum position, Dad realized he had been used. This unfortunate event resulted in my father spending the next three years in charge of the National Sanatorium in Johnson City, Tennessee, instead of enjoying the good life in his beloved New England. However, the sojourn in the south proved to be memorable. They made many friends who later came north to visit them when they moved back to Massachusetts. Their departure because of homesickness was not the primary reason for their return to New England. Though Dad was an inveterate gun enthusiast, he feared for his family's safety when he encountered gunfights as an almost daily occurrence on the city streets. Sad to say, Johnson City in the early 1920's resembled Dodge City in the 1840's.

The trek north by Chevrolet touring car ended at the farm in Norfolk where I imagine the hospitality offered by the sisters left much to be desired. I'm sure that Alma held her own in conversation, backed by her Maine heritage and independent demeanor. The only incident that I recall during this medical hiatus, other than the deer incident, was being butted by a Billy goat. According to witnesses, after the animal had the audacity to knock me down, I got up and landed a good punch on its nose. I'm sure that Dad expected me to be the next John L. Sullivan when this occurred.

CHAPTER TWO

Moving to Needham

Using the farm in Norfolk as an operational base, my parents searched for an ideal location not only suitable for the practice of general medicine, but also for the public school education of their only child. Initially, they chose Medford where Tufts University is located, assuming that the local schools would be superior because of their proximity to a great University. However, they failed to recognize the heavy density of physicians in the area and the extreme difficulty of starting a new practice. After re-evaluating the Boston area for educational quality, physician density, and economic status of the community, they found a 15-room, three-story Victorian mansion in Needham which, fortuitously, was even larger than the local (Glover Memorial) hospital. And it had an added advantage as a potential doctor's residence: it was located near the commercial district adjacent to a large field (Greene's Field), which served as a playground weekdays for youngsters and weekends as a ballpark for adolescent youths. At its distant end, the stately Stephen Palmer elementary school, soon to be expanded to accommodate the burgeoning population, stood.

To purchase the house, mother emptied her piggy bank and Dad borrowed enough for a down payment from his father and took out an $8000 mortgage. To start a cash flow at once, the patients refused by the overcrowded Glover hospital occupied the high-ceilinged rooms in the house at 921 Great Plain Avenue and mingled with the occasional transient who wished to stay a week or so. I should emphasize that these patients were not seriously ill but recovering from minor operations or broken bones. As a 3-year-old, I used to drop in on these bed-ridden people. I have a photograph of the future Chief of the Fire Department and me in bed when he was recuperating from an automobile accident. Worse, my mother discovered me in bed with a woman who had just expired. Fortunately, no camera was available to record the incident. Occasionally, a patient would be admitted who had mental problems. There was one woman who repeatedly climbed out on the porch roof in her pajamas, which created quite a scene as my mother tried to coax her back to safety with a stern voice. This spectacle, incidentally, took place only a few yards from the main thoroughfare into Needham. Imagine the impression this must have made on those entering the town from Boston to see a woman sometimes clothed and sometimes scantily clad bickering with a nurse on a porch roof.

(Today, the Stephen Palmer school is an apartment complex and my house is Needham's YMCA, complete with a large swimming-pool and health complex facility embodied in a brick structure in the rear with a large parking area in the front).

Grandpa Mann played an important role in the selection of Needham as an alternative location to Medford, besides loaning Dad money for the down payment on the property. Because of minor farming injuries, he had become acquainted with a physician named Mitchell who lived in Medfield, a village near the slightly larger town of Needham (in population if not in land area). Dr. Mitchell knew another physician named Mitchell (no relation) who lived in Needham and was highly regarded there as the dean of physicians. Indeed, Dr. William Mitchell was beyond doubt the most beloved doctor in the town's history, and the perfect person to serve as a mentor and friend of a neophyte physician. He was also a well-known Dickens' scholar who, each Christmas, opened his home to the local folk to whom he recited "The Christmas Carol." It occurs to me now that the elder physician's interest in my father had much to do with Dad's English heritage. When Dr. Mitchell's only son was killed in a N.Y. highway accident while changing a tire, the entire town mourned his loss. Years later, when I was a young professor of pharmacology at Temple University, an older professor named Frank Eby (a professor of botany and pharmacognosy) seemed to be the reincarnation of Dr. William Mitchell in appearance, sense of humor, manner, and intellectual honesty. We soon became close friends and occasionally appeared on the same lecture circuit that introduced professors to pharmaceutical societies in the small towns of Pennsylvania. Dr. Eby's prime interest was establishing a chapter of *Kappa Psi,* the great pharmaceutical fraternity, at Temple. In this capacity he and I (a member) spent many hours searching for a suitable home in northeast Philadelphia for our members. In his later years, he became a mainstay of the National Chapter and received many honors.

In the early nineteen twenties, Needham was a town of 8,000 whose inhabitants could travel to Boston by trolley. The first electric car I remember was an open-air type that began its journey beside the Town Hall. I never rode on it as a child because it was considered too dangerous for a pre-teen kid. When I visited Boston to see the Sells Floto circus in Sullivan Square and watch Tom Mix, the legendary cowboy, do his stuff for $10,000 *per week,* I went by bus to the Charles River interchange, and then took a conventional trolley to Boston. I was always accompanied by Mom because Dad's patient roster was beginning to swell. Aside from attending circuses, the only other visits I made to the big city were to enjoy the Cab Calloway and Glenn Miller orchestras, the latter introducing "Elmer's Tune" on the occasion when the maestro himself announced it was the premier performance of that catchy tune. The Big Band era ended soon after Glenn Miller's plane disappeared over the English Channel during WW II. Although Cab Calloway and Benny Goodman were also superb bandleaders, the financial pressures of changing times ended their ongoing contributions to music history. During my junior year in high school, I also traveled to Boston to see the 1939 films: *Snow White and the Seven Dwarfs, Stagecoach,* and *Gone With The Wind.*

Speaking of school, every kid remembers the first day at the steps of the educational treadmill when the transition from freedom to captivity occurs, usually in the presence of a teary mother, whose tears are misread by the child as chagrin at the momentary departure from longterm togetherness. My first day was memorable in that it was rainy and consequently I decided to start the educational process alone, sneaking out the back door clad in a yellow raincoat which Gloucester fishermen wore, the kind that smelled of fish even when it was hanging new on a hook in a store. A few moments later, I tripped in the mud of Greene's field and had to return home to scrape the goo off my clothes and receive an embarrassing rebuke from mother. After a change of clothes, I took the sidewalk this time and arrived at Ms. McDougall's class reasonably dry and ready to absorb knowledge. The highlight of the year's learning process was coincidentally reported by a former kindergarten teacher from New York in a letter to the *New York Times* in 1995:

"The children sat in a circle while the teacher poured a pint of heavy cream into a jar and then handed it to a child who would shake it vigorously before passing it on to the next." Mary Schreiber of Ossining, N.Y., wrote: "What excitement when the butter began to appear." And over 65 years later I recall this event as the most interesting thing that happened to me during my first year at school. Why? Because I was able to take something distasteful and transform it into something that was delicious!

The fact that Miss McDougall was unmarried emphasized an unfortunate circumstance that strained the qualifications of elementary school teachers in those days. Married teachers were not allowed to work full time in the educational system. Therefore single teachers were theoretically more capable of instructing their younger charges without the encumbrances of children and husbands. Although I'm sure that some violated this mandate, discovery by higher authorities meant immediate dismissal. This edict, however, did not apply to elementary school administrators.

In the third grade I was fortunate to have won a spelling contest where the prize was a dictionary, which I still cherish. I didn't win anything else until my senior year when, having written the class song, I received a music medal at graduation and was given the privilege of conducting the orchestra, whose sound improved noticeably with the absence of one amateur musician. My grandmother Rose was in the audience and was shocked to see one of the teachers in the wings thumbing her nose at me. "Oh, For Heaven's Sake," said my mother. "That's Eva Churchill. She's trying to get David to relax at the podium!" Although Eva was in her fifties, she owned a Harley-Davidson and knew how to fly a Piper Cub.

CHAPTER THREE

The Sea Student

My elementary school education began while the country was experiencing a terrible economic downturn known today as the GREAT DEPRESSION. It started in 1929 when I was in Miss Doughty's first grade class where my only goal was to be taller than she (slightly under five feet). I was unaware that many people were standing in breadlines, visiting soup kitchens, and scrounging in rubbish containers to survive, while those who were more affluent and had played the stock market were leaping out of skyscraper windows as their portfolios vanished in worth. But Needham in the early thirties seemed to be immune to these horrors. What happened in Boston mimicked the situation in many other cities: skilled workers suddenly found that their jobs no longer existed; professional people, including college professors, who had enjoyed the fruit of the bountiful twenties, were forced to do menial tasks such as chopping wood and sweeping streets to survive. Youngsters in the early grades, however, were oblivious to these events unless their families were directly affected, for children didn't read newspapers and rarely listened to radio newscasts.

There began for me in April 1931, a most unusual happenstance that was entirely out of context with what was occurring in the U.S. A patient had informed Dad that if he had any funds available, now was a most opportune time to go on a cruise to the Caribbean because the rates were extremely low as shipping companies struggled to fill their passenger rosters in the face of a growing financial crunch. As a lover of the sea and ships (and what person of British origin isn't?), Dad arranged for the three of us to sail to Bermuda, Nassau, and Havana aboard the Anchor Line's S.S. Caledonia, a three-funnel ship that departed from Boston. I had no trouble getting permission to miss a couple of weeks of school because the trip was considered a great educational opportunity.

To step aboard a ship for the first time is a truly wonderful experience. Building a behemoth that has all the comforts of home, country club, theatre, library, boutique and hospital, that is capable of plowing through all kinds of ocean at 15 or 20 knots, is certainly a remarkable human achievement. Seeing the lines cast off from the pier, feeling the shuddering of the hull as the propellers start to revolve, and observing the valiant efforts of tugs in moving the ship, are experiences not easily forgotten. Then, there is the excitement associated with the boarding of the ship by the harbor pilot, who scampers up a hastily thrown (over the side) wooden ladder to the bridge, where he guides the vessel to the open sea. Later, a whistle blows heralding his job is over and signaling the crew of the waiting craft, bouncing in the waves, to pick him up from the same wooden ladder on which he displays the skills of a seasoned gymnast. Finally, the belching of black smoke from the Caledonia's center stack (the other two are dummies) indicates that the engines are primed and the ship is now ready to head for Bermuda at full speed.

The two days sailing to Bermuda were spent exploring the ship from its spartan bridge to its cluttered hot engine room where the stainless steel railings, slippery with oil, prevented the crew from falling into an array of asbestos-covered boilers, pipes of all sizes, and complex instrumentation, all devoted to the turning of the glistening propeller shaft. Where the shaft disappeared through the stern into the ocean beyond, a steady stream of salt water poured into the bilge, requiring a pump to control the volume. All ships have this problem, for it is impossible to make a seal tight enough to prevent this inflow of ocean, otherwise the shaft wouldn't turn at all. Also, passengers would be appalled if they knew how frequently (daily) small fires ignited in the holds of these ships. Usually they're easily extinguished in a large vessel, in contrast to the fire that started in a narrow passageway of the French liner Normandie on February 9th, 1942, from an unguarded welder's torch in the passenger section.

I don't remember the details of debarking in Bermuda, Nassau, and Havana in 1931. I do recall, however, that Bermuda was a strip of land dotted with brightly-colored dwellings, set like gems in a matrix of white coral outcroppings and tropical vegetation. The glare was awful and gave me a headache. Cars were not permitted on the dusty coral roads in the thirties because, if rain fell, sometimes for only a few minutes, the surface became too slippery. Horse-drawn carriages and bicycles were the most common modes of transportation, although an electric train carried tourists from Hamilton harbor to the other side of the main island, where they could visit the aquarium and see a giant green turtle and playful penguins.

Nassau was hotter than Bermuda and had three things not shared by the latter: a straw market, where you could buy anything made of straw that was usable and readily transported; glass bottom boats that allowed one to peer at myriads of colorful fish in crystal-clear water; and DIRTY DICK'S, a bar and steel drum orchestra, that boasted a floor show of talented native dancers.

When we finally reached Havana, I became ill with dysentery, possibly from drinking tap water collected off the roof of a Bermuda hotel. Or maybe I got the bug at a beer baron's exotic showplace in Havana where I drank water, while the folks had alcoholic beverages. Regardless, I spent the return trip in the Caledonia's modest infirmary, nervously lifting myself up to the porthole overhead hoping to glimpse the welcome sight of Boston harbor. The cruise ended with my departure from the ship on a stretcher with sore biceps and a coy smile.

The following year (1932), we went on another cruise in August, this time to the North Cape on the Anchor Line's California, a single-stacker that departed from New York. The shift in geographical direction probably had something to do with my physical condition upon arriving in Boston from Havana the previous year. But, as in April, 1931, the ship, after exploring the beautiful Saguenay River, sailed to Bermuda before returning to New York. Theoretically, if my dysentery had begun in Bermuda, I would be sick again after visiting the island the second time. However, if I arrived in New York in good health, it would imply that Havana had been the origin of my intestinal problem the previous year. The culprits eventually proved to be Havana and its drinking water.

The trip up the Saguenay River, a tributary of the mighty St. Lawrence, was enjoyed under blue skies part of the way and in fog the rest, necessitating the ringing of a huge bell and the doleful moaning of a foghorn. Although the river is located in a remote area of Canada and bordered on each side by huge granite cliffs, it is a busy watery highway for small craft, fishing boats, and a few sightseeing vessels which, if involved in a collision, would sink a mile before reaching the bottom. Finding this knowledge a bit unsettling, I was relieved when at last the ship turned around and headed for Bermuda.

Bermuda, with its 365 islands of varying sizes, was still without automobiles a year later. I did spot a truck which belonged to a highway crew who were busily repairing a section of the road. A carriage ride to the aquarium in the center of the main island and a visit to the crystal cave were exciting. Only the Mammoth Cave in Kentucky was somewhat comparable to the crystal cavern, while the first marineland aquarium in St. Augustine, Florida, was still in the planning stage.

As we entered the cave, past a steel gate, the path descended into a large chamber where stalactites hung down from the ceiling and stalagmites protruded up from the floor. These formations consisted of crystals of calcium carbonate (calcite) which required thousands of years to reach their present form. "Watch your head," someone yelled, as Dad inadvertently hit his head on a stalactite, shattering fragments in all directions. A few moments later, we walked across a wooden bridge that floated in an area of cathedral-like proportions where one could clearly observe the bottom, although the water was at least fifty feet deep. The level of the water rose and fell with the tide, but, oddly enough, there were no fish present and the bottom seemed to be devoid of sea life in any form. The water filtered into the cave from the ocean nearby, entering through tiny apertures which removed fauna and flora visible to the naked eye.

What was going on with the educational experience during these years? At the end of the first grade (1929), I was still shorter than my teacher, Miss Doughty. The second grade was taught by Mrs. Blake (her husband had died, thus she was single and qualified to teach), who was the nicest person I'd ever met and never scolded me for dipping pigtails of girls sitting in front of me into inkwells. I did have to come in early each morning to prepare the ink, however. Because the Stephen Palmer school was getting a new addition the following year, my third grade class was moved to the Kimball school, a three-story wooden structure that was the town's foremost firetrap. Fortunately, Mrs. Blake was my teacher again with a new set of tasks to replace my ink-preparing duties. She made us rehearse fire-drills, which were easily accomplished from classrooms on the first and second floors, but almost impossible to do from the third floor where, strangely, gym classes were held. The stairways were wide enough on the lower floors, but extremely narrow on the third, for this area was reserved for storage of school paraphernalia, not for holding large gym classes. Luckily, no emergencies occurred during this one year hiatus from Stephen Palmer, but severe vibrations from gym classes weakened the floor structure. The Kimball school was, at best, an example of bad architecture and, at worst, a miserable excuse for an educational facility. Presently, at its former Chestnut Street location near the new Glover Memorial Hospital, a large brick building stands which serves the dual purpose of housing fire and police departments.

Back at Stephen Palmer, I entered the 4th grade of Miss Cornish, a tall, thin, pretty brunette who was pleased with her brand new classroom that had sliding doors at the rear of the room behind which raincoats, rubber boots, and snowshoes were stored. There was also a new, 48-star, silk flag at the front of the room to which we directed our pledge of allegiance, standing with our right hand touching our right eyebrow and extending the arm forward upon reciting the word "flag." After WW II, Hitler's Nazi salute, which closely mimicked the salutation to our flag, abolished this part of the pledge, with the right hand now being placed over the heart. Miss Cornish also read from the Bible, but the class never recited a prayer aloud or observed one in silence. Multiplication tables were learned during the year up to 12 X 12=144. I don't recall learning anything else of importance. That academic hiatus would be filled in the fifth grade during the reign of Miss Miller.

Ah, Miss Miller, the most innovative teacher I encountered in elementary school! Today, students sit on movable plastic chairs with writing platforms affixed to the right or left sides to accommodate righties and lefties. In the early thirties, the classroom had a large wooden desk where the teacher sat facing six or seven rows of fixed desks, each with an attached chair, totaling 7 or 8 students per row. Thanks to Ms. Miller's genius, the occupancy of each chair depended entirely upon the mathematical skills of the student. Those sitting nearest the desk in each row were the ones who had achieved perfection on the math quizzes over a period of several weeks. If you missed an answer, with the grade now in the nineties, you dropped back a seat. Mathematical dunces thus occupied the last row.

At the front of the room was a large table on which an authentic Japanese village, meticulously duplicated in cardboard, plaster of Paris, and eskimo pie sticks, surrounded a large *papier mache* Buddha with a peculiar bulge sticking out of its forehead. We learned that the bulge was a blob of pure silver with religious significance. Every student in the class desired to visit Japan someday, not realizing that there would soon be a strong possibility that a few might get a bird's eye view as they dropped firebombs on Tokyo prior to Enola Gay's visit to Hiroshima.

When it came time to study Canada, Ms. Miller assigned a province to each student and, after some practice, we marched single file to adjacent classrooms where the provinces were uttered one after another as follows: student number one would say British Columbia; student number two would say: British Columbia, Alberta; student number three would say: British Columbia, Alberta, Saskatchewan, etc. Unhappily, I was Nova Scotia.

Before the school year had ended, it was time to go on another cruise. March 18th, 1933, we departed from New York on one of the most famous ships in the world, the Cunard Line's Mauretania. For 22 years she had held the famous Blue Riband as the fastest ocean liner to cross the Atlantic from New York to Southampton, England. For years she was a member of Cunard's distinguished triumvirate of transatlantic ships that included the Berengaria and Aquitania, bringing immigrants to this country and tourists to Europe. Now, in the twilight of an unparalleled career, the formerly black hull was painted white to prepare her for Caribbean cruises on which we were about to embark. On the "White Queen", we were soon to visit Trinidad, La Guaira (S. America), Curacao, Colombia, Panama, and Havana.

When the Mauretania was built in Tyne, England and launched in 1906, she was the largest ship ever built: 790 feet long, 88 feet wide, displacing 31,938 tons. Her public rooms, staircases, and first class staterooms boasted the rarest woods, most exquisite marble carvings, most beautiful crystal chandeliers, and most remarkable examples of gold leaf inlays outside of British and French Royal residences. Unlike today's ships, however, she did not have outside on-deck swimming pools. These were considered unnecessary by Cunard because even during the summer months, the air was often cold and the water temperature was more attuned to raising lobsters than offering recreational swimming. When "Maury", as she was affectionately called, was adapted for Caribbean routes, a large canvas bag suspended by thick ropes and filled with sea water served as the pool. Considerable sloshing occurred during rough weather, and diving was impossible, even with smooth sailing, for the ship's movement made the depth 5 feet one moment and 4 inches the next.

As soon as one boards a ship a peculiar feeling comes over the body as the feet begin to explore the new home. First of all, the floor or deck is not flat but slightly rounded, as if one were standing on a giant cucumber. The width of the ship is noted and the large public rooms are invaded while the parents are more concerned with finding the stateroom.

"I'll be down to join you in a few minutes. Just want to look around," says the kid who is trying to avoid carrying anything heavy, or perhaps casing the oncoming passengers for kids of comparable age. Perhaps this isn't the reaction of a kid who is about to go on his or her first voyage, but for me, the experienced sailor, this was number 3.

After finding the stateroom, usually on C deck, we visited the purser to obtain the necessary dinner cards without which you have as much chance of getting fed as a stowaway. The dinner card not only reveals the table position by number but, more importantly, which sitting you can attend. With 2300 passengers to feed and a dining area that can accommodate only several hundred at a time, several sittings are required. There are, however, at least 2 large dining areas: one for first class passengers and one for tourist class, harking to the days when the ship carried three classes, the lowest being steerage where the immigrant grabbed a wooden bench in the bowels of the ship and hastily ate a meagre meal. Now, on a Caribbean cruise, there is only one class, but accommodations vary considerably according to the price paid. The food, though, is similar, and the passengers enjoy a string quartet whose dulcet tones apparently aid digestion.

At the first dinner, we found ourselves located on the starboard side of the ship beneath a porthole with a table set with fine china and glasses stenciled with the Cunard Line insignia. A hard bun was sitting on a small plate next to a paper cup filled with salted peanuts. At once, I hollowed out the bun with my index finger and filled it with peanuts, chomping down on a new taste treat. After the meal, I became queasy and soon was seasick. But I discovered a remedy: chew radishes and celery, then go to bed.

It took only a day to meet Sid, the middle-aged sailor, whose work area was the equipment-laden sun deck that extended several hundred feet aft of the bridge. Strung near four huge smokestacks were nets for the games of ring-toss and badminton, which required setting up each morning. Outlined in white paint on the teak deck, were several games of shuffleboard, reminding passengers of the chalk-marked, sidewalk game of hopscotch of their childhood. I soon made friends with Sid, who hailed from England and had been a sailor with Cunard for twenty years. He enjoyed my companionship, while arranging the athletic equipment each morning. One morning Dad recorded Sid's acrobatics with his new *16-mm.* Cine Kodak movie camera. Recently, I had the black and white film transferred to video tape and was pleased with the quality of the recording. In 1971, when my young family sailed to Southampton on the QE-2, I asked a ship's officer if Sid were still alive and was pleased to learn that he was enjoying retirement in England.

Our first port of call on this cruise was Trinidad, at a harbor known as Port of Spain. This island is the wealthiest in the entire Caribbean group because it has oil, asphalt, and a populace that has benefitted from these natural resources. We visited the famous "pitch lake", a large tar pit in Brighton where one can walk on wooden boards covering asphalt, the principal ingredient of macadam roads. With each excavation, more asphalt is replaced, oozing up from a seemingly inexhaustible source. As with the LaBrea tar pits of Los Angeles, strange things come to the surface from time to time, providing a bonanza of fossil treasures for scientists. Pitch lake is conveniently located slightly beyond a pebble's throw from the ocean where freighters load this valuable cargo that becomes road surfaces and roofing material throughout the world.

A car trip (Packard touring car) around the island revealed bamboo forests, fields of sugar cane, and grapefruit trees. On our return to the ship, we stopped at a small gift shop and examined a wire-topped box containing a deadly fer-de-lance snake. This serpent, native to Central and South America, was introduced to Trinidad to prevent overworked natives, laboring on sugar cane plantations, from escaping their slave-like existences. A similar situation existed in Martinique in the old days and, in both Trinidad and Martinique, the poisonous snake flourished. The mongoose, supposedly its arch enemy, was introduced to eradicate the serpent, but the animal preferred chicken and soon reduced the fowl population.

Our next stop was the continent of South America or, more specifically, the harbor of La Guaira in the country of Venezuela. It was a beautiful day when I scampered on deck and, open-mouthed, gaped at the highest mountains I had ever seen, the foothills of the Andes. Because the harbor was shallow (only small freighters such as those belonging to the United Fruit Company and the Grace Line could dock), the Mauretania had to anchor a considerable distance offshore where the wind had whipped up large waves. To transport passengers safely to shore, motorized tenders from the ship served as ferries and, to convey them from the ship to the small boats, a large wooden platform was lowered to the waterline that was reached by descending on wooden stairs. On the platform a number of husky sailors watched the wave motion and, when the crest of the waves matched the height of the platform, a passenger was lifted up and then down into the waiting tender. What made this operation especially exciting was the fact that some of the largest sharks I had ever seen were gliding a few yards from the white hull. Cunard's enviable record of never having lost a passenger remained intact thanks to the strenuous efforts of its well-trained seamen.

Once on land, a retinue of taxi drivers picked up their human cargo and headed for Caracas, only a few miles away by crow but because of the terrain, approximately 20 miles by tortuous road laboriously carved from the barren slopes of the Andes. Halfway there, we stopped at a weird monument consisting of a severely compressed, black (Henry Ford era?) car perched atop a concrete piling. This tragedy was a gruesome reminder of the first highway accident to occur shortly after the ribbon cutting ceremony. There had been many similar accidents since, but no survivors because going off the road meant falling several thousand feet.

Soon we arrived in Caracas, a large, bustling city of predominately white buildings where one of the world's largest bullrings reminded visitors of the Spanish kinship of its citizenry.

A must-see in Caracas is the tomb of Simon Bolivar, the George Washington of South America. The first time we visited it, he was entombed in a beautiful marble work of art located in a government building. Years later, in the sixties, when I was again in Caracas, he had been moved to a cathedral on the outskirts of the city. The tone of the city seemed to be serene in the thirties, but I found out how mistaken I was when I read in *Reader's Digest* in the fifties that Dictator Gomez was a horrible person who subjected his enemies to terrible human rights violations.

An overnight sail from La Guaira brought us to the Dutch island of Curacao, a dry, lizard-ridden, coral-tipped, extinct volcano that lies 30 miles off the coast of Venezuela. It is one of the A, B, C islands, of which Aruba and Bonaire comprise the rest. Ships are more likely to stop at Curacao because it is a major refueling station, which explains why the pier we had tied up to was a network of pipes leading to storage tanks on the horizon. In contrast to Trinidad's bonanza, Curacao imports its oil from the mainland where huge deposits were discovered under the tranquil waters of Lake Maracaibo in Venezuela.

We took a taxi to the island's major city of Willemstad, which is divided by a large canal that is actually an inlet from the ocean that ends at a protective harbor. Spanning the canal, bordered on either side by brightly colored buildings of Dutch architecture, is the world-famous pontoon bridge known as "Queen Emma." When a small ship approaches, the bridge swings aside like a giant door. Tourists entering the city usually have two destinations: those searching for bargains head for the jewelry stores such as Spritzer and Furman (Jewelers of the Caribbean), or venture to the nearby wharfs, where schooners arrive daily from the mainland to display their wares of produce, crafts, and exotic animals (sea turtles, macaws, pigs, and pythons). On this trip, one of the Mauretania's passengers bought a python which he said was going to live in a specially-constructed cage next to his telephone to discourage the neighbors.

In the evening, returning to the ship, our taxi passed several posted swimming areas reserved for tourists in one place and the crew of cruise ships in another. I noticed that a steel sharknet was visible a few hundred yards offshore in the sailor's swimming locale, but none was visible in the tourist's section. Is one to deduce from this that the local sharks prefer those smelling of oil to those reeking of cologne and Chanel #5?

Our next stop was the ancient city of Cartagena (Cartahena), Colombia, where our ship's draft required anchoring some distance from the shore. Instead of using lifeboats as tenders, the Colombian harbormaster had arranged for vintage sternwheelers, formerly plying the Mississippi in the Huckleberry Finn era, to ferry passengers ashore. These riverboats have a very shallow draft of several feet and thus are not easily controlled when a stiff wind is blowing. As one of them approached the ship, a gust of wind surprised the pilot, who was unable to control his vessel, with the result that it slammed violently against the massive hull of the Mauretania. Dad captured the collision on film, but after it was developed, all we saw were a few boards flying through the air.

Visiting Cartagena for the first time, one can't help but compare it to the modern city of Caracas. What a difference the discovery of oil makes! Before the days when Colombia was associated with cocaine, emeralds were the chief export, hardly a source of great wealth for the people and a rejuvenator of their architectural heritage. But Cartagena does possess something that Caracas lacks, an aura of pirate history with its plundering of terrified victims who once felt secure behind their stone forts and great sea walls bordering the harbor. Realizing that such defenses were useless, engineers constructed a tunnel that enabled the citizenry to sneak out of the city and head for the hills without being observed. Tourists, for a fee, were encouraged to walk several miles through the tunnel, noting the cavities in the walls where prisoners were occasionally chained until they died of thirst, starvation, or physical abuse. Two years later, when we visited the city again, a guide informed us as we entered the tunnel that a cache of gold coins had been dug out of the wall after someone had found a treasure map in Spain which revealed its location. The presence of several doubloons scattered in the dirt below the hole suggested that the treasure seekers had worked hastily to remove the loot before someone discovered them.

One of the tourist attractions that everyone seems to visit is the church of Pedro Claver, who converted the African slaves (for a modest fee) to Christianity and then gave them work building his house of God. This was during the Spanish Inquisition, when the alternative to conversion was hardly a viable choice. Inside the dank, candle-lit sanctuary, reposing in a glass case mounted on an altar, are the skull and long bones of Pedro, partially wrapped in a velvet cloth. The sight is somewhat unsettling to a kid who, as a result, lost his appetite for chicken soup for several months.

The most interesting aspect of Cartagena, to one who admires the writings of Ernest Hemingway, is the presence of a small bullring in the center of the city where, according to the season, fights were held. March apparently was during the off-season, for there were no posters proclaiming forthcoming fights, and the only activity I saw in the ring was a kid with a wheelbarrow, on which were mounted a couple of large horns, who was haphazardly pursuing another kid swirling a cape in an awkward display of incompetence. I felt as though I were looking at another Manolete or, more likely, a soon-to-be basket case at the local hospital.

As the ship departed that evening, I noticed a monastery atop a high cliff from which, it was said, nuns leaped into the sea below to escape being captured by pirates. The ocean, however, was no longer at the foot of the cliff, because several centuries of earthquakes had rearranged the sea floor.

In the thirties, before Castro wiped out their holdings in Cuba, the United Fruit Company bolstered the economy of many countries and islands in the Caribbean community. In Colon, Panama, a city on the Atlantic side of the canal, this aid was manifested in part by construction of piers to accommodate not only the freighters of the U.F. Company, but also large vessels, such as the Mauretania, whose owners, the Cunard Line, had shifted from the unprofitable Atlantic route to the Caribbean, where white-hulled ships brought tourists from the United States to the gambling casinos and bordellos of Havana, Haiti, and Panama.

The city of Colon reminded me of a smaller version of Havana with its stucco-fronted shops, arranged like rowhouses, displaying their wares of leather goods, perfumes, silks, linens, carpets and, astonishingly, shrunken human heads! When Dad spotted these horrendous so-called trophies hanging in a drugstore window, he remarked to mother: "I'd like to see how much they cost." They were selling for 50 dollars each, but mother was ahead of the situation. "I'm not staying in a stateroom with that thing!" she exclaimed.

Perpetual scholar that he was, Dad learned that the "trophies" came from Ecuador and were gathered by the Jivaro Indians from vanquished enemies. Once in awhile, when Caucasian missionaries or explorers got in the way of local skirmishes, their heads, sometimes sporting long red hair, appeared in the stores of Panama, quite recognizable as human but the size of baseballs. Of course, the main attraction in Panama was the canal, but several decades later, when I viewed the great ditch with Canadian friends, I was miffed when they joked about the antiquity of the Canal and invited me to see the modern Wellington Canal in their country. In retrospect, they completely disregarded the heroism of thousands of workers who lost their lives to disease (yellow fever) and cave-ins from dynamite blasts, and the era of Teddy Roosevelt, when engineering of this type was in its infancy. One should also mention the frustration of Ferdinand deLesseps, the builder of the Suez canal, who attempted to duplicate, with sand-moving equipment, his Egyptian success in Panamanian mud. His failure is seen today as a water-filled gully located several miles from the Caribbean Sea. In the thirties, Richard Halliburton, a Princeton graduate and explorer, swam the Canal at a cost, imposed by the government, of 36 cents (according to the tonnage fees at that time). In 1939, he disappeared in the Pacific while sailing a Chinese junk from the Asian mainland to San Francisco to commemorate the opening of the World's Fair on the West coast at Treasure Island. There are those who believe he was killed by Japanese military personnel who were constructing defenses for the forthcoming war. It is also likely that Amelia Earhart may have suffered a similar fate in the same area of the Pacific.

Arrival in Havana, the last port, came with cautionary advice: don't drink the water and slake thirst with plenty of fruit. At the harbor's entrance, Morro Castle, perched on a rocky promontory and resembling a truncated Hadrian's tomb more than a castle, gave no hint of its ugly past. Centuries earlier, political and religious dissidents, chained in dungeon cubbyholes, had awaited the rising tide and eternity. On the opposite shore, cars were moving along the palm-fringed esplanade known as the Malecon. Approaching its pier, the Mauretania pointed her bow toward the stately, black-hulled Holland-America liner, Statendam. Sadly, before the decade ended, this beautiful ship, built in 1929, was destined to be blown to smithereens by the Nazis while tied up at her home port of Rotterdam in the Netherlands.

Among those meeting the disembarking passengers was a man with leprosy. He was well-dressed and middle-aged, with gnarled hands so ravaged by the disease that chalk white bones were exposed.

While hailing a taxi, I spotted a teenage girl, clad in torn clothing and shoeless, hobbling with a severe limp down an alley. With each step, the exposed heelbone of a clubfoot touched the filth left by donkeys, dogs, and human derelicts. These sights made me wince in contemplation of the pain she must have endured.

As she disappeared in the distance, a taxi arrived and we drove down the mosaic-encrusted Malecon, slowly enough to enjoy the beauty of the harbor. Shortly, we arrived at a beer garden where giant oak trees provided shade for tourists busily drinking free beer at little tables. The trees looked natural, but close inspection of a trunk revealed a wire mesh where the cement covering had fallen off. A brick mansion belonging to one of the beer barons attracted our attention and, at our driver's request, we walked over to the building and looked through a window. The floor, sparkling in the morning sun, had been meticulously inlaid with chips of mother-of-pearl. Beneath a giant artificial oak tree we enjoyed a beverage: parents sipping free beer; a quarter's worth of coke for me.

Ever wonder why the first stop in any Caribbean port is also a place where rum or beer is available? The unwary tourist is soon put in the proper mood to buy something that'll end up in the attic. Returning to the city, we saw the most famous bar in the world with a counterpart today in Key West—Sloppy Joe's. The redwood bar seemed hundreds of feet long, a wooden Great Wall of China, in a cigar-smoke-filled room with a decibel level of a college library at exam time. Leaving the bar after contemplating its Cuban atmosphere, we arrived at the famous Upmann Cigar factory where workers convert leaf tobacco into the world's finest cigars. The skilled artisans wrap the leaves into cylindrical shapes and trim the excess with triangular knives, while they listen intently to a man perched high on a tennis referee's wooden chair reading a novel in Spanish.

The final stop of the tour was a visit to Havana's capitol, which resembles the Washington structure, but is smaller with maroon-colored, rubber floors, massive bronze doors depicting battle scenes, and a nickel-sized diamond embedded in the white marble floor directly beneath the apex of the dome. Thick, bullet-proof glass protected the flawless gem from tourist thievery. During the upheaval of Castro's revolution, someone removed the stone from its resting place. Recently, it was discovered in a desk drawer of a Cuban general.

Before returning to the ship, we passed a memorial to the battleship Maine, whose explosion in Havana harbor precipitated the Spanish-American War. The monument to this ocean and earth-shaking event was a tall, concrete column on top of which was placed an anchor from the ship. At its base, a bronze plaque, polished daily and bearing the names of those who lost their lives in the tragedy, honored the memory of their sacrifice. When Castro took over the country in the fifties, the monument was depicted in the local newspapers undergoing the same fate as its *raison d'etre.*

Back aboard ship, two interesting things were taking place: Betty Carstairs, the British millionairess, had dispatched a mahogany launch from her huge schooner, anchored at the harbor entrance, to pick up cartons of liquor from the R.M.S. (Royal Mail Ship) Mauretania. Off the latter's bow, a large wooden raft, equipped with hard-hat diving equipment, bobbed on the incoming tide. Presently, a copper helmet broke the surface, followed by a person in a canvas suit, holding an air-hose and hawser. One might assume that the diver had been inspecting the creosote-covered pilings of the pier for signs of rot, but in reality he was searching the muddy harbor bottom for an anchor the Aquitania had lost on a previous trip.

CHAPTER FOUR

Down on the Farm

In the fall of 1933 I said goodbye to the Stephen Palmer School and trudged to the Emory Grover school, a yellow-brick edifice on Highland Avenue, to begin the sixth grade. My teacher, a very attractive young lady in her early thirties named Mary Kett, was a welcome change from the staid Ms. Miller. Instead of the mathematical chairs game introduced by the latter, Miss Kett exposed the class to classical music by moving the entire student body once a week to a special room where we listened to a symphony orchestra conducted by Walter Damrosch on the radio. I have no recollection of learning anything of significance except the difference between a tuba and a sousaphone.

What I do remember, however, were three male friends who were destined to achieve different degrees of notoriety because of the forthcoming war and a crime wave of bank robberies.

Bob Collamore, a heavy set, athletic type with a pleasing personality and ever-present smile, sat next to me in an adjacent row. Two tragedies struck his family within a few years: his father succumbed to anesthetic overdosage during a routine appendectomy. The attending physician told Dad that "he just slipped away and I couldn't get him back!" Such deaths, though relatively rare, can happen to anyone during general anesthesia. Once tonsillectomies were performed in a sitting-up position. Then it was discovered that the heart is more apt to stop beating if the patient is sitting up than lying down. The reason for this discrepancy is that a vascular readjustment to the presence of the anesthetic in the circulation is easier to accomplish if the patient is lying down. If sitting up, this reflex activity is hampered and the heart stops. The second tragedy occurred when Bob stepped on a landmine in Europe and was killed.

Charles Perry, an ebullient young man, who resembled a young version of Noah Beery (Rockford's TV Dad), like Bob joined the Army and saw service in Europe. Last year, while watching Charles Kuralt in a TV segment in Albertville, I was shocked when he mentioned that he was standing next to a memorial to 6 paratroopers who had lost their lives during an invasion of Nazi-held territory. He read the names and one was Charles Perry, who died when his parachute failed to open. My mind wandered back to those happy, carefree days in the thirties when Charles and Bob had lifetimes of accomplishment ahead of them. No one in Miss Kett's class could have envisioned that Charles would shortly be buried in a winter Olympic town in France while still a youth.

Although he didn't attend our 50th high school reunion, my third friend, Arnold Mackintosh, was in good health. Back in the thirties, he had been in class when a call to report to the principal's office was heard over the loudspeaker. Minutes before, the screaming of sirens was heard, presumably made by fire trucks responding to a routine alarm. But the cause of the fracas was far more sinister than firemen simply answering an alarm. Police in patrol cars were pursuing the infamous Millen-Faber gangsters, who were fleeing to Boston after robbing a bank and mortally wounding Forbes McLeod, a beloved policeman, as he crossed the tracks and headed to the crime scene beside the railroad station on Great Plain Avenue.

To protect themselves while speeding away, the bank robbers grabbed Arnold Mackintosh (senior), the bank's treasurer, and held him on the running board as a human shield, where he hung on for dear life, with a Thompson submachine gun pointing out a window over his shoulder. As their car careened through Needham Heights and passed the fire station, they fired a burst at police officer Francis Haddock, who happened to be standing on the cement runway of the firehouse, wondering what the commotion was all about. Mr. Haddock was also mortally wounded and died at the Glover Memorial Hospital the next day. A 0.45 calibre slug from the submachine gun had ripped through his back, severing a ureter and causing urine mixed with blood to drip from the gaping wound.

Dad vividly described the fatal wounds over dinner and sadly remarked: "They didn't have a chance!" Ironically, Dad and I had seen the weapons involved at the Sportsman's Show, held annually in February at Mechanics Hall on Huntington Avenue in Boston. In addition to the usual displays of canoes, camping equipment, rowboats, and fishing gear, the State Police had assembled a fascinating exhibit of machine guns, ammunition, and macabre photographs of homicidal victims' eyeballs with images of their attackers imprinted on the lenses. Dad and I were at the Show on a Sunday and the following day the Millen-Faber gang robbed the exhibit of its weaponry.

The call to the principal's office was a terrible ordeal for young Arnold. However, the gangsters, brothers Irving and Murton Millen and M.I.T. graduate Abraham Faber, were captured a few months later and, in contrast to today's stalling with endless appeals, were electrocuted at Charlestown prison before the year had ended. Officer Haddock's daughter, Mary, who was a child at the time of the bank robbery, became a nurse and contributed many years of devoted health care to the Glover Hospital. She eventually married a young man named Barnicle, which brought a smile to those who read their wedding announcement heading—Haddock-Barnicle—in the Boston Globe.

Like most men, Dad loved automobiles, so it didn't take much coaxing by two of his patients, Al Tocci and Phil Mays, who were members of an auto racing association, to join their organization as a judge. I saw my first auto race at Rockingham Motor Speedway in Salem, N.H., on Columbus Day, October 12th, 1926. My parents paid $5.50 each for Grand Stand tickets (4-yr-olds were free) and witnessed two 25-mile dashes and a 200-mile race. The participants were Bennie Hill, Harry Hartz, Frank Lockhart, Leon Duray, and a Peter somebody. Dad was smitten by the racing bug that day and, during the next decade, attended dozens of events in Keene and Newmarket, N.H., and Weymouth, Mass. Tragically, many drivers were seriously injured, and several were killed, while racing on mile long dirt tracks originally designed for horse racing. After our favorite driver, Armand Ferrande, was killed when his race car left the track and exploded in a grove of trees at Spofford Park in Keene, N.H., Mom refused to go to other races, leaving Dad and me to occupy the judge's booth, She relinquished her boycott, however, to attend the Indy 500 in 1950 while I was a graduate student at Purdue. The three of us were enjoying a great race when the skies opened up and a cloudburst stopped the event. Johnny Parsons won with a cracked engine block that would have spelled disaster after several more laps. The drivers I met in the thirties were dedicated, fearless athletes who were completely devoted to the sport and gave their all for the spectators.

When our National League baseball team was about to win a pennant, I took my young son, *Scott, to a game. He wanted an autograph from one of his heroes so, after the game, Scott approached the star pitcher and made the request. "Out of my way, kid," came the abrupt reply. By contrast, I took Scott to Langhorne to see a 200-mile auto race in the seventies. Bobby Unser won and my son approached him for an autograph. "Sure," smiled the tired, dirt-covered man, and he signed his name on a race program which is now a cherished possession of the family and a reminder of the type of men that race car drivers are.

It was also in the fall of 1933 that I began a precedent that continued almost to the start of my college days in the forties: I spent each weekend during the fall and spring school year at grandfather Jesse's farm in Norfolk, Mass., where the only other legitimate creatures living there were a Holstein cow named Nancy, a dog named Towser, and a flock of nervous Rhode Island Red chickens. When school ended each Friday at 3:30, I changed into overalls, had dinner, then got in the car with Dad and anxiously awaited our arrival at "The Cedars", as the farm was called. Jesse's wife, Marion, had died in her seventies, and the children had grown up, married, and only occasionally visited the farm on special occasions such as birthdays and holidays. So Jesse was alone with his livestock most of the time which was the reason for my weekly visits.

*Scott Elliot, our youngest child, born on his paternal grandmother's 70th birthday, May 29th, weighing 8 pounds 10 ounces.

When I arrived at "The Cedars", a once productive farm that had seen its heyday decades ago, the 17th-century farmhouse, barn, and outbuildings were fragile-looking, like 82-year-old Jesse himself, yet the prospect of fun and freedom from school for a while put the reality of what I saw in a happy perspective. Not that grandpa Jesse was sickly and bedridden. On the contrary, he was quite active, fun to be with, and endowed with a British sense of humor. His humorous outlook always peaked when his daughters arrived with their husbands, for they always came fortified with arguments about how awful FDR was and why the country was headed for socialism. No mention of what was going on in Europe, as though Hitler were a fly-by-night dictator soon to be replaced by a stable Kaiser, or of Stalin

29

and Mussolini who seemed to be bit players on the world's political stage. The truth was that Americans didn't know much about foreign politics for the Great Depression had usurped everything in the news that dealt with the suffering of people an ocean apart. Where the next paycheck or mouthful of food was coming from was of far greater importance to them than hearing about youth movements in Germany and Italy and what mischief they were considering. Even the warning from Charles Lindbergh, that the Nazis were assembling the greatest air force the world had ever seen, was shrugged off with a disdainful—so what?

Thus, the weekend gathering did little to educate me of world politics. Instead, the last person to personify the word "presidential" in manner and speech in this century was vilified as an opportunist who was taking the country down the road to ruin with his stopgap policies such as the CCC, NRA, and WPA. To their credit, however, my relatives never mentioned FDR's polio-induced handicap, which required assistance in walking and frequent rides in a wheelchair. Everyone was aware of his ambulating problem, but so great were his personality and stamina, that this infirmity was overlooked.

Daughter Agnes and her husband, Wallace, were the most frequent visitors to the farm because they had established an apiary near the barn, consisting of 6 or 7 multitiered hives that required constant attention. This hobby almost resulted in disaster when Wallace, a high school physics teacher, installed screens in his Franklin car and then transported several hives on the rear seat to his home in West Newton. As he rounded a curve, the hives separated, releasing thousands of angry bees. He somehow managed to drive the buzzing horde to West Newton without being stung or running over anyone.

Aside from arguing with "Thin," her affectionate name for Jesse, Agnes spent her free time at the farm playing solitaire, cribbage, or packing her oil paints and easel for a brief session of landscape painting in the apple orchard behind the barn.

Youngest daughter Anna and her husband, Harry, who lived two hours away in Hampton, N.H., came to the farm monthly with their only child, Marion. Upon arriving at the farm, Uncle Harry, like Uncle Wallace, left the women to gab about politics while he pursued his hobby assembling radios from the hundreds of parts he kept stashed away in cardboard cartons in a nearby shed. Uncle Wallace, on the other hand, clad in sting-proof clothing and wearing an old Stetson hat with a wire screen protecting his face and neck, pursued his hobby of distracting bees with an aluminum gadget that emitted puffs of acrid smoke as a tranquilizer, while he raided their stores of honey. Because the house lacked electricity, Uncle Harry used large red batteries to test the assembled results, one radio of which carried the reassuring voice of FDR delivering his inaugural address in March, 1933: "We have nothing to fear but fear itself."

While the adults were occupied with their assorted interests, Marion and I explored the apple orchard behind the barn, creeping like Indians to the pond beyond a stonewall where we tried not to startle turtles sunning on partially submerged boulders. Their perception, however, was exceptionally keen and we only managed to get a glimpse of their dark green shells as they plopped into the water. Sometimes we saw a large black water snake coiled on a log awaiting an unwary frog or fish. One of these harmless serpents scared my father when, as a kid, he was swimming in the Stop river near the Medfield-Norfolk line and it slid across his shoulder. That was the last time he swam anywhere, claiming he was susceptible to ear infections. In reality, he was susceptible to the sudden presence of limbless swimmers. The oldest surviving daughter, Carol, and her husband Wilbur rarely visited the farm with their two children, Carolyn and Willard. The reason for this was that Uncle Wilbur owned and operated a fuel company in Stoneham, Mass., that required his constant attention. Incidentally, with Dad smoking Chesterfields, Mom smoking Raleighs, and Uncle Wilbur puffing on a stogie (sometimes in his car), grandpa Jesse must have yearned for fresh air during these get-togethers. Long ago, he had given up the weed when he dropped his meerschaum pipe with the amber stem while ploughing. Of all the cousins, Willard was to become the most beloved for a reason to be divulged much later.

The farm was typical of those in the countryside of 17th century Massachusetts that were carved from the deciduous forests under the suspicious eyes of Indians. Thousands of years earlier, the glacier had deposited boulders as it moved down from the north, gouging depressions that became the lakes and ponds of New England. After the land was cleared for crop planting, rocks hidden beneath the surface wrecked havoc on the plows of those tilling the land. These rocks, however, also supplied the building materials for the walls demarking the fields and the farm itself. Although the boulders were not in the same league as those used to build the pyramids, many of them which provided a base to support the walls approached their dimensions and required comparable ingenuity to position them.

The two-story farmhouse was built in 1675 in the proximity of a sycamore tree that had reached such monstrous proportions a person could hide inside its hollow trunk. The largest area of the first floor was the kitchen with a brick-encased Dutch oven, wood-burning stove, and a handpump that drew water from an outside well. The remaining three rooms were the parlor, dining-room, and a small bedroom. Upstairs, the bedrooms were denoted by color. The room over the dining-room was called the Green Room where Dad was born on the longest day of the year-June 21st (1892). Two other bedrooms and a windowless room in the center of the house, presently used for storage of Uncle Wallace's SatEvePosts, once offered a safe haven for the occupants during Indian attacks. Inside the walls of the second floor were slots through which muskets were inserted to ward off attacks by the Indian warrior, King Philip, whose encampment on Noon Hill was only a half mile away. The only plumbing in the house was a handpump in the kitchen. An outhouse next to the woodshed offered toilet facilities in the company of spiders. During the winter months, outside temperatures often registered 35 degrees or more below zero (Fahrenheit), so the fireplaces in each bedroom and parlor were welcome amenities. Kerosene lamps, which required frequent cleaning to remove carbon from the glass chimneys, provided an eerie glow that enhanced the spookiness of the detective magazines I enjoyed perusing. One got the impression from them that a walk in the woods anywhere in the U.S. would lead to the discovery of a body resembling the out-of-focus, rotogravure-toned corpse depicted in each article. A reminder of the growing prevalence of crime in the country was re-enforced by the many trucks laden with sand that passed the house each day on the way to Walpole where a huge prison was being built. In later years, criminals such as the Boston Strangler would call the place home.

The barn was originally designed for horses at one end and dairy cattle at the other, with a vast area above the livestock quarters where loose hay was stored. Wide doors in the middle of the barn gave access for the unloading of hay wagons, a process usually accompanied by the awakening and frantic scurrying of bats leaving the security of the large beams and flying outside to become stupefied by the sun's rays. Now there was only one resident, Nancy the Holstein cow, who would answer with a loud MOO from a distant field when her name was called. I know cows rather well: Jerseys and Guernseys are small, temperamental, and provide milk rich in cream; Ayrshires are big and clumsy with low intelligence and large teats with warty skin that discourages hand milking, yet yield copious amounts of fat-laden milk; Holsteins are intelligent and lovable and give large amounts of rich milk.

Two chicken coops, one the size of a small bungalow with a dirt floor and the other a small structure covered with tar paper, were located between a carriage house and a large grapevine entwined in an elm tree within a sloping, grass-covered field behind a small rock formation several feet high, where the mongrel Towser usually could be found on sunny days, presumably guarding several dozen Rhode Island Reds. The actual guards were two Plymouth Rock (Barred Rock) roosters who strutted around the yard like members of the Queen's Scots Guards ready to squawk if predators approached.

The approach of the fall hunting season meant that Dad remained with me at the farm on Fridays instead of returning to Needham for a weekend of card games or a trip to New York with Mom for gourmet dining and theatre. He was a handsome man in his early forties with a resemblance to FDR, even possessing the president's penchant for cigarettes. His weight, however, was approaching 240 pounds, far exceeding what the health charts recommended for a trim, six-foot male. Therefore, the arrival of the hunting season was a godsend, for he was able to walk a few miles in the woods and, away from rich food, at the same time greatly reduce his intake of fatty food.

For weapons, he brought along a 12-gauge, double-barrel shotgun for himself and a smaller gauge, single shot gun for me. Our quarry was diverse and bountiful: red and gray squirrel, rabbit, fox, woodchuck, quail, pheasant, wild duck, Canadian goose, woodcock (timberdoodle), partridge (Ruffed grouse), and whitetail deer. (Today, despite the inroads of civilization and the disappearance of habitat, one can add to this group the coyote, cougar, black bear, and even moose). Dad was an expert hunter. Partridge, quail, woodcock, and goose succumbed to his skill. On the other hand, I never shot anything. I believed that the single barrel and smaller bore handicapped my skill. Regardless, I didn't have to cook and consume the game I killed. But Dad did and enjoyed it. He ate everything except woodchuck, fox, and chicken hawk. I recall once when we were driving to Norfolk he spotted a fox several hundred yards away in a field. He stopped the car (Nash), stepped out, and shot it with a 0.22 rifle right in the eye. The mounted trophy stood guard on top of a curio cabinet in his waiting room until its beautiful red bushy tail fell off and it had to be removed because it looked like someone's dog. Dad had instructed me in the proper use of firearms: 1-Never point a gun at anyone unless in self-defense; 2-Assume that all guns are loaded until proven otherwise; 3-Use the safety until ready to fire; 4-Never fire blindly into the air; 5-Always remove ammunition when traveling in a car; 6-Never use the gun as a cane; 7-When climbing over a fence, be sure the safety is on and gun is pointed toward the ground; 8-Only shoot when quarry has been identified; 9-Never stick the muzzle of the gun into the ground; 10-Always clean your weapon immediately after use.

I learned from a relative that when he was a teenager he once fired a bullet through his mother's dress while testing the trigger of a new rifle. Fortunately, she wasn't in it.

Enjoyable as hunting with Dad was, and it continued after WWII when I shot a 340-pound whitetail deer in New Brunswick, Canada in the forties which had a huge asymmetric rack, those rare times when we went fishing are most vividly recalled. There was the Nantasket Beach episode when, using clam bait, we cast our lines off rocks at the edge of the ocean and hooked sculpins, the ugliest fish in the sea. Then there was the Medfield incident when we fished a small pond in the shadow of Noon Hill and hooked hornpout, the northern version of catfish. They were about 5 inches long and resembled a miniature piscatorial version of a killer whale, black on top and white below. They were horrible to debone with very little edible meat. One summer we stayed on a Maine island near Mount Katahdin and caught smallmouth bass from a rowboat. But the most memorable occasion was when we sailed in an 18-foot bass boat from Woods Hole, Massachusetts, to Nantucket and returned hours later with hundreds of bluefish. Using eelskin bait, each time we entered a riptide off the island, a bluefish attacked the hook like a hungry piranha. The boat, incidentally, belonged to a Mattapoisett millionaire, Eugene Boardman, who once lent money to a young man whose dream was to own a tanker fleet. A few years later, a magnificent white yacht appeared in the bay near his home and discharged a gleaming brass-rimmed mahogany launch. It was the young man returning the borrowed funds with interest. His name was Daniel Ludwig, the American counterpart of the Grecian tanker king, Onassis. I have discussed this phenomenon—the vivid retention of parental fishing experiences—with others, and find that they are in agreement, although also lacking an explanation (perhaps another example of fish influencing the brain).

In March, 1934, we again boarded the Mauretania in New York for a cruise to Trinidad, LaGuaira, Curacao, Haiti, and Havana. March is the time of year when thunderstorm activity causes bodies lying on the bottom of New York harbor to rise to the surface as floaters. Fortunately, I didn't see any as the ship plowed through the flotsam and headed for the open sea. During the voyage, however, a despondent passenger leaped off the ship into four swirling propellers. His jump was not witnessed by anyone, and his body was never recovered (he probably jumped into the Gulf Stream, also called shark alley), but a suicide note tucked in some discarded clothing confirmed his act. Cunard's record of never having lost a passenger or crew member remained intact because the line was not responsible for suicides.

Few, if any passengers were aware that within 6 months this great ship would be sailing on her last voyage to England and, by July, she would be heading for the shipbreakers in Scotland to be converted into shells and bombs in retaliation for the sinking of her sister, Lusitania, by a German submarine in 1915 with the loss of many innocent lives. The Mauretania was President Roosevelt's favorite ship. By a remarkable coincidence, an 18-foot model presented to him by Cunard was being installed at the Smithsonian on the very day the famous ship left England for Rosyth, Scotland.

What made the cruise unique, other than the eventual realization that it was one of the last for a maritime legend, was the opportunity to visit Haiti for the first time. As a 12-year-old schoolboy, I knew nothing of the country save for its reputation in horror movies as the land of voodoo and zombies. It was so insignificant to Miss Miller that she had failed to assign a study session about the island to her fifth grade class. Perhaps the zombie aspect of the place reminded her of someone on the school committee.

Our arrival in the harbor of Port-au-Prince revealed a lush, mountainous land whose beauty resembled that of Tahiti as depicted in the National Geographic Magazine. Ashore, the concept of paradise quickly dissipated as the extreme poverty of its people became evident. We had seen flagrant begging in Trinidad involving young children, but not the rampant begging of older people who obviously had adopted this behavior as a full-time occupation. These were victims of the economic travail responsible for a per capita annual income of 60 dollars. (This figure is in contrast to the per capita income of 1,820 dollars per year for a U.S. citizen). Haiti relies heavily on tourism for income because, aside from art (oil paintings and wood carvings), nothing of significance is produced. Speaking of wood carvings, as we went by taxi to a hotel at the top of a mountain for lunch, and some gambling, a fellow in his thirties followed us on a bicycle carrying a set of mahogany tables that unfolded one atop the other when fully extended. At first his price was 10 dollars. A few miles later, it was down to 5 dollars. To get rid of him we bought the contraption for 2 dollars. While visiting the so-called Mahogany Factory, we purchased statuary presumed to be of mahogany. Later, aboard ship we discovered the wood was white pine stained with oxblood shoe polish.

The most intriguing feature about Haiti was our discovery of a tiny shop on a side street that sold American and Canadian silver and gold coins, with dates almost a hundred years old, whose individual worth exceeded a year's salary of the citizenry. There were California gold coins, large octagonal pieces the size of silver dollars, that sold for several hundred dollars. I bought two ancient American dimes which became the basis of a fascinating collection that ended out of boredom about fifty years later. We spent only a day in Haiti and, regrettably, did not have time to attend a voodoo ceremony. As the ship was leaving the harbor, passing only a few hundred yards from shore, one could hear the rhythm of distant drums heralding a religious ceremony. We left Haiti not realizing that during the year

(1934), the U.S. Army, in control of the government there for 19 years, was soon to leave. The era of Papa and Baby Doc was about to unfold.

A few comments about my third visit to Havana: at age 12 I was, in Cuban eyes, old enough to enjoy the nightlife, which allowed entry to the exclusive Tropicana, with its famous dance floor of inlaid brightly-colored glass rectangles in the style of Mondrian, on which the most naughty shows in Cuba were performed, and observing the wealthy champagne and caviar crowd being suckered in the gambling casino of Sans Souci. At the Tropicana, we sat at the edge of the dance floor opposite 12-year-old Norma, of Belmont, Massachusetts, and her young, fortyish parents. The show, consisting of a dozen scantily dressed girls dancing to the staccato beat of a steel drum band, was exciting, yet the highlight of the performance was the appearance of a stunningly beautiful young girl who did acrobatics while removing her brief costume. After the last garment dropped to the floor, she smiled coyly at the audience and disappeared through a door. This act, in retrospect, was probably babyish when compared to that being offered to the Harvard, Tufts, and B.U. students at the Old Howard theatre in Scollay Square. However, back aboard ship, the prudish puritanical passengers severely rebuked my parents for allowing an innocent youth to be exposed to such depravity. Norma's parents were likewise verbally chastised with scathing comments from these virtuous idiots.

At the entrance to the Sans Souci casino was a beautiful marble fountain in which larger-than-life statues of the Three Graces, in all their nakedness, were coyly standing amid the gushing waters (as three-dimensional equivalents of the cartoon depicting the loser departing from Las Vegas clad in a barrel). Inside, the gambling rooms were dimly-lit, tastefully glitzy, and crowded with well-dressed people who appeared to be from the Caribbean area and South America, with only a handful of familiar faces from the Mauretania present. I was intrigued by the behavior of a middle-aged lady, clothed in an expensive gown, bedecked in jewels, and topped off with a sable neckpiece, who was wagering a rouleau of 1000-dollar chips at a roulette table. In the blink of an eye her wealth vanished like a pearl in a glass of vinegar. Oddly, one-armed bandits seemed to be missing. The reason for this may have been the availability of this type of gambling device throughout the United States, whose ready access in this exclusive casino would have cheapened its exotic aura.

When I returned to my classroom in Needham a few days later, I didn't tell my teacher about the most exciting aspects of my latest Caribbean tour.

CHAPTER FIVE

The Junior High Years

Upon completion of Miss Kett's class in June, 1934, at the yellow brick Emory Grover school, I awaited the arrival of fall and entering Junior High School, a nearby two-story redbrick building with the architectural charm of a civil war fort. Although just a coin's throw from Emory Grover, the Junior High attendees experienced a great leap forward, at least in terms of academic stress. Advanced student feedback annually emphasizes the difficulties associated with pupil-teacher relationships, but never more dramatically than when the naive student enters Junior High from sixth grade.

"You're not dealing with one teacher in Junior High; you're at the mercy of many," says the know-it-all advanced student (by one year) who, by now, is probably a retired politician living on a canal in Florida. "Furthermore," he or she continues, "your exposure time with each is drastically shortened, unlike the situation in elementary school when you get to know the quirks of the teacher better than those of your relatives."

In July, a wonderful thing happened: Dad bought an old, beat-up Studebaker coupe from Les Cook that was rugged as an Army tank. He broke the news while I was sitting in a bathtub soothing the sores from a bad case of chicken pox.

"You're going to learn to drive and use the car as a tractor on the farm. We'll drive to Norfolk after sundown," he added.

It was a clever move for, if he were spotted by a patient driving this wreck in the daytime, his reputation as a successful physician might be questioned. At the Cedars the old car performed numerous tractor duties as well as teaching me how to drive on dirt roads at the farm in all kinds of weather. To start required the release of the hand brake, turning on the ignition, and allowing it to roll slowly down a hill while releasing the clutch. There was no power steering, no automatic transmission, just a solid, well-built vehicle that could pass through an opening in a stone wall and dislodge a boulder without showing any visible damage to the fender. After WWII, when grandpa Jesse was long gone and the farm had been sold, the old green Studebaker was sold to a junk dealer for 35 dollars.

The three years of Junior High were eventful: Miss Goering, the math teacher, was an excellent instructor who, because of her name and appearance, was destined, in the 1940's, to be investigated by the FBI as a possible relative of Field Marshal Herman Goering, head of the Luftwaffe; Miss Russell, the English teacher, who wore men's clothing, including a shirt and tie, had a nervous breakdown after her class performed with tiny bells at an assembly; Mrs. Hatch, my fortyish civics teacher, gave birth to her first child in the presence of the Needham Fire Department, who graciously lent searchlights for the historic event when a hurricane, originating in the Caribbean, turned Needham into a vast swamp and destroyed power lines to the Glover Memorial Hospital; the Nazi zeppelin Hindenburg interrupted our baseball games on Fridays by passing overhead as a gigantic silver cigar on its way to eventual ignition at Lakehurst, New Jersey.

In the 7th grade, I enjoyed a subject called MANUAL TRAINING, which was taught by a Mr. Pelletier (pronounced Pelcher). Part of the attraction of this course was the delightful aroma of freshly sawed wood mixed with the odor of fish glue that permeated the classroom. The remainder of the attraction was the feeling of accomplishment that creating something useful would bring as a hands-on experience. Most of the subjects in Junior and Senior High were nothing more than regurgitation exercises that the curve of forgetting could quickly eliminate from disuse, leaving the student with secondary ignorance. Aside from his primary duty of seeing that students finished the year with the same number of fingers they started with, Mr. Pelletier proudly ingrained in his charges a love for tools and craftsmanship.

Each student was expected to work on a project, however small, that would be useful in the home. I made an ashtray that consisted of a little wooden man crouched in a fishing position whose head was a hollowed out cocoanut. He even held a small fishing rod in one of his carefully sawed out hands. It took only a week to finish and lasted about that long at home where it ended up in the attic. Next, I bought cedar boards, drilled holes for dowels, and glued them together creating what appeared to be an aromatic raft. After I had a dozen or so boards glued together, which bore no resemblance to a cedar chest

I was trying to construct, I brought them home because the course had ended. Dad split them apart with a hatchet and put a board in each closet to thwart moth activity.

My history teacher, Nellie T. O'Brian, straight as a fencepost, white as a piece of chalk, and capped with a powdered gray wig, imbued students on the first day they met with the impression that she had acquired her historical knowledge directly from the actual participants. She proved to be an outstanding teacher who was loved by even the most recalcitrant student. Although by no means young, she attended many of the athletic events and even had a car named for her in the soapbox derby. Most of the townspeople who were educated in Needham fondly remembered Nellie T. as a legend and one of their cornerstone educators.

My greatest turnoff in Junior High was gym which required doing silly exercises on flying rings, bouncing around on dusty mattresses, and standing in cold rooms awaiting the chance for basketball tryouts. Then, stripping, showering, and placing one's feet in that smelly green solution to kill Athlete's foot fungi I found revolting.

At the time my parents decided I should spend weekends in Norfolk with grandpa Mann, Dad fulfilled a longheld ambition to purchase a canine companion to improve his hunting skills. Thus, one day he drove home and got out of his Pierce Arrow with a silly grin on his face. Opening the rear door, he reached in and produced a pointer puppy wrapped in a blanket which wriggled uncomfortably as Mom and I ran out of the house to greet him. One look at his coloration and the pudgy little fellow was named Specky. Several months before my arrival in Johnson City, Dad had given Mom a Manchester terrier named Tootie as a Christmas present to get her mind off the rigors and stresses of my approaching birth. Also, during her pregnancy while she became more and more uncomfortable with nausea and other physiologic disturbances, he was being entertained by nurses at cocktail parties. The unfairness of this situation was the sole reason why I came into the world destined to be an only child. There is always some ignorant person who feels sorry for you because, unlike them, you had no siblings to play with while growing up. I always answer them by saying, "on the other hand, unlike your situation, when my parents have passed on, I'll get everything!"

Back to Tootie. When the folks came north after their three-year stint in Tennessee, Mom, Dad, Tootie and I comprised the passengers in the Chevy touring car which was headed for "The Cedars." I'm sure that the dog resented my presence although it never tried to show its feelings outwardly. Yet, it obviously had a stress problem, for it ballooned in size from overeating and underexercising. It was the personification of a lap or boudoir dog and so introverted that Dad bought me a little brown mongrel named Betty as a companion. Tootie lived to be 12 years old and Betty reached age 5 as both departed to the great doghouse in the sky at about the same time. This pet void in our lives was temporarily filled with a litter of white mice, a parrot, a box turtle, rabbit and duck until Dad arrived with Specky.

When our male pointer reached maturity and weighed 80 pounds, he was ready to join me at the farm where he remained with grandpa Mann as a companion for him and the aging Towser, who by now couldn't care less if chicken hawks or foxes devoured the chickens. When Specky's hormone levels peaked occasionally, he would leave the farm and explore the canine inhabitants of Norfolk causing much distress for Mom who would search for him in her Studebaker coupe. Her greatest fear was that Specky might be dognapped for someone's pet or a laboratory experiment. Usually, she would come upon the dog wearing a foolish look on his face as he danced around another dog not necessarily of the same breed. With considerable difficulty, she would get him into the car and back to grandpa who would see that he got a nourishing meal and a good night's sleep before Specky repeated the process of disappearing sometime during the following day.

Into every (Christian) child's life there should appear, at least once a year, that jolly embodiment of the Christmas spirit—Santa Claus. I was lucky—Santa Claus was a relative! It was one of those serendipitous happenings attributed to war. John Burke, a sergeant at Fort Devens, Mass., was about to be sent overseas during WWI when the Armistice was signed ending the conflict. His acquaintance with the saintly presence of nurse Bertha Reed, R.N., Grandma Rose's oldest daughter, who ministered to his violent case of food poisoning, continued on a less formal basis for a few months before they were married.

Uncle John, as I remember him in the 30's and 40's, was an Archie Bunker look-alike who dressed in vestless grey suits and wore white shirts and socks (as did the TV personality), retired each evening promptly at 10, and was overweight by at least 30 pounds. He differed sharply from Carroll O'Connor's portrayal of the famous bigot by showing a great love for humanity, especially children.

During much of their lives, Uncle John and Aunt Bertha lived in a black-stained, wood-shingled house at 34 Upland Road in Quincy, Mass., about a quarter of a mile from the purple-painted, barn-like dwelling of John Quincy Adams. John left each weekday morning (and occasionally on Saturdays) by car to work as a novelty jobber, buying toys, candy, and bubble gum from manufacturers and selling these items retail through a local store owned by Louis Pearl, the "toy king" of Lawrence, Mass. Each December, Uncle John would don his Santa Claus outfit and host the annual Christmas show at Lawrence's Palace and Broadway theatres for the benefit of the Eagle-Tribune Santa Claus Fund which was partly sponsored by Mr. Pearl. The chief recipient of the funds collected from the show were the Catholic charities without whose dedicated work many of the residents of Lawrence would either have starved or been deprived of warm clothing.

In 1932 and 1933, Uncle John invited me to participate in the show, along with my closest pal, James Davidson, who agreed to pair up with me as an eskimo named "Inkie" (derived loosely from his middle name Ingalls), while mine was "Boysie" (arrived at for no particular reason other than Dad's love of Tarzan movies). Jimmy had been my friend ever since the day he dropped by my house while I was feeding my pet chickens in back of the carriage house. The eventual day that I asked him about joining me on the stage of a Lawrence theatre as an eskimo, he readily accepted, and then waited anxiously for the arrival of December 2nd, 1932. On that date, Mom drove the two of us to Quincy where we picked up "Santa" and continued on to Lawrence. There, in astonishment, we witnessed an enormous line of kids stretching for several blocks outside the two companion theatres. The first year we made a mistake by wearing clothing that resembled what the kids in the audience were wearing, which diminished our authenticity. However, the following year we were better organized, donning white eskimo parkas and black boots.

Once on stage we responded to Santa's cue as to what we had in a brown paper bag by yanking out a huge codfish which we said, with a loud grunt, was our breakfast! The audience responded in kind with an even louder "UGH!"

Just before the stroke of midnight on Christmas Eve, the sound of crunching snow heralded the approach of Uncle John and Aunt Bertha in a Pontiac loaded with gifts. After downing an eggnog or two, we gathered around the Christmas tree and opened presents.

The most memorable Christmas was the one when a Buddy-L red firetruck equipped with rubber tires, a water tank, tiny hose, and a ladder occupied center stage. Perhaps it was the euphoria induced by a spiked eggnog or a throwback to youth that prompted Dad and John to pile wrapping paper in front of the living-room fireplace, light a match to it, and then douse the blaze with water pumped by hand from the little tank behind the driver's seat. Everything worked perfectly except for a slightly singed border to the Sarouk oriental lying just beyond the tiles of the fireplace. Twenty years later, after Dad had died, Mom donated the firetruck and several other Buddy-L trucks to the fire department for distribution to local charities. Along the same vein, she gave a magnificent moose head that graced the wall in the hallway next to the waiting room to the Quincy chapter of the Loyal Order of Moose, but due to an oversight never received a thankyou note from them. It is strange how altruistic one becomes after the loss of a loved one. As if donating prized possessions will somehow bring a person back from the dead.

Jimmy Davidson and I enjoyed a happy childhood that included building a treehouse that was reached by climbing a rope, and an underground house dug in the garden plot behind our carriage house garage. The trench-like structure was covered by several boards to make a roof and there was room for three kids if they just sat without moving. One October I removed the roof and heavy rains filled in the 4-foot deep excavation with muddy water. A neighbor's kid came by and started to step into the "puddle" even though he was warned that the "puddle" was several feet deep. Not heeding my warning, he stepped forward and disappeared for a moment into the coffee-colored water. Resurfacing, he struggled to get out of the hole before the mud walls collapsed and ran home, a firm believer in the spoken word.

During my second year of college, I saw Jimmy alive for the last time standing in the foyer of our home. Dad called me to chat with him because he was going to have his appendix removed the following day. I told him I'd visit at the hospital when I got home from college that weekend. Jimmy died shortly after the operation from an infection of gas bacilli due to contamination of something in the operating room. Every instrument and piece of equipment in the areas occupied by Jimmy were discarded and the whole place was disinfected. Thus, Jimmy Davidson, my closest childhood friend, had died suddenly just before the introduction of antibiotics into medicine that would have saved his life.

What irony that was! Several years later, during a graduate course in mycology (the study of fungi), I found a way to induce *Penicillium notatum* to produce the antibiotic penicillin more quickly by using carrot agar in place of potato agar. More on that later.

After Jimmy's funeral, which was held at his home in the presence of several hundred friends, classmates, and relatives, I reflected on some of the memories which occasionally entered my consciousness that dealt with our relationship as youngsters. Once, during the Saturday matinee at the local Paramount theatre after the cowboy film had ended and the amateur contest began, I was shocked to see Jimmy take to the stage and sing a song that was popular at the time: "Tiptoe through the tulips," which became popular again years later when TINY TIM sang it on THE TONIGHT SHOW. I didn't know that Jimmy could sing. I should have known he had talent for his sister Jean was an excellent musician and occasionally played the organ at the Congregational Church. And his introduction to show business as an eskimo at Uncle John's Christmas Show in Lawrence had removed any fear of stage-fright from his singing performance.

I remember the time his family invited me for a weekend at Dennisport on Cape Cod when we engorged ourselves on fried clam rolls and strawberry floats and then dove into the cold waters of the Atlantic. A giant wave threw us both upside down and, when he stepped out of the water, blood was streaming down his face where a stone had scratched his forehead. Jokingly I had told him, "Don't worry. When you appear as an eskimo next Christmas, you can point to the scar where the polar bear attacked you!" He didn't think it was amusing at the time.

And then there was the time when, standing in line at gym, some kid had made a crude remark that was overheard by the teacher. Thinking Jimmy was the culprit, the teacher stepped in front of him and slapped him in the face. I didn't see what happened because I wasn't in the class at the time, but the slap left an impression on his psyche. The teacher happened to be one of Dad's patients, so I never mentioned the incident. The bully boy was one of those he-man, macho-acting guys with bulging muscles and a commanding voice, who was well-suited to be a sergeant in the Army. When I see Mel Gibson, the Academy Award-winning actor and director, in a movie or on TV, I'm reminded of bully boy, for there's a slight physical resemblance. Imagine in today's litigious climate what would happen to a teacher who displays such behavior to a student?

Every time I visit Needham, which is rarely now, I go to the cemetery where my parents and friends are interred. Not far away, on the slope of a hill, lies Jimmy's grave, barely a stone's throw from Blackie's shallow pond across the street where we used to skate. When Blackie's was overcrowded, we'd take our hockey gear to Newton Upper Falls and continue our game on the more dangerous Charles River.

"Of all the ships that sail the sea, the one for me is the Kungsholm. More than just an ocean liner, no other ship could be any finer!"

This was the silly ditty that set the tone for our next cruise from New York in April, 1935, on the Swedish-American motor liner, Kungsholm, to Haiti, Cartagena, Panama, and Jamaica. Painted hospital white, immaculate inside as a royal palace, and crewed by Scandinavian athletes, this cruise ship was one of the most memorable for the great fun everyone seemed to have, so much so, that with the arrival at each port, a groan emanated from the passengers because their good times were about to be interrupted by the onerous task of getting up early, dining quickly at breakfast, and then getting into a lifeboat for the trip ashore where the natives were waiting to relieve their victims of money. This was accomplished in subtle ways such as pretending not to understand English when going by taxi to some remote place where the journey was twice as long as it should have been. Another ploy was to charge exorbitant prices for drinking water. This so infuriated my father that on one cruise he pocketed a silver pepper grinder (which we still use today) from a swanky country club in Caracas. Of course, there are legal ways to relieve tourists of their money, such as a head tax for those arriving by plane, or the port tax for those stepping ashore on some dismal island.

The Kungsholm was the cleanest ship I'd ever seen. I'm willing to bet that it was the only ship in the world that was rodent-free, unless the chefs served rabbit. I mentioned that the crew were athletes. Being of Swedish origin, this ship had a sauna and offered massages to passengers. The cost of a massage was ten dollars, which was more than I wanted to pay, so I gave the masseur a quarter and had my big toe massaged. Not toes, just a toe that I had stubbed traversing a passageway.

Our arrival at the first port of call, Port-au-Prince, Haiti, was greeted by a well-dressed native who lost no time in showing us a large diamond ring that was for sale for a couple of hundred dollars. One look at the ring and you knew if he'd dropped it on a rock, it would have shattered into a thousand pieces. A day spent in Haiti is more than enough time for one to appreciate how well off the citizens of the United States are, even during a depression. To watch aged women, who are probably only in their thirties, walking slowly downhill to the market with live turkeys on their heads, and smoking pipes held between toothless jaws, is a sight long remembered.

I wasn't mature or astute enough in those days to appreciate art, otherwise I would have insisted that my parents search for the studios of the local Rembrandts whose works, if purchased in the thirties, would have been worth more today than the baseball cards of Ruth, Hubbell, Dean, and Cobb combined. Although the style is primitive and vibrant with dazzlingly gaudy colors, some connoisseurs of the fine arts would rank them in the rarefied atmosphere of Blue Period Picassos. As usual, we left Haiti with sadness for its people. The police, known as the tonton macoute or bogey men, exerted tight control over the populace. It was the only Caribbean island where smiling faces were never observed.

When we were leaving Cartagena, after having seen all of the usual sights (except for Saint Pedro Claver's cathedral), Dad, from the stern of the Kungsholm, shouted to a couple of natives sitting in a dugout canoe that he'd like to buy their tiny boat, which was only about 5 feet long. Before they could answer, the ship started its propellers and both had to jump to safety as the little craft turned to splinters. Fortunately, no one was hurt, but the request was ill-timed, and Dad was remorseful that he couldn't somehow remunerate the swimmers who, by now, were being picked up by friends in a larger craft.

Our most memorable adventure took place in Panama where we were able to dock alongside a gigantic concrete pier in Colon. Because we had neither reached the Pacific side of the canal, nor even seen the Pacific ocean on our previous trips, it was decided that a train ride would be exciting. We boarded a narrow-gauge freight car filled with native women and chugged behind a belching locomotive moving parallel to the great ditch, past the Gatun locks and the namesake lake with tree stumps sticking out of the water, to Culebra cut (now called Gaillard cut), Pedro Miguel locks, Miraflores lake and locks, and finally to Balboa and Panama City, sitting on the edge of the Pacific. The chief attraction in Panama City was the church whose altar, now gleaming in 24-carat gold leaf, fooled the pirate, Sir Francis Drake, for the metallic surface had been painted gray in anticipation of the infamous pirate's anticipated plundering. When it was time to board the ship, Dad decided to return to Colon by a more exciting route than rails. He always wanted to fly and the sight of a seaplane floating at a pier in the harbor caught his eye and imagination. So the three of us walked down a narrow plank to the aircraft and boarded, along with six other passengers, with Dad occupying the seat next to the pilot. As the propeller spun, oil splattered over the windshield, obscuring Dad's view. Soon we were speeding over the waves and then soaring over the canal, fascinated by the dense greenery of the jungle on either side. Dad, in the meantime, was watching the pilot's every move in case he suddenly had a heart attack. Half an hour later, we arrived at Colon on the Atlantic side. We had gone 50 miles at 100 m.p.h.

If one follows closely the itinerary of these annual cruises, a frequent repetition of the ports-of-call is noted. In the depression-tainted thirties, only a few ports were deemed tourist friendly, and the stellar attraction among these was unquestionably Havana-the sun and sin city that was unequalled in its exotic appeal. When Castro and the communists took over Cuba in the fifties, Havana was quickly replaced on the cruise itinerary by San Juan, Puerto Rico, a fortress city whose history reeked of buccaneer deeds and Spanish conquests.

On this cruise, our new port was Kingston, Jamaica, where the Kungsholm docked beside a bustling concrete pier laden with crates of bananas, grapefruit, mangoes, and barrels of cocoa pods and coffee beans interspersed with stalks of sugar cane awaiting shipment to Great Britain and the States. As the gangplank was being positioned to disembark passengers, a strange sight was unfolding: a small retinue of stevedores carrying canvas stretchers were waiting to come aboard. Each stretcher held a giant sea turtle, positioned on its back with feet waving in the air, blissfully unaware that it was shortly to become the main ingredient of delicious soup.

Kingston is a beautiful city that represents a rebirth of Port Royal, the ancient pirate stronghold that slid into the sea in 1692, when an earthquake jolted the area. Some of the passengers elected to fly over the harbor in small aircraft to observe the streets and historic buildings preserved in salt water for almost three centuries. The trick is to fly on a windless day, otherwise the waves obliterate the underwater view.

In the few hours available, we were more interested in exploring the island in the traditional Packard taxi, so off we went to the east and the mountains, stopping briefly to enjoy a botanical garden fringed with luxuriant bamboo and exhibiting tropical species native to the Caribbean. In a somewhat reckless mood, I mistakenly patted the head of a water buffalo, who responded by trying to gore me. Later, I learned that they detest Caucasians (it was time to get a tan). On the return trip, we drove along a coastal road where sandy beaches were fringed with gracefully contorted cocoanut palms that looked out at an azure sea that was devoid of bathers.

"Sharks," said our driver, to answer my obvious question. "Tiger sharks are man-eaters!"

The intense heat of the afternoon sun forced us to depart for the Myrtle Beach hotel where a large outside swimming pool awaited.

The following June, Dad decided to visit Mom's relatives in Canada, although he was greatly handicapped in conversing with them, for they knew little English and he little French. He had an ulterior motive, though, for he had just purchased a new car, a maroon Pierce Arrow the dealer said could do at least 120 m.p.h., and he was

itching to try it out in Canada, having read in the Boston Herald that a
new six-lane highway had just opened outside of Montreal. Also, he
decided to take Uncle John and Aunt Bertha along with Mom and me.
The relatives, especially John, were lots of fun and their presence
would take the attention off his romance language deficiency, for they
knew fewer words than he. After swigging a couple of rounds of
Canadian ale, who'd care?

We left Needham at daybreak for our first stop overnight in
Burlington, Vermont, at the old hotel Van Ness, which was across the
street from another ancient dwelling known as the hotel Vermont. On
that street, however, were three great attractions for Dad and me: the
Flynn theatre, which was showing "Earthworm Tractors," a film
starring my favorite comedian, Joe E. Brown; a pawnshop which Dad
loved to peruse looking for weaponry and old musical instruments;
and, at the western end of the street, a pier from which the ferry
Ticonderoga sailed across Lake Champlain daily to Plattsburgh, New
York, where the geologic attraction, Ausable Chasm, with its rugged
Potsdam sandstone walls, appeared as a sort of mini-Grand Canyon.
Also, in the city square, there was a large cast iron fountain brimming
with lake trout the length of baseball bats. Add to these unusual
features Coggin's restaurant, which served a delicious breakfast, the
Black Cat Nightclub restaurant, specializing in seafood (lobster
stuffed with lobster not crabmeat), and two other theatres showing
cowboy films exclusively, and one will understand why, until I saw
the bay area beauty of San Francisco, Burlington was my favorite city
in the United States.

It consumed most of the day to reach Burlington, an urban oasis in
the midst of the agricultural solitude engendered by the presence of
miles of hayfields and the occasional dairy barn with its resident cows
blissfully munching cud (when most of them were lying down, rain
was in the offing). The weather, however, was sunny and the day was
particularly long because it was approaching the summer solstice
(June 22nd). We arrived after 6:00 p.m., washed, and headed for the
Black Cat Cafe. After a sumptuous meal of lobster, we headed for the
Flynn theatre as the curtain was ascending on Joe E. Brown's
hilarious comedy about Alexander Botts, a Caterpillar tractor
salesman, whose adventures appeared regularly in the SatEvePost. By
11:30 we climbed into our beds and shortly were dreaming about

ribbons of highways. In the morning, we had breakfast at Coggin's and then ventured down to the pier where a half-sunken gunboat, the *Philadelphia* from the Revolutionary War, was being salvaged. It was filled with mud along with black cannonballs and an occasional white human skull. An hour later, we left for Montreal. *Now at the Smithsonian.*

A few minutes after leaving the city limits of Burlington, we passed through the small town of Winooski on a road that was elevated because of occasional floods. On the left side of the road, I noticed a barn-like structure with a huge dusty window through which a number of rectangular shapes could be seen that were obviously caskets piled haphazardly. A village this small must have an unusually high death rate with so many caskets available. Soon, we arrived at the beautiful town of St. Albans where long ago Grandpa Jesse had married Grandma Marion. We had lunch in a restaurant owned by a Mr. Pryor, a gentleman whom we met a few years later on a cruise accompanied by his wife and daughter, Frances. Mr. Pryor, at the time of our lunch, was nowhere to be seen. The pleasure of his acquaintance awaited several more years.

Now came the thrill of the trip. Having crossed the border and passed through Customs, we were soon speeding along a 6-lane superhighway on our way to Chateauguay where Mom's relatives resided on several farms. Suddenly, Dad stopped the car and asked: "Would you like to take the wheel?"

"Wow!" I jumped out of the car and climbed into the driver's seat. Then, carefully releasing the emergency brake, I pressed the accelerator pedal and the car took off, soon reaching a speed of 85 mph. The relatives in the rear seat along with Mom were beginning to get nervous, so I slowed to the fifties and continued on for a few minutes before stopping and turning the wheel over to Dad. At the next gas station everyone went to the bathroom while the tank was being filled with Imperial gallons of fuel. It was an exciting experience and certainly explains why some people are enamored enough with horsepower to become race car drivers. It may also explain why the nearby town of Winooski had so many caskets in abeyance at the local warehouse! A few hours later, we arrived at the Caughnawaga Indian reservation beside the Lachine rapids, where Mom and Aunt Bertha purchased several handmade baskets woven by the women. Without the construction skills of their fearless husbands, there'd be no Woolworth, Chrysler, or Empire State skyscrapers. These native Americans were the foremost high-altitude steelworkers in the world, scampering hundreds of feet above city streets to assemble the world's tallest buildings. Yet, they lived during the winter months in single-story homes on a Canadian prairie near the St. Lawrence river. In the off season, the women-folk wove exotic-looking baskets, which they sold to tourists in the summer. On the return trip, having disposed of several boxes of gifts to the relatives, enough space remained in the car's trunk for a basket the size of a Sinbad amphora, to be used as a laundry hamper.

Without experiencing the formalities of passing through Customs or noting the messages on highway signs, one would still be aware of traveling in French Canada if only for two striking influences of Catholicism evident in the elaborate architecture of the cathedral-like houses of worship, which differed sharply from the clapboard and brick and mortar simplicity of the churches of rural New England, and the presence of outdoor chapels at the crossroads near small villages. Fifty years later, although these features were still present, the first signs that we'd left the United States were roadside stands selling firecrackers and speed limits posted in kilometers per hour.

As we approached the village of Chateauguay, I thought about the people I was about to meet for the first time. Mother's father, Napoleon Reed, who lived in Canada as a child, had a brother, Adolf, who remained in the Province of Quebec, retaining the Scottish surname Reid. He had five boys named Felix, Eubald, Alfred, Albert, and Eugene, all of whom followed their father's farming profession except for Felix, who hated working in dirt and chose a career in the fur industry instead. To learn this vocation, he came to the States where he stayed at Grandma Rose's home in Quincy, Mass., and commuted to Boston to learn how to manufacture fur apparel. During this period of several years, he learned English and taught French to Alma and Lillian, while Bertha was away studying nursing at Quincy City hospital. These were the sad days after his Uncle Napoleon had been summoned to the granite quarry to repair equipment on that fateful New Year's Day when an explosion changed their lives.

Returning to Canada after completing his apprenticeship, Felix settled in Montreal and started a fur business in a building that had a storage vault in the basement, a show room on the first floor, and two floors above where coats and hats were manufactured (92 handsewn pelts of mink were required to make one full-sized coat). While his business was growing, Felix met and married Yvonne, a beautiful young lady, who spoke French fluently and understood spoken English, but could not converse in the latter. Although childless, Felix and Yvonne adopted one of Eugene's children when he died of tuberculosis while in his forties, and Felix helped financially when his brothers required help during the lean years of farming. The farm we were about to visit belonged to Alfred and Philomena and their three daughters, Cecilia, Denise, and Juliette. It was a treeless expanse of several hundred acres of gently sloping terrain next to a brook and included, besides the two-story, wood-shingled farmhouse, a shed containing milk cans and a DeLaval cream separator, a large wooden barn with stalls at one end for four horses, stanchions for 30 cows at the opposite end, and a thick-walled pen where a bull paced nervously awaiting work duty. To this day, I will rank the odor of that barn as being far superior to that of any French perfume I have sniffed.

It was early evening when we arrived at the little white farmhouse in Chateauguay. Several cars were parked near the milk shed, and the house seemed to be bustling with people as Dad parked the Pierce Arrow beside the other vehicles. Suddenly, the door swung open and half a dozen people came out waving their arms excitedly and shouting something in French. Then, we heard some words of greeting spoken in English.

"Felix and Yvonne," cried mother. And then a torrent of French followed this greeting while Dad, John, Bertha, and I smiled and shook hands with the eager group of relatives. Having met everyone, John lit a cigar and went to the car to bring gifts for the happy crowd who by now were settling down in the large kitchen and pouring Canadian Black Horse ale for everyone. When I tasted my glass, the bitterness ignited my tongue and I decided to add a little sugar, whereupon the resulting chemical reaction caused the drink to explode and hit the ceiling. No one noticed this indiscretion, for Uncle John was handing out little Japanese toys and games to the children and Mom and Dad were giving the adults clothing from the States. I studied the girls, Cecilia, Denise, and Juliette, trying to determine their relative ages, finally deciding that Cecilia was the oldest and

Juliette the youngest. Denise was probably a year or two younger than Cecilia who looked like a teenage version of her mother, Philomena. Alfred was a tall, raw-boned, balding man who looked every inch a farmer with large hands and calloused fingers from years of hand-milking cows and holding leather straps while plowing and harvesting with his team of Belgian stallions. Philomena was tiny, wore thick glasses, and had a white cap over a head of hair that sat on her head like another cap. She seemed to smile all the time as though it were a defense against not understanding English.

Felix was well-dressed, wore glasses, sported an Adolphe Menjou mustache, and stood a little shorter than his wife. She had a pretty face, nice complexion, and a pleasant smile which, unlike Philomena's, reflected an understanding of the English language. The other guests were Albert, Eubald, Eugene, and their wives and children who came from the nearby farms with the exception of Eubald, who had forsaken agriculture to live in Montreal and work as a furrier with brother Felix. The surprise of the evening occurred when the party ended and the folks got in the car, along with John and Bertha, and Dad said to me: "We're going to spend a couple of days with Felix and then take Cecilia and Denise back to Needham with us for a week. How would you like to spend a week on the farm?"

"Oui," I replied.

While the folks were spending two days visiting Felix and Yvonne along with John and Bertha, I was enjoying the hospitality offered by my Canadian relatives, which involved awakening at 5:00 a.m., eating a hearty breakfast, milking several cows (my wrists got tired after the first one), and watching Juliette pour fresh milk into the DeLaval cream separator, whose two spigots spewed forth cream from one and milk from the other. I was also bemused by the family collie which understood French and responded to the command *"a la vache,"* by chasing aberrant cows, or acting likewise with horses at the order of *"a les chevaux."*

When the folks finally arrived to pick up Cecilia and Denise for the long trip back to Needham, I learned that John and Bertha had decided to take the train to Boston because of sudden business demands. (Actually, they were afraid I would do another driving stint at Winooski). As the girls were climbing into the rear seat, Dad glanced at the front of the house and observed that the shades were drawn. Turning to Mom, he remarked offhandedly: "Tell Alfred it would be healthier if sun were allowed into the kitchen." Mom nodded her approval and conveyed the message to Philomena without going into details as to why it was a good idea. Dad's specialty in pulmonary medicine had kicked in as he recalled that some of his TB patients in Rutland and Tennessee had come from situations where dampness and darkness were partly responsible for their condition. Sadly, this warning proved to be too late to save Juliette, who succumbed to the inroads of this disease before she attained thirty.

In the years that followed my week-long adventure in Canada on that wonderfully exciting farm, Felix and his fur business prospered and he soon became a millionaire when such a financial accomplishment was as rare as seeing a brown polar bear. By the 1970's, he owned the second and third largest fur stores in Montreal, lived in a beautiful home in a posh section of Montreal, and vacationed abroad, enjoying the scenery of Europe in a chauffeured limousine. In 1956, when Dad and Mom drove my Scottish-born wife, Mary, son David, daughter Caryn and me to Montreal on our way to Quebec City for a transatlantic crossing to Great Britain on the Home Line's Homeric, we visited Felix and Yvonne for several hours. They had just returned from a European trip which included a visit with the Pope. On the mantle of the fireplace was a photograph of Felix and Yvonne standing with the Pope, obviously having a private audience with the prelate. The little lad who had commuted from Quincy to Boston years ago had attained a level of business success that was a tribute to the opportunities afforded in the United States and Canada to those with the intelligence and diligence to pursue them. We were fortunate to have visited Felix and Yvonne in the 70's, a few months before she was stricken with a fatal disease. At that time, although in his mid-eighties, Felix still worked six days a week in his office to supervise his staff of salespeople, designers, artisans, and TV advertisers.

In September, 1935, I entered the 8th grade at the brick fort on Highland Avenue, the escape route taken by the Millen-Faber gangsters in February, 1934, after killing Needham's popular police officer, Forbes McLeod. Another shooting had been reported that year in the spring and again in the fall, but it was recorded only in the local newspapers, not in the Boston papers, although the town boasted among its residents two outstanding journalists, Joseph F. Dinneen, who worked for the Boston Globe, and the other, Bigelow Thompson, for the Christian Science Monitor. Mr. Dinneen had authored an article about the Needham bank robbery entitled *Murder in Massachusetts,* which appeared in Harpers Magazine. Another article, about the sensational Brink's robbery, was made into a movie entitled: "Six Bridges to Cross," starring Tony Curtis. Who was involved in these later shootings and why were they covered up?

It seems they were relatively innocuous. As kids crossed old lady Ryan's land near Bird's Hill on the way to school, their raucous behavior disturbed her to such an extent that she filled her double-barrel shotgun with rock salt and let them have it as they crossed her property. She never hit anyone, partly because her eyesight was on a par with that of a rhinoceros and, deep down, she probably liked the kids and pointed the gun at the sky, which explained why her backyard was covered with brown vegetation due to salt contamination of the soil. She was also mentally unbalanced. A number of years after these initial incidents Dad, as chairman of the Board of Health, had to investigate the sanitation of her property and was shocked, when he opened the door of the chicken coop behind her house, and frightened half a dozen hens pecking at the decomposed body of her supposedly long lost brother. The undertaker, who up until that time had been a friend of the family, never forgave Dad for the indiscretion of calling him to pick up the corpse.

In the 8th grade I encountered two great teachers: Nellie T. O'Brian, the historian, and Mrs. Hatch, the civics instructor. It was the latter who held up a piece of paper one morning and asked the class to identify it. No one did, so Mrs. Hatch informed the students that this was the new look of the dollar bill with a pyramid whose summit displayed an eye and a background of colorful threads to thwart counterfeiters. Also, the size was a bit smaller than the one in current circulation with a shade of green that appeared to be lighter. I had read somewhere that there were only 12 engravers in the United States who were skilled enough to produce new dollar bills and of these 10 were in jail. As I looked around the classroom, I wondered if my generation would have the skill to engrave in such fine detail the magnificent scroll work the new dollar bill required. A few years later, I saw an FBI exhibit of counterfeit bills and was appalled to see one of a dollar bill laboriously handdrawn in ink. It must have taken the idiot many hours to "make" a dollar.

In the February, 1936, edition of the weekly Needham Chronicle, the following item was reported:

"Dr. and Mrs. David E. Mann and son David sailed Wednesday from New York on the Empress of Australia for a sixteen-day Central American cruise, their ports of stop including Bermuda, Martinique (St. Pierre), Port-au-Prince (Haiti), Jamaica, Havana and Nassau. Before leaving, Mrs. Mann entertained her contract (bridge) group at a luncheon on Monday."

Also, during the same month the Chronicle listed those students who had made the 8th grade honor roll for the January-February marking period. The recipients of the 85+ average grade in each subject were 9 girls and 7 boys, one of whom was a peripatetic youth named David Mann.

A close friend of the family, Mr. Lester (Les) Cook, also of Needham, greeted us at Boston's South Station when we boarded the train of the New York, New Haven and Hartford railroad, waving as he climbed into the cab of the locomotive where his job required stoking coal into the oven until the train reached New Haven, whereupon the electrified system of the Pennsylvania railroad took over for the remaining miles to New York City. Les Cook was a tall, handsome man who was often mistaken for Dad's brother, in his later years, when he worked for the Chrysler Corporation at its Needham dealership. He took great pride in his appearance and always dressed in fashionable clothes. Amazingly, when we reached New Haven, he was still immaculate, although he had stoked several tons of coal into the insatiable oven of a 4-6-2 Baldwin locomotive. Les Cook played a leading role in the lives of the Mann family and even accompanied me on my first date with my future wife when we drove to Dedham to see a midget auto race. He delighted in telling about the time he drove the film star, Dorothy Lamour, from Logan Airport to Needham in a canary-yellow Dodge convertible for a bond drive during WWII. "My closest proximity to greatness," he'd proclaim with a hearty laugh.

David E. Mann, Jr.

The "Empress of Australia", clad in white like the old Mauretania, was launched in Germany in 1913 as the black-hulled "Admiral von Tirpitz," for the Hamburg-America Line. After WWI, she became part of reparations to Great Britain, eventually plying the Pacific from Vancouver to the Orient as one of Canadian-Pacific's magnificent Empress ships. Eventually, she was transferred to duty in the more lucrative Caribbean area where Canadians, Australians, Welsh, English and Scots comprised her splendid crew.

I mention their countries of origin because I found these people, some of whom were less than a decade older than I, more interesting than many of the passengers. I recall one passenger, a lady of obvious wealth, bejeweled ostentatiously and in her sixties, who had no idea why the ship moved, and rudely ignored my effort to explain how propellers functioned On the other hand, the sailors had stories to tell, some of them unprintable, but all tremendously interesting. For example, I was surprised that piracy still existed in certain parts of the world. Bias Bay in the China Sea was a good example of pirate-infested territory. Yachtsmen sailing in that area disappeared never to be seen again. Anyone reading a modern yachting magazine realizes that certain parts of the Caribbean are extremely dangerous as pirate strongholds. And yacht stealing is commonplace in the waterways surrounding Florida, where multimillion dollar boats are stolen by drug traffickers who, when their job is done, scuttle the vessels to get rid of incriminating evidence.

On one occasion, I was wandering among the huge anchor chain links on the bow deck when I spotted a small door at the base of the foremast. I glanced around to see if anyone were watching, then, finding the coast clear, I lifted up the metal latch and slowly opened the steel door. There was a loud sound of rushing air followed by a booming voice from high above: "Either close the door and go away, or shut it and come up!"

I chose the latter and proceeded to climb nimbly up the ladder to a large opening for entry into the crow's nest.

"Welcome to my penthouse!" cracked a loud voice in an Aussie twang. And then an arm reached out and pulled me into a crescent-shaped structure with a swinging seat upon which a sailor in his thirties sat rocking with the motion of the ship.

"I'm Roger Sicklin," said the occupant, shaking my hand while looking straight ahead at the open sea. "You'll excuse my inhospitality, but if you bear with me, I'll show you what I do up here."

During the next few minutes, Mr. Sicklin showed me how he alerted the officers on the bridge behind him to ships on the horizon. A ship seen on the left or port side of the bow required a single pull of the cord attached to a large brass bell positioned over the canvas canopy of the crow's nest. A ship on the right or starboard side required two yanks, and one straight ahead off the bow was three pulls. With each yank a loud clang rang out followed by a similar but fainter clang from the officers on the bridge in acknowledgment. When I told Mr. Sicklin that I was president of the Sharp Eyes Club in 2nd grade he said: "Go ahead, see if you can spot one before me."

After a few minutes, I saw something on the port horizon and called his attention to it. "Right on, matey," he cried, and gave me an enthusiastic pat on the shoulder. He told me how, during WWI, submarine captains aimed their torpedoes at a ship's hull where the mast was positioned. An explosion there sinks the vessel more quickly because the hull's strength is dependent upon steel beams that provide support for the outside plates by radiating outward from the base of the mast. He also said that when such an explosion occurred at that location, the poor soul inside the crow's nest would be blown hundreds of feet into the air from the concussion. (Like Hugo Zacchini's cannon act at the Ringling Bros. and Barnum & Bailey circus in the thirties. The last time I attended a circus in Boston Garden I sat in an upper row beside a man dressed in white. I asked him if he were in the circus and he replied, "I'm in the last act." It was Hugo).

Just before WWII broke out in Europe in the fall of 1939, my family entertained Roger Sicklin with a home-cooked meal when, as a crew member of a British warship docked in Boston, he telephoned that he had a few hours to spare. We were thrilled to see him at our doorstep an hour or so later. Grateful for our hospitality, he gave us a Union Jack the size of a tablecloth from a British destroyer. As we said goodbye, he promised to send me an authentic boomerang when he returned to Australia. We never heard from him again, probably the victim of a Nazi wolf pack lurking off the East coast.

Every cruise I've experienced, the Line has always invited passengers to tour the bridge and engine room. On this trip, Canadian-Pacific went a step further and asked me if I would like to steer the ship. The helm was a large mahogany wheel set next to a brass binnacle containing a large compass. In front of the helm was a teak floor mat upon which the helmsman (or quartermaster in the Navy) stood peering through thick glass windows usually specked with salt from the sea spray. I said, "no thank you," and have regretted being so stupid ever since. Today's ships do not have helms of that size, for steering is accomplished by a computer linked to satellites. Exact ship position can be determined in fractions of a second. Regardless, once a week the ship's officers are required to "shoot the sun" with sextants as part of maritime tradition. The new port on this cruise was St. Pierre, Martinique, where the magnificent volcano, Mt. Pelee, dominated a harbor filled with jelly-fish. On May 8th, 1902, it erupted, killing over 30,000 inhabitants. This historic event was undoubtedly the first proof that crime actually pays in some instances, because the only survivor was a prisoner confined to a basement cell, the location of which protected him from deadly gases.

CHAPTER SIX

Dad's Patients

In a town the size of Needham, one would hardly expect the people who made up Dad's practice to be famous personages, many of whom would make significant contributions to humanity. Perhaps the most outstanding of these people was an MIT dropout named Westcott, who lived in a two-story, brown-shingled bungalow in the pine woods of Dover, across the Charles River from Needham. On Saturdays and school holidays, I often accompanied Dad on house-calls, which I found very interesting and sometimes sorrowful. Once he emerged from a home with tears in his eyes to relate that his patient, a young girl, was dying from lockjaw (tetanus). If the patient's prognosis were favorable, I would sometimes be invited into the home for cookies or cake. On the occasion I met Mr. Westcott, I was not only invited into his charming bungalow, but also explored his laboratory on the second floor. After he showed me his collection of black and white nude photographs (his hobby), he told Dad and me about one of his recent projects. He had been commissioned by the Kalmuses of Hollywood to improve a new kind of film called technicolor. The entire process was extremely intricate, for it entailed the use of three cameras which filmed the same scene in each of the three primary colors; red, yellow, and blue. The underlying weakness of the film was its failure to pick up each basic color accurately, which meant that the light did not diffuse properly onto the film. He related how he had used bits of aluminum foil pasted on a glass lens in different shapes, such as squares, stars, circles, etc., without success. Then, he had inadvertently dropped the lens and, having tried everything else, picked up the jagged pieces containing the aluminum strips and rearranged them haphazardly on a fresh lens. Lo and behold, they worked perfectly. Natalie Kalmus and her husband, Herbert, were ecstatic and sponsored Mr. Westcott's tour across the country in a freight car that processed the technicolor film shot experimentally in different parts of the country. Most astonishingly of all, Mr. Westcott told a hitherto unrevealed secret: "Mrs. Kalmus, although she became the technicolor director on many films, was actually color-blind! And color-blindness is presumably rare in the human female." I thoroughly enjoyed meeting Mr. Westcott, a stocky, balding man in his fifties who, like many geniuses before him, had been helped by serendipity.

One of Dad's patients was an eccentric young widow who, having inherited wealth, bought a small farm in Dover (not far from Mr. Westcott's laboratory) and populated it with several children, Sicilian donkeys, riding horses, and an assortment of exotic fowl. On one of his frequent visits to this mini-estate, he was in a second-floor bedroom treating one of her kids when he happened to look out the window and saw the mother siphoning gas from his Buick. He said nothing, but Mom, who did the billing, adjusted the fee accordingly. In fact, the young widow paid a premium price in the days before premium gas existed.

To counter this event, a chaffeured car entered the yard one day and out stepped the driver bearing a Belgian-made shotgun in a wool-lined leather carrying case. Handing it to Dad, he said: "This is from my boss, Bancroft Davis, who doesn't think you charge him enough for house calls!" The gun was valued at $400 in 1940 dollars.

A few years after these experiences, a woman confronted Dad during a routine office visit and broke down in tears. The reason for the emotional outburst was the forthcoming marriage of her daughter to a man she despised. Although the solution to this dilemma belonged in the hands of a marriage counselor, Dad tried to dissuade her from doing something irrational. Disregarding his advice, she tossed her antique silver heirlooms into a well. Somewhere in rural Needham, Dover, Medfield, Wellesley, Westwood, Walpole, or Dedham there is a well containing a small fortune in silver. And drinking from this well could cause argyria or silver poisoning! Hiyo Silver!

I've already mentioned Mr. Joseph F. Dinneen, the Boston Globe's star reporter, who lived in Needham. Mr. Dinneen was born in Boston's old North End at a time when the Irish there were becoming politically viable. One of his several best-selling novels, "Ward Eight," reflects many aspects of this phenomenon. He gave Dad a first edition of the book inscribed as follows:

"To Dr. David E. Mann, who can deliver babies even better than I can deliver books. He brought some of mine into the world."

With the best wishes of the author.

Signed Joseph F. Dinneen

Boston. Oct. 14, 1936

Mr. Dinneen accompanied another famous Needham resident, Walter Queen, chief engineer of the ship "Bear of Oakland", which carried personnel and supplies to the South Pole for Admiral Richard E. Byrd. It was Mr. Dinneen's responsibility to report on the suitability of a newfangled vehicle, equipped with huge balloon tires, to maneuver through the polar ice fields. This invention was the prototype of today's monster trucks.

Needham's most famous citizen was N.C. Wyeth, the illustrator, artist, and patriarch of the Wyeth dynasty, who was born and lived on South Street before moving to Chadds Ford, Pennsylvania, where his brilliant career was ended by a train which also killed his grandson. N.C. was not a patient of Dad, unlike his brother Stimson who was a frequent visitor to the office at 921. I never saw N.C., for he had moved to Pennsylvania prior to our arrival in Needham. But having viewed photographs of him, Stimson could well have been his twin. He taught French at Wentworth Institute on Huntington Avenue in Boston (a stone's throw from the Museum of Fine Arts) and had a son, John, who was several years younger than I. N. C. Wyeth immortalized Stimson in some of his illustrations for Treasure Island, one of which depicts his brother as the pirate, Blind Pew, one arm outstretched the other grasping a cane, leaving the Admiral Benbow Inn.

In the 1940s, Needham's economy was partly dependent upon the sale of gravel which, in the early part of the century, had been used to fill in the swamps of Back Bay and the marshes of Cambridge and, during the war years, for highway construction. Industry, such as Carter's factory, which manufactured underwear and pajamas, and the Old Trusty dogfood company, provided a substantial boost to the economy while truck farms, vineyards, and dairy farms afforded the least. But a dairy farm on the outskirts of Needham, the *Walker-Gordon facility,* pioneers in the use of the rotolactor, was world renowned. The rotolactor was a slow-moving, round platform, a kind of merry-go-round, on which cows were gently bathed with warm water, dried, and milking machines were attached to their udders. When the milking was completed, the metal suction units were automatically released from the teats, sterilized, and ready for the next cow as the milked one left the platform to return to her stanchion.

Walker-Gordon's acidophilus milk was the favorite non-alcoholic beverage of President Franklin Delano Roosevelt and was so popular throughout the country that the dairy had to expand by purchasing acreage in New Jersey and constructing barns with the square footage of airplane hangars. At the New York World's Fair in 1939-1940, the rotolactor was one of the most popular exhibits, eclipsed only by RCA's presentation of television which proved to be a more effective "milker" of the public.

Weeks after I was graduated from high school in 1940, I accompanied Dad who was making a house-call at a stately mansion that seemed out of place next to a large dairy farm. After a half-hour visit, he came out of the mansion with a big smile and jumped into the car.

"Do you know who lives here?" he asked.

"No, I have no idea. But he must be a big shot from the size of his house."

"It's Irene Walker. I told her about your graduation from high school and plans for college and she gave me this check for you."

It was for 25 dollars. That evening I wrote her a thank you note. I never met Mrs. Walker in the few years I remained in Needham. Her (deceased) husband had helped to make agricultural history with his innovative approach to dairy farming, and yet his accomplishments appeared to go unrecognized. I wished I could have met them both.

Two people I did meet, and within a week of each other, were Joe Dobson and Mal Hallett. Each was standing in the foyer chatting with Dad who summoned me from the adjacent living-room where I was reading the Boston Globe.

"I'd like you to meet Mr. Dobson of the Boston Red Sox.," said Dad. "This is my son, David."

I shook his hand firmly (athletes hate to shake a hand that feels like a rubber glove filled with cold cream) and replied: "I'm pleased to meet you." I had seen Mr. Dobson pitch at Fenway Park and knew he was one of the better players on a staff dominated by lefty Mel Parnell. I was at a loss for words, however, to continue a meaningful conversation.

"Pleased to meet you," I muttered again and quickly retreated to the living-room.

Joe Dobson was a member of that great Red Sox team that included Birdie Tebbetts (catcher), Billy Goodman (first base), Bobby Doerr (2nd base), junior Stephens (shortstop), Johnny Pesky (3rd base), Ted Williams (left field), Dominic DiMaggio (center field), and Rudy York (right field), who suffocated at the Miles Standish hotel in Boston while smoking in bed. The pitching roster also included Dave (Boo) Ferris, Mel Parnell, and Jack Kramer, whose throwing skills, along with a handsome visage, packed the stands with femininity on Ladies Day. The first time I saw the Red Sox in action was at an exhibition game at the Bainbridge, Maryland, Naval Base, in the late forties, when I was a sailor in boot camp recuperating from cat fever. From a wheelchair I watched them play the Washington Senators. The Senators won and "Boo" Ferris was the losing pitcher.

The other celebrity, Mal Hallett, was New England's most famous band leader. His band had a reputation for being in existence longer than any other in the country. It boasted in its organization such outstanding musicians as pianist Frankie Carle, trombonist Jack Teagarden, vocalist Teddy Grace, and drummer Gene Krupa, who played in other bands before starting their own. Mr. Hallett was a dashing individual who could have become a stage or screen star. As a successful bandleader, he commanded the respect of his peers as an innovative and competent musician; the band, however, generated so many decibels that bookings were restricted to large dance halls.

It was a great moment to shake the hand of this dynamic man. I was secretly proud of Dad that these famous people had entrusted their health to him.

The other end of the spectrum, with respect to human accomplishment, was also demonstrated by several of his patients. One evening there was violent knocking and shaking of the front doors as a patient tried to get into the house after office hours. Dad arose from his comfortable chair in the living-room, where he had been reading the Boston Herald, and stepped into the foyer. As the shaking continued, he unlocked the doors as I was coming down the stairs to find out what the ruckus was about. A well-dressed man with a red face entered and I got my first look at what Dad later described as a paraldehyde addict. His most outstanding attributes were nervousness and a terrible odor to his breath. Years later, when I discussed sedative-hypnotic agents with my pharmacy and dental students, I mentioned chloral hydrate and paraldehyde, the old-timers that are still used occasionally, and thought about the paraldehyde addict with the red face. According to Dad, the man had started out as an alcoholic, but his wife had carefully monitored his access to supplies of wine, beer, and liquor. Complaining of insomnia, he had somehow convinced a pharmacist that he needed paraldehyde to get a good night's sleep. In short order his addiction to alcohol was transferred to a psychic and physiological dependence to paraldehyde. Somewhere along the way he had obtained prescription blanks from a doctor's office. That was his subtle reason for coming late to Dad's office, hoping that he would find blanks lying around he could steal when no one was looking. Aware of this tactic, also employed by the morphine addict, Dad opened the front doors and bid him *adieu.*

"You've just seen a rare bird," he remarked. "A paraldehyde addict."

"How can anyone take a drug with an odor like that? It smells like embalming liquid."

"The body's need for the drug, whether it's alcohol or paraldehyde, is so overwhelming that the disagreeable taste and smell are not enough of a deterrent for the addict to stop this behavior. And, at this stage, it's only a matter of a few months before he departs from this world. The other old-time sedative-hypnotic agent, chloral hydrate, is the infamous Mickey Finn. When added to an alcoholic beverage, it forms chloral alcoholate,—knock-out drops."

Although I didn't realize it at the time, my interest in a career in pharmacology was in its embryonic stage.

There was another memorable patient whose behavior I recall from Dad's actions. At 3:00 a.m., when the house was quiet except for the humming sound of the basement gas heater, the telephone would ring and shortly I'd hear Dad getting out of bed and dressing.

The next morning I'd ask him what happened and he'd reply: "It's Mrs. so-and-so. She drank too much and fell down the stairs like a wet towel. She didn't hurt herself except for a few bruises."

This event occurred once a month until Mrs. so-and-so was finally sent to a rehab hospital where she eventually succumbed to a severe disulfiram (Antabuse) reaction. This drug is given to addicts whose systems have been purged of ethyl alcohol for at least 12 hours. Usually the regimen entails administering several hundred milligrams of Antabuse at least 12 hours before the anticipated imbibition of alcohol. Ten minutes after consuming an alcoholic beverage, a severe reaction develops of nausea, vomiting, flushing, sweating, rapid heart beat and respiration and rarely cardio-vascular collapse and death. Antabuse blocks the oxidation of ethyl alcohol at the acetaldehyde stage. With the accumulation of this metabolite in the blood stream, a series of toxic reactions occur which serve as a Pavlovian reflex to thwart any additional drinking by the addict. If, during Antabuse therapy, a nurse walks by with alcohol vapors emanating from a tray and they are inhaled by the patient, a typical reaction occurs even though the alcohol was not taken orally. This is what happened to Mrs. so-and-so who received a back rub with alcohol while on Antabuse therapy. The sulfur-containing oral anti-diabetic drugs can also cause a similar violent episode in the presence of alcohol. Antabuse, which chemically is tetraethylthiuram disulfide, is a relative of ANTU, a thiourea-containing agent used as a rodenticide (rat poison). When added to bait, this drug can wipe out a rat colony by causing pulmonary edema (fluid in the lungs).

The town's self-styled comedian was Frank Quinlan, co-owner (with his father) of a Texaco service station on Dedham Avenue. Frankie, as he was called by those older than he, had the perpetual demeanor of a winning football coach. Filling a tank with gas at his station served a dual result: the vehicle was energized and so were its occupants with his refreshing humor and manner. The nearness of his station to an Italian community, whose farms and knitting mills were barely a stone's throw away, enabled Frankie to acquire their native tongue with such proficiency that, although he was of Irish heritage, he could easily have fooled the immigration officials at Ellis Island as a native of Naples. This skill was put to practical use by such organizations as Rotary whose impresario hired Frankie as an impersonator of famous Italians. The act went something like this: the chairman of the entertainment committee of a local Rotary Club would introduce the uniformed Frankie as a pilot who flew around the world in a squadron led by General Balboa to publicize the opening of the Chicago World's Fair in the thirties. Applause followed and Frankie would bow graciously and begin to extol the greatness of the United States in a cultured Italian accent. After the audience had settled down with satisfying looks on their faces, *Giuseppe Baldini,* as Frankie was called, would say: "But Italy under fascism leaves this country in the dust! Mussolini has brought a renaissance to my country and made the trains run on time!"

The audience would suddenly gasp and soon become hostile, with booing becoming louder and louder. When the situation seemed desperate, the chairman would stand up and smile saying: "I'd like you all to meet Frank Quinlan from Needham!"

Thunderous applause would follow as Frankie thanked the audience for their patriotism and invited them to drop by his station on Dedham Avenue.

Frank Quinlan will be remembered for his beaming countenance, charm, unusual comedic talent, and for unknowingly making available his parking place in front of the Needham Trust Bank for the Millen-Faber bandits on that fateful day in 1934. So precise was the planning of the robbery that the eventual nefarious activity was obscured by a local train that arrived just as officer Forbes McLeod, summoned by the bank's silent alarm, crossed the tracks and was killed in a barrage of machine-gun fire. My recollection of the tragic event, apart from the student incident in Miss Kett's class, was looking through the grocery store window across from the bank the next day where, standing alone on a shelf, was a box of Wheaties with a bullet hole in its side.

Mr. Quinlan was rarely a patient of Dad. His comedic talent, however, may have been responsible for a disorder that brought him to the Victorian doors at 921 Great Plain Avenue. Known as "quinsy sore throat," the condition is characterized by severe inflammation of the throat and may result from excessive vocal activity.

The vast majority of Dad's patients were decent people who did not abuse their demands of him. A few, however, were nothing more than parasites and four-flushers who left their bills unpaid, yet drove around town in the latest model cars.

CHAPTER SEVEN

The High School Years

Back in frigid Needham, after the Empress of Australia cruise in February, 1936, my Caribbean tan quickly dissipated upon exposure to the New England weather. Leaving New York's ice-filled harbor for the Caribbean in winter requires a day and a half to reach the blue waters and flying fish of the Gulf Stream and climatic conditions warm enough to don summer apparel. Today, cruise ships headed for similar ports in winter depart from Florida because precious time and fuel are saved thanks to the clement weather. A consequence of this change in departure locale is a cruise length from Florida of 4-7 days in contrast to that from New York of 14-16. Another advantage of southern departure is the avoidance of Cape Hatteras whose turbulent waters dramatically increase the incidence of seasickness. Of course, there are several disadvantages for those intrepid travelers from the New England area who choose to leave from Florida: the flight down and back during unpredictable weather conditions, and worse, for drivers, over 3000 miles of roads to traverse.

As my days in the 8th grade slowly ground to a halt in mid-June, we visited the Forands in Wrentham, Massachusetts, who had recently purchased a summer cottage on a promontory at Mirror Lake. Dick Forand, his wife and daughter lived in Needham where he was a member, along with Dad, of the Rotary Club. Like Dad, he was also an avid outdoorsman who enjoyed hunting and fishing, especially fishing, for this was the prime reason he had selected Mirror Lake as a vacation spot. The lake was exactly a mile in length, approximately a half mile wide, and only 12 feet deep where a diving raft floated adjacent to a sluice-way that drained surplus water from the lake into a stream below. In contrast to nearby Lake Pearl, a natural lake deeply gouged out by a glacier thousands of years ago, Mirror Lake was created in the twentieth century by damming a small stream. Of the two, the latter was far safer than its prehistoric counterpart because of its shallowness. Consequently, each summer Lake Pearl claimed the lives of several swimmers (as famed Walden Pond does in Concord to this day), whereas Mirror Lake lacked comparable statistics. The presence of the King Philip dance hall at Lake Pearl, with its boisterous crowd of inebriated kids, accounted for some of the deaths among those whose dancing in a building without air-conditioning required a momentary dip in the cool waters of the lake. The staid, older folk who frequented the shores of Mirror Lake were accurately represented by the Forand adults who were in their early forties, yet would seem ancient to the vivacious youths who enjoyed the fun and social aspects of the King Philip ballroom. This perception of age reminded me of my vacation days spent on Martha's Vineyard with the family of a high school classmate in the early fall of 1939. When I asked my teenage friend about Nantucket as a vacation resort, he answered: "Nantucket is for old fogies!"

Returning home after a delightful afternoon spent at the Forand's summer cottage, Dad said nothing about acquiring property at Mirror Lake, although it was in the back of his mind. Shortly after, while driving to Quincy to visit John and Bertha, he took the long route through Reading, an important railroad terminal. On the right side of the road, as far as the eye could see, was a vast collection of outdated, rusting steam engines and tenders awaiting shipment as scrap to Japan. Just beyond this yard, and before the road went under a massive stone bridge, there was a display of cottages of different sizes, ranging in price from several hundred to several thousand dollars. Dad stopped the car and the three of us inspected the various sample dwellings. In a few minutes, a decision was made: the eleven hundred dollar cottage was selected, comprising a kitchen, bedroom, living-room, porch, and brick fireplace. Next, Dad purchased lakeside property ($400) and arranged for the assembly of the delivered wood, bricks, and asphalt shingles. Within a few weeks, all of this was accomplished and the cottage was ready for occupancy except for a lack of plumbing facilities in the kitchen and toilet, located in a small room adjacent to the kitchen and opposite the side door.

The next step was to purchase a large wooden mallet, thirty feet of pipe, and fashion a divining rod from an apple tree bough to search for water. The cottage was only a few feet from the edge of the lake, so water was available almost anywhere a pipe was driven into the ground. Now the main action took place in the kitchen with several men taking turns pounding the pipe with the heavy mallet until gurgling was heard and a line dropped inside the pipe revealed moisture present. A Sears-Roebuck pump was attached to the pipe and water gushed out after the pump was primed by pouring a glass of water onto the plunger. There was a distinct taste to the water which therefore could not be used for drinking, but for washing dishes and silverware. A very large wooden barrel was buried at the rear of the dwelling as a septic tank attached by clay pipe to the toilet adjacent to the kitchen. Now the place was habitable and ready for entertaining our Needham friends who fished, swam, boated and played bridge (Mom's women friends). Ironically, Dad rarely visited the place preferring to spend his free moments, while we were at the lake, with his father in Norfolk.

During my high school years, and before my senior year in 1940, I enjoyed boating and sailing on the lake during the summer months, and ice fishing in the winter. Dad bought a new rowboat at a Cape Cod boatyard for eighty dollars, an Evinrude outboard motor, and a sailboat called a Cape Cod tern with a blue hull 12 feet long, a mahogany tiller and centerboard, and an 8-foot mast equipped with a bright red mainsail and jib. As an indication of how shallow the water was, one day I was in the middle of the lake when a gust of wind flipped the sailboat over. After I had righted it, a glob of mud stuck to the top of the mast plopped onto the white deck. Dad also bought a beautiful canoe, complete with paddles and seat cushions, which was kept beneath the cottage in an unfinished basement along with the sailboat and rowboat. One winter, while visiting Mirror Lake to check on frozen pipes, we discovered our canoe had been stolen. The robbers were obviously kids because they left the rowboat and sailboat behind. If they had taken these across the frozen lake's surface, as they did the canoe, they undoubtedly would have suffered from hypothermia as the whole kit and kaboodle crashed through the ice. Speaking of thievery, when the farm had first been abandoned with all of the livestock gone, Dad and I drove to the "Cedars" and surprised a fat kid siphoning gas from the old Studebaker. Dad always carried a handgun since the time years ago when, driving to a house call in nearby Dover late at night, he came across a gang of kids forming a barrier across the road obviously attempting to stop the car and rob or even kill him. Instead of stopping, he blew his horn and increased speed, and the little bastards barely got out of his way. I asked Dad why he didn't fire a shot over the head of the rapidly retreating kid to teach him a lesson. "When you fire a gun," he replied, "you aren't always sure where the bullet will end up." The kid could have been the arsonist who burned down the barn a few months later when his heart beat returned to normal. Fortunately, he didn't set fire to the historic farmhouse and the other outbuildings. The farm, with its hundred or so acres, was eventually sold to Mr. Harold Hillyard, the manager of the Boston office of American Airlines, for 5500 dollars! About 25 years later, the farmhouse and five acres surrounding it sold for $38,000! This devaluation of our currency is attributed to a number of factors, not the least of which is inflation. During WWII, admirals and four-star generals were paid the princely sum of $8500 per annum! Destroyers, such as those constructed at the Bath Iron Works in Maine, cost the taxpayers 5 million dollars. An aircraft carrier, such as the Lexington or Enterprise, had a construction cost that approached 200 million. Tanks built by the Chrysler Corporation cost the government 10,000 dollars. Fighter aircraft, some with folding wings for carrier duty, cost slightly less than what a tank cost. These were the days when dimes, quarters, and halves contained 90% silver (the rest copper), and, prior to FDR's election, gold certificates could be exchanged for gold coins of exquisite beauty. In the late thirties and early forties, any salary above 35 dollars per week was considered sufficient to provide for a family of four with a wife raising two children at home while her husband was paying off a mortgage on a 4500 dollar dwelling. Postmen, who delivered mail twice a day on foot, lugged huge leather bags filled with letters and bills, and rarely a catalogue unless Sears-Roebuck's annual edition was due. School busing was unheard of, for no buses were available. Schools were built where children could easily attend them by walking or riding bicycles. The family doctor, once called a general practitioner and now referred to as a family medicine practitioner, charged 2 dollars for an office visit and 5 dollars for a house call. To deliver a baby, he or she charged 35 dollars or a week's pay for the more affluent. A major operation could cost as much as several hundred dollars! A Studebaker Champion cost 550 dollars; a Cadillac, 4500 to 7500, depending upon whether one wanted a coupe or a hearse. A year at Harvard cost $770. When the new Glover Memorial Hospital was built on Chestnut Street, as a large rectangular-shaped building, it cost $50,000. Dad was able to convince one of his patients, Charlie Haskell, a farmer who lived near the Wyeths on South Street and one of their closest friends, to contribute $10,000 towards its realization. The old wooden structure, formerly located on the site of the new facility, became the nurses' home. When Aunt Lillian was pregnant and expecting her first child, she heard a peculiar sound in a nearby room at the old hospital. It was eventually revealed that it was Dad killing flies with a newspaper while awaiting her delivery! Being a relative and a minister's wife, the bill for hospital services, including delivery, was zilch.

In the ninth grade, a happy event occurred which reminds me of one of the funniest scenes in the British film, "Hope and Glory." The Junior High School building, showing its age, began to disintegrate. For safety's sake, all of the students were sent home a month early for a lengthened summer vacation. Huge bolts were fastened in the walls and the building was used for several additional years until a bulldozer ended its usefulness and left no trace of its existence. In the British film of WWII, children about to enter the schoolyard were surprised to see their building had been completely demolished by a German blockbuster. The joy of a few months of freedom from education was appreciated regardless of the cause.

At this time, when I was about to enter 10th grade by walking a mile and up a hill to reach the stately High School building, I had a companion to accompany me in rain, snow, and fair weather. His name was Specky II, another pointer replacing the first one that died at the age of 6 at the "Cedars." The new Specky seemed to be even more intelligent than his predecessor, for he knew exactly when high school classes ended and would be outside the exit, wagging his tail and awaiting the mile walk home. It was Specky who barked when the Baptist minister's sons set fire to our garage, saving the structure from total incineration. It was Specky who accompanied mother downtown, often stopping by Charles Stevens' ice cream, grocery, and novelty store on Great Plain Avenue, where Charles once caught him lapping strawberry ice cream from a large cylindrical container. He said nothing but carefully scraped the top off the ice cream and sold the rest. Charles Stevens, an heroic figure as a volunteer fireman, was well known for his prowess in hopping on the rear platform of a firetruck that would stop in front of his store on the way to a fire. I wish I still had the baseball cards I purchased from him! The miniature WWI tank I bought from him to entertain my father, while he was a patient at Glover hospital after a hemorrhoid operation, almost drove Dad up the wall when I wound the diminutive military toy and placed it on the sheet of his hospital bed.

Before delving into my experiences in high school where the IQ of my canine companion, if he'd been human, would have guaranteed him entry into the 10th grade, I must mention the last Caribbean cruise with my parents. On my 15th birthday, I celebrated the occasion aboard the Cunard White Star liner, M.V. Georgic, as it sailed toward the Venezuelan coast. This is how the card marking the affair looked;

Birthday Greetings   *Menu*

to

David   Caviar Glace

Clear Green Turtle

Poached Salmon, Balmoral

————

Plank Steak, McAlpine

Salade Princesse

Crepes Suzette

————

M.V. "Georgic" February 13th, 1937 Coffee   Dessert

The trip was especially memorable for we traveled with my parents' closest friends, Clifford and Floss Locke of Needham. The Locke's held the highest social status in the town and belonged to a distinguished New England family that had accumulated considerable wealth by investing in lumber and real estate. They were also the nicest people anyone could have for friends. Their two daughters, Doris and Eleanor, went to the finest schools (i.e., Wellesley College), and lived successful and fulfilling lives. Sadly, Clifford's sister was murdered by her husband who also killed their children before sticking a shotgun in his mouth and blowing his head off. Frank Kettlety had been employed at the Locke's lumber yard in Needham and was a devoted husband, but something went awry in his brain that caused him to go on a murder rampage. It was Clifford who discovered the bodies in the bungalow in Dover, which was purchased several years later by the widow who siphoned gas from Dad's Buick. I remember seeing one of the children, a thin, pretty girl of nine, when I accompanied Dad on a house call to the white bungalow on the banks of the Charles River.

The M.V. Georgic (called motor vessel because its engines were diesel instead of steam turbines) and her sister-ship Britannic were two-funnel vessels, the latter being the last ship to bear the White Star colors on the Atlantic run. The former was converted to a British troopship at the outbreak of WWII and its forward funnel removed to confuse Nazi submariners. I don't recall the itinerary of this cruise, but I do remember the botanical displays on the rattan-carpeted promenade deck and the superb cuisine. As a result of this trip, the Locke's treated me as they would have their own son. On weekends, when I came home from college, they often invited me to share *hors d'oeuvres* at their home while they played bridge with my parents. During my Purdue graduate school days, when I flew from Indianapolis to Miami on a vacation break to meet my folks at the Everglades Hotel, Clifford and Floss were there to greet me before my parents arrived hours later by train.

There were three people who had a profound influence on my life during my years at high school. Two were people I knew as close friends; the third was a person I never met, but admired immensely. He was the comedian Fred Allen. I admired his wit and ingenuity in creating his own radio scripts which were generally amusing and often hilarious. His real surname was Sullivan which, as a competitor in show business, partly explains his dislike for the famous newspaper columnist, Ed Sullivan, who later became famous as the popular TV variety show host.

"Ed Sullivan will be around as long as someone else has talent," Allen was often quoted as saying in his distinctive nasal twang. There was another reason why he disliked his namesake: Allen, a Roman Catholic was married to a Jewish lady named Portland Hoffa; Ed Sullivan's marriage was similarly mixed. Perhaps Fred thought that Ed was a sociologic copycat. Regardless of these human frailties, Allen provided the nation with wholesome entertainment in radio (and occasionally in movies with Jack Benny) for many years. In the fifties, however, competition from a radio show called STOP THE MUSIC and live TV comedy shows caused his ratings to nosedive. Commenting on the inanity of television shows, it was Fred Allen who once said: "Television is bubble-gum for the eyes," a statement often attributed incorrectly to the world-renowned architect, Frank Lloyd Wright.

As an example of the prudishness of TV in those days, when Elvis first appeared on the Ed Sullivan Show, TV cameras were not permitted to exploit the King's gyrations. Yet, in a different setting and time frame, the Harvard freshman smoker I attended in the fall of 1940, Ed Sullivan entertained the all-male audience with jokes and introduced several stripteasers whom everyone had seen at the Harvard-owned Old Howard Burlesque Theatre in Scollay Square.

In my High School Graduation Yearbook of 1940, juxtapositioned next to my name, was that of Fred Allen under the title—*the person I most admire.*

The two remaining persons, my close friends, were as diverse in their occupations as any two persons could be. One, Bill Patton, was a salesman for the Globe Ticket Company and the other, Bill Fisher, was a high school music teacher and writer who published a series of music appreciation books that were well received nationally. Yet, they had much in common: both loved nature, had a wonderful sense of humor, married remarkable loving wives, and were respected as active citizens in the community. I first became acquainted with Bill Patton when I met him at a Congregational church supper and we shared a table, eating oyster stew which, in New England, is the traditional dish of such gatherings. During our conversation, I happened to mention my interest in aviation by asking him if he had ever seen a Republic amphibian. I was intrigued with the aircraft because it had a Franklin engine set in the center of the single wing to which was attached a "pusher" propeller. The craft was actually a flying boat designed to carry 4 passengers to hard-to-reach fishing spots and was equipped with a tiny Danforth anchor that the pilot tossed from a window upon landing.

Bill had seen the amphibian land and take-off at nearby Wiggin Airport in Canton, Mass., and confirmed my suspicions that it flew like a Greyhound bus. When he invited me to go flying with him, I got my parents' permission and one Saturday we drove to the airport and climbed into an Aeronca, a red-colored monoplane with side-by-side seating. After a few minutes of instruction, he started the engine and proceeded to maneuver the plane while I looked out the window at the ground below. Suddenly, he revved the engine and we sped down the concrete runway with the tail assuming a horizontal position. As he slowly pulled the crescent-shaped half-wheel toward his stomach, we climbed into the air and were soon leveling off a few hundred feet above the ground. Reiterating his instructions he said: "Remember your feet steer, the wheel takes you up or down by pulling it toward you or away, and by turning the wheel and pressing your feet, you go right or left. The feet steer by moving the rudder and turning the wheel moves the ailerons to aid the steering." I took over and Bill, after a few moments, said: "You're acting as though you're driving a car! Relax, you have nothing to worry about unless you slow the engine in which case you'll go into a spin!" After a few minutes, he took over and we landed safely.

Following thirteen hours of instruction from Bill, I was ready to fly from Canton to Needham, which only took a few minutes. I flew over my house, its slate roof shining in the sun's rays. Seeing no one around, I headed for Plymouth. At all times, I was aware of appropriate surfaces to land on in case the single engine failed. At Plymouth, there were beautiful stretches of beaches where this could easily be accomplished. In landing at Wiggin about a half hour later, I flew in a little too low and hit the branches of a pine tree but managed to land safely. That was the last time I flew a small plane and, after that experience, I disliked flying in any aircraft.

Bill Fisher was Dr. William Randolph Fisher, a native of New York, who had been hired to teach music at Needham High School. He was in his early thirties, freshly graduated with a doctorate degree in music from N.Y.U., and married to a brilliant mathematician, Doris

Nufer, also of New York City. I first met Bill when he entered the high school gymnasium one morning to introduce himself as the new music teacher who was recruiting members for the school orchestra and also planning to organize the first band the school had ever had. He became an instant friend when I left the room for a few minutes and returned with a folding chair. Recalling the incident a few years later, he remarked: "You were the only student among all those seated that gave a damn I'd be standing for several hours."

His gratefulness was acknowledged at graduation three years later when I, as a lousy clarinetist, received the music medal. It's true that I composed the class song and conducted it before my beaming mother and grandmother, but my musical skills were minimal. In fact, as a member of the orchestra which played at Wednesday school assemblies, I once forgot to clean the mouthpiece after practice the day before and had to run out of the auditorium before throwing up outside the principal's office. This had to be my most embarrassing moment in high school.

Bill Fisher, Doris, and I spent many happy hours together in the early days of their Needham arrival when they occupied a rented house near the Glover Hospital. In the summer of June, 1940, as a graduation present from my parents, I visited them at their parents' home in New York City and spent five days at the World's Fair. The imposing white trylon and perisphere dominated acres of interesting exhibits such as the General Motors' Futurama with pothole-free superhighways, Billy Rose's Aquacade starring Buster Crabbe and hundreds of young ladies cavorting in a gigantic swimming pool, and RCA, where a demonstration of a newfangled gadget called television transferred images of pedestrians onto a big screen. Remnants of the Russian Pavilion, a huge step-like pyramid on which torrents of water gushed the year before, now stood rotting at the mercy of international politics. A week after I left the Fair, a bomb was detonated at the British exhibit in protest of their entry into the war. By far the most dramatic exhibit was the show in the United States pavilion that reviewed the colorful history of our country and ended with two giant locomotives, entering from opposite ends of the stage, symbolically commemorating the cross-country union of two railroads in Utah in the 19th century.

One evening, the three of us were in the audience at NBC studios where Major Bowes hosted his famous amateur hour. A young announcer stepped to the microphone and, with crisp enunciation, intoned: "Four great cars, Plymouth, Dodge, DeSoto and Chrysler, present Major Edward Bowes and his original amateur hour!" It was Ralph Edwards who became even more famous than the host of the radio show, the major himself! The show never reached the airways, though, for it was preempted by a political announcement for presidential candidate Wendell Wilkie.

During the week of my 16th birthday in February, 1938, my mother drove me to Waltham in her maroon Studebaker coupe for my driver's examination. There was little trepidation about either the oral or driving test, for I had memorized the rules of the road and had driven my old green Studebaker at the farm under a variety of field conditions.

The examiner was an officer in his forties who had a gruff manner tempered by years of evaluating all kinds of people who put his life in jeopardy hundreds of times a week throughout the year.

"I'll ask you questions while directing you where to drive," he growled, as he crawled into the adjacent seat. I recall only one question which stuck in my mind because of its uniqueness.

"What does a flashing green traffic signal mean?" he asked with a slight smile on his face (secretly hoping I'd give the wrong answer and free up some time to drink coffee).

In any other state in the nation this electronic aberration would mean the traffic control people should correct the problem as soon as possible. But in Massachusetts it means something else.

"Slow down as you would for a constant yellow light."

"Correct. Now pull over and park next to that telephone pole."

I drove past the pole and continued on. "Didn't I tell you to stop?"

"Yes, but the rule book says keep several car lengths away from fire apparatus and that pole has a fire callbox on it!"

"You're right," he answered. "I wasn't trying to trick you. I didn't notice the callbox."

I got my license and he had a few minutes left over from my shortened instructions to have a cup of coffee.

I mentioned earlier that for a graduation present my folks provided me with funds to attend the New York World's Fair in 1940.

That was only part of my reward for not getting into trouble in high school and passing all of the courses. A year before my graduation in June, Les Cook and Dad drove into the yard in a black Plymouth convertible and handed me the keys. "It's yours - and safe driving!"

There was one restriction imposed on my driving: Never drive the car to school.

During my sophomore year, I had the good fortune to meet Mrs. Gillespie, a woman of Scottish ancestry who was in her fifties and had a voice that sounded like that of Sir Harry Lauder, the famous British singer/comedian. Mrs. G. was the high school dietitian who, aside from preparing the daily fare of hot dogs, sauerkraut, and vegetable soup, made sandwiches of ham and cheese and egg salad that required wrapping in cellophane. Mrs. G. was my first non-relative employer, for my folks thought it would be a good idea for me to earn pocket money while working for someone else. Incidentally, the pocket money I earned didn't actually exist, for my task of wrapping sandwiches in cellophane was rewarded with coupons that entitled me to daily free meals at the school. However, I was able to pocket part of my weekly allowance as money I would otherwise have spent on milk and a sandwich of ham and cheese or egg salad. Also, the time I spent each day with Mrs. G. amounted to only an hour or so at mid-day. Measured as today's minimum wage, I probably earned two dollars an hour.

My association with Mrs. Gillespie was memorable, although her Scottish accent was at first impenetrable, leading her to believe I was hard of hearing. As a result, her directions to me during the lunch hour were heard by most of the students munching away in the dining-hall. I always saved a portion of my sandwich for Specky, who was anxiously awaiting his snack outside the appropriate door, drooling copiously and wagging his tail.

When Dad bought me a Plymouth convertible with red leather upholstery and a rumble seat during my junior year, I raised funds for gas and general maintenance of this black beauty by chauffeuring elderly ladies to the downtown stores in Needham. During the summer months, I drove to Tillotson's rubber factory near Rosemary Lake, behind the beautiful brick library, where I made ink sacks for fountain pens by dipping a stick with metal molds into a thick brew of melted latex. Occasionally, I would repeat the process by substituting cotton gloves for the metal molds in order to produce gloves with a coating of black rubber. Although the work was accomplished in a hot, odoriferous environment, the experience was well worth it, for the opportunity to study human behavior (women discussing their marital problems; men bragging of their sexual prowess) was rather reminiscent of Charles Dickens' exposure to similar human frailties while he labored in a factory on the banks of the Thames bottling shoe polish. The 10 weeks of work were fortified nutritionally by a daily intake of one ham sandwich and a soft drink without the presence of Specky to share part of the meal.

On weekends, Specky and I would drive to Norfolk to see if the buildings were still intact. On one occasion we ran into cousin Martha walking toward her home, so I stopped the car and invited her to join us.

"Martha, you'll have to crawl into the rumble seat," I requested, "because Specky will only sit alongside me."

Driving through Norfolk with this arrangement must have brought a few smiles to the faces of the townsfolk as we sped along. Specky, who was thinking of his next meal, was spewing long strands of saliva while Martha, in the rumble seat, valiantly tried to avoid the onslaught. These humorous moments remain forever entrenched in my brain.

During the first two years of high school, two teachers who also remain entrenched in my mind were Miss Fessenden, the geometry instructor, and Miss Steele, the English teacher. The former was a taskmaster who threatened her students with such comments as: "You'll graduate from here and be a truck driver if you don't master my subject!" Or, perhaps to the girls: "Expect to be a failure at whatever you attempt,' cause you're stupid, stupid, stupid."

Miss Steele, on the other hand, was encouraging in her attempts to teach the King's English to her charges who learned how to pronounce words such as airplane as aeroplane, and the name of author Ernest Thompson Seton as Seeton not Seton. The irony of this latter rebuke was the simultaneous appearance of an article I wrote entitled: How to Hunt Whitetail Deer in *Hunting & Fishing Magazine* in the same issue with one by the great woodsman, Ernest Thompson Seton, in the late forties!

Both Miss Fessenden and Miss Steele were not appreciated for their perseverance in demanding scholastic perfection in their respective areas at the time, but years later their efforts paid off when former students became successful in their chosen fields partly from the knowledge these ladies imparted.

When the summer of 1938 arrived, it was time to go on another trip, but the days of Caribbean cruising were over. Instead, our attention was focused on San Francisco and the Rotary Convention to be held there in June. Dad, as president of the local Rotary Club, was expected to attend, so the three of us traveled by train (The New England States) from Boston to Chicago and then, via the Santa Fe railroad, through Illinois, Iowa, Missouri, Kansas, Colorado, New Mexico, Arizona, and California. We visited the southern rim of the Grand Canyon in Arizona and descended on the Bright Angel trail for a mile or so. Looking up at the rim behind us, we were surprised to see we were only a few hundred feet from our point of departure. The Colorado River below looked like a thin gray ribbon surrounded by green grass that was actually tall trees. At the El Tovar we watched the Hopi Indians perform a ritual dance and, after a meal at the hotel, boarded the train for Los Angeles. While enjoying a steak dinner that evening the train stopped for water (steam engine) and I peered out the dining-car window to gaze upon the decaying carcass of a longdead steer.

I was fortunate to have two companions, also from Massachusetts, whom I met at the first sitting in the dining-car as the train departed from Boston's South Station late on a June afternoon. They were the Marshall twins from Lexington who, like myself, were traveling west for the first time along with their parents who were also attending the Rotary Convention in San Francisco. Although the boys resembled each other, they had facial differences that made it easy for me to distinguish Roland from Richard upon first making their acquaintance.

Their parents were very personable and obviously fun-loving, for Dad and Mom took to them immediately and enjoyed their companionship throughout the trip, which for them, ended in Portland, Oregon, on the homeward trek, where one of the boys came down with appendicitis and had to have the vestigial organ removed at once. Prior to that incident, however, we explored the city of Los

Angeles, which seemed to be a hodge-podge of buildings attempting to be skyscrapers amid a landscape dotted with oil derricks. We stopped in Chinatown to enjoy a dinner but were told that Chinatown had been destroyed by fire the day before and would require several months for rebuilding. And I thought the smokey odor in the air was due to chinese cooking!

We went by taxi to a pier where we boarded a small ferry for the island of Catalina thirty miles away for a day of sight-seeing. Also aboard was the actress Ann Sothern, a pretty blond, who was going to meet her husband, bandleader Roger Pryor, whose orchestra was playing at the casino in Avalon. When we arrived several hours later, Mom spied one of her favorite actors sitting on a bench on the pier and ran over to give him a big kiss. He was Frank Morgan, taking a day off from playing a role as the wizard in the classic film, The Wizard of Oz. Prior to the kiss, Dad recorded on 16-mm film a fan chatting with the famous character actor and comedian and one can sense that the actor is beginning to feel uncomfortable at having his space invaded. In viewing the film recently, I also felt uncomfortable seeing his distress. Thank Heavens the kissing incident was not recorded.

Catalina Island once belonged to William Wrigley, the chewing gum tycoon. After his death, the island was sold to individuals with considerable wealth, one of whom was the western novelist and University of Pennsylvania Dental School graduate—*Zane Grey.* We saw his home, a lavendar-colored adobe structure perched on a cliff overlooking Avalon harbor. Only recently did I realize why purple was chosen as the color of his home. Although he was a novelist, he was a dentist before he became famous as a writer of great western stories, and purple is the color of the cape that a graduate dentist wears at commencement.

Before boarding the train for the trip to San Francisco, we toured Hollywood, gawking at the beautiful mansions of the stars. As we passed the residence of Joan Blondell and Dick Powell, she the comic actress and he the perpetual college boy crooner, we saw Joan leaving in a hurry. (The next day we read in the L.A. Times the reason why. She had given birth to her first child.) The large mocha-colored stucco mansion of my favorite movie comedian, Joe E. Brown, was magnificent, reposing in a setting of tall palms surrounded by tropical plants. We visited the Hollywood Bowl, an updated version of a Grecian amphitheatre, Sid Grauman's Chinese Theatre (now-called Mann's Chinese Theatre), stood in the footprints of Clark Gable, Jimmy Durante, and William Powell, and gazed at the mouth-print of Joe E. Brown who had actually kissed the cement to make the impression.

At Riverside, we toured the famed Mission Inn whose cheery atmosphere was heightened by the sound of the hit song of the day, "Tipi-Tipi-Tin." At the California Institute of Technology in Pasadena, we saw the world's largest telescopic lens, 200-inches in diameter, that was painstakingly being polished before its installation at Mt. Palomar. Finally, in San Bernardino, I climbed aboard a huge Santa Fe steam locomotive and Dad took my picture as I waved from the cab's window.

Soon, it was time to board our waiting train for the long trip to San Francisco, to view the Bay area attractions of Alcatraz, the Golden Gate and Oakland-Bay bridges, and Japanese tea gardens, while Dad attended the Rotary Convention. I mentioned earlier that Burlington, Vermont, was my favorite city among those I had seen: Bangor, Boston, Buffalo, Burlington, Chicago, Havana, Montreal, and New York. But after seeing San Francisco (when the fog lifts), the Vermont city took second place. It is not hard to imagine what the city looked like in the days of the gold rush when the'49ers arrived in windjammers and stage-coaches. Just replace the present-day skyscrapers with two-story wooden structures and you have captured the look. The streets are still narrow and afford wondrous views of the Bay, for the terrain borders on being mountainous. But geologically the harsh, granite look of the Rocky Mountains has been replaced by the smooth, rounded look of the Scottish Highlands. The vegetation, however, resembles that of Italy rather than the flora of the British Isles. Yet, it is the Bay that makes the city unique and constantly anointed in fog which arises from the warm water of the Sacramento River meeting the cold water of the Pacific Ocean. In this churning marine environment, a host of edible species proliferates and provides sustenance for the human population somewhat comparable to the situation that exists in Vancouver.

We stayed at the Plaza hotel in San Francisco near the world famous St. Francis Hotel. Market Street, the main route to the Ferry

Building on the edge of the Bay was becoming crowded with arriving conventioneers, which implied that the restaurants nearby would require long waits for a decent lunch, so the folks decided to enter the Ferry Building and see the exhibits extolling the riches of California. And what exhibits they were! The theme was the *California Gold Rush,* which increased the population of that State far more than did its agricultural prospects, proximity to trade with the Far East, or fantastic climate.

I had never seen so many forms of the base metal: nuggets, ranging from pea-to hen's egg-size, gold as filamentous as dental floss, or in glistening sheets resembling the delicate filigree of butterfly wings. Indeed, the Spanish explorers, Cortez and Coronado, would have lost their cool upon viewing the geologic splendor I beheld. What an exciting place this was! Dad was profoundly bitten by the gold bug. Later, during a motor trip to Sacramento, we stopped at Sutter Creek, where a man named Marshall discovered gold nuggets in 1846 that launched the Gold Rush.

Dad photographed me standing beside a large sign designed as an open book over which the upper torso and bearded visage of John Sutter loomed like a gigantic bookmark. Across his chest was the word WELCOME, and below, the words SUTTER CREEK, founded in 1846 by Capt. John A. Sutter, Population 1370, Elevation 1300, Good Hunting & Fishing, Principal Industry Gold Mining, Production in County 350 Million and Gold Rush August 13-14, 1938, Boosters Club-Inc.

In Jackson, we stopped at a working gold mine and collected a few discarded rocks which Dad and I carefully examined. Sure enough, back in our hotel room over a sink with considerable brushing, we saw tiny flecks of gold.

A boat ride around the Bay allowed us to see the stark buildings of Alcatraz, America's penal colony counterpart to France's Devil's Island off the coast of South America, and similarly guarded by man-eating sharks and treacherous currents. In the distance, at the Bay's entrance, the newly-constructed Golden Gate Bridge shimmered in the sun's rays while a ship of the famous President Line passed beneath its orange-colored span on its long voyage to Hong Kong. In the opposite direction, the Oakland-Bay Bridge could be seen with an immense white concrete buttress, located near the central part of the span, supporting the structure as it approached Treasure Island. This man-made island, sucked up from the bottom of the Bay, was soon to be the site of the San Francisco World's Fair of 1939. At this stage, the Tower of the Sun, a white cathedral-like structure rising above a number of flat white buildings, could be seen in the distant mist. As we passed beneath the Bay bridge near the concrete abutment, someone remarked that "a ton of sugar was added to the concrete to hasten the hardening process."

Although Dad did attend a few meetings at the Rotary Convention, he found most of the sessions boring and preferred instead to clip highlights of the daily sessions reported in the San Francisco Examiner. Thus, he was able to see the sights with us during the day and spend an hour or so each evening, culling the events he believed of interest to the members back home.

After his trans-isthmus flight from the Pacific to the Caribbean-Atlantic side of the Panama Canal in a vintage seaplane, he had to see the fabled China Clipper take off for the Orient from the Bay at Alameda. Like the seaplane, the Clipper was constructed of aluminum with a single wing, but instead of a single engine located at the front of the fuselage, there were four mounted in the wing to ensure safe passage across the broad expanse of the Pacific. In place of floats, the hull was designed as a huge boat with portholes through which staterooms and a dining area could be seen. In retrospect, with four engines booming in unison, sleep must have been impossible and the intense, constant vibrations must have rattled everything in the galley that wasn't bolted to the hull. A Delta Airlines walk-thru exhibit at Disney World, displayed several years ago, showed a mid-section of the China Clipper's hull. I was surprised at the small size of the aircraft contrasted with that of today's mammoth Boeing 747. Dad would have been astonished, too. The ancient aircraft he had captured on 16-mm. Kodak film, skimming over the waves of San Francisco Bay like a metal dragonfly, today would be mistaken for a toy at a model airplane meeting.

Mother's enjoyment of the Bay area was enhanced when we visited the Japanese Tea garden, a botanical wonderland with lush plants, brightly lacquered wood pagodas, and bridges arching over Monet-like lilypad pools. She also marveled at the world's largest outdoor swimming pool at Fleishecker's and the occupants of the nearby zoo. San Francisco has something for everybody and thank God it doesn't have frost and snow or it would soon become the car insurance capital of the world.

With our California sojourn coming to a close, an unusual incident occurred at the hotel desk. The clerk gave Dad a container filled with dice and asked if he would care to gamble for his week-long hotel bill.

"Sure," he replied. "What do I have to do?"

"Just throw any combination adding up to seven and your stay is free! However, it will cost you 50 bucks for the throw."

Dad shook the cylindrical container and out popped the dice. The four ivory cubes added up to seven; we left with a hotel bill of only 50 dollars instead of one four times that amount! The only time Dad and Mom gambled was during their winter vacation in February with the Lockes, who were similarly frugal in betting only a few dollars per race at the Hialeah track in Miami. They were more intrigued with the awkward behavior of hundreds of flamingoes in the pond within the track than the skills of the jockeys.

Reflecting on the events of the past two weeks, as the train headed toward Portland, Oregon, for the first stop on its homeward trek, the highlights of California, in addition to those already mentioned, were: visiting Robert Louis Stevenson's home on the coast of Carmel, a two-story New England-style colonial set among gnarled cypress trees, inhaling the cool oceanic breezes while gazing at Seal Rock and the famous Cliff House near San Francisco, and experiencing an ant's view of the gigantic redwood trees that were seedlings several centuries after the birth of Christ. How lucky I was to live in a country that was so diversified in its topography and citizenry! Many years later, when a professorial acquaintance was bragging of her imminent trip to Europe, another professor, in disgust, asked: "Have you ever seen a cream separator in Nebraska?"

Arriving at Portland (actually named for the city on Maine's rugged seacoast), one of the Marshall twins became nauseous and complained of pain in the lower right abdomen. When the pain subsided momentarily, slight pressure over the appendix area caused greater discomfort. Pain symptoms can be tricky, for patients have been known to experience pain in a big toe (right or left) and have an appendix ready to burst because of the phenomenon of referred pain.

Richard was rushed to a hospital and his parents and twin brother made arrangements to stay in a hotel room for a week before returning to Lexington with the convalescing sibling.

During our one-day stay in Portland, we boarded a tour bus and, of all places, visited a cemetery that was elevated as a plateau surrounded by a high stone wall on all four sides. What the attraction was, I don't recall, but it must have had some unique feature that escapes me, unlike what will never happen to the occupants. Climbing aboard the train, again, we passed through Washington and Idaho at night, awakening to see Missoula, Montana, and the Anaconda copper mine at its namesake city before reaching Billings at nightfall. We arrived in time to see a movie there and it was the first time I had ever attended a cinema where an Indian (native American) with a single feather in his headband sat in front of me. Our tour by train was momentarily terminated, so we stayed overnight in Billings before heading to Cook County, Montana, where the northern entrance to Yellowstone National Park is located. Before entering the Park, the bus driver warned tourists not to stick arms out windows where a ravenous bear could chew off a limb in their search for snacks. And these were not the 200-hundred pound black bears that populated the Park, but much larger grizzlies that could amputate an arm in seconds. When we arrived at the Yellowstone Inn, the world's largest log structure, park guards stationed at the entrances warned visitors that a black bear had wandered into the dining-room. Not wishing to wrestle with a hungry bruin, we walked a few yards to the viewing area and sat down on the logs that surrounded the clay mound from which Old Faithful was about to spout hot water and gases at 65-minute intervals.

After watching Old Faithful discharge hundreds of gallons of steamy water, we took a tour of the Park to Yellowstone Lake, where a small crater several feet offshore enabled a fisherman to catch a fish and cook it without removing it from the hook by dropping it into the boiling water. Native Americans probably did that, but such shenanigans were now prohibited. Not so, however, at another location where one could drop a soiled handkerchief into swirling hot water at the mouth of a cave and have it laundered and discharged at another opening. Why not throw a freshly caught fish into this natural cauldron and have it reappear cooked?

Throughout the Park there were geysers with signs indicating their own spouting schedules. I observed one that meant the visitor had to wait a year before it shot a spray several hundred feet into the air! Unfortunately, its discharge date was during the winter months when the Park was closed! At the Morning Glory pool, so-called because it resembled the flower from a plane, the water temperature was 212 degrees Fahrenheit, according to the Park ranger, yet, amazingly, it contained primitive life in the form of heat-resistant bacteria! Speaking of Park rangers, how did a forest ranger distinguish a geyser's discharge from the smoke of a beginning forest fire? Of course I asked that question, and the ranger answered: "Each look-out tower in the Park is equipped with a map that designates the location of the geysers, so a campfire that is illegal can easily be detected."

At Artist's Point, we gazed down on the Grand Canyon of The Yellowstone and observed the spectacular falls in the distance. One would expect to see many animals in the Park, but few were observed: a black bear, a bison, a few antelope and, in the evening, while sitting in a fenced area near the Yellowstone Inn, we watched half a dozen grizzlies loping slowly out of the woods to feast on the garbage dump placed there for the entertainment of the guests.

We boarded our train at Cheyenne (Wyoming) and ended our cross-country tour two days later in Boston. What a wonderful trip and what a great country! In 1971, we sailed to England on the QE-2 and visited Scotland, France, Italy, Switzerland, Germany, and the Netherlands. Five weeks later, we boarded the Nieuw Amsterdam in Rotterdam and sailed down the Rhine and across the Atlantic on a five-day voyage to New York. One great advantage of traveling extensively in the United States, in addition to encountering people everywhere who know the English language, is the requirement of only a single currency here. What a bother it is to keep changing currency with each new country and trying to keep up with the rate of exchange. And there is also less danger of offending anyone due to ignorance of ethnic customs.

In the fall of 1938, I started the junior year at high school and also began taking piano lessons from Mr. Lewis Bray, who was known professionally on radio as "Flying Fingers." I had taken classical music lessons for several years from Helen Moulton, a charming lady who, along with her husband Charles, were brilliant pianists and teachers. Charles had a studio in Boston where the elite of Massachusetts society paid 80 dollars an hour for lessons, while Helen, who taught at home in Needham, charged 35. They lived in a large, beige stucco house filled with expensive antiques that included a Japanese warrior's suit of armor which stood next to their grand piano awaiting a sour note (in my imagination) ready to lop off the hands of the offender with a samurai sword. The only distasteful aspect of these lessons was the annual recital that brought forth the Boston superstars from Charles' studio to bolster the poor souls with less talent from the Needham area. It was sheer terror to listen to their musical skills while trying to remember how my rendition of "The Happy Farmer" started on the nearest flat surface backstage that resembled a keyboard! And the recitals were always held in the Unitarian Church in the center of town which meant that every seat was occupied with parents, worshippers, and vagrants due to the accessibility.

"Flying Fingers" did not hold annual recitals and his music was from the Hit Parade, the popular radio show with singers such as Snooky Lanson who sang the hits of the week. Mr. Bray introduced an innovative approach to playing that enabled a student to concentrate on the right hand while playing the appropriate chords automatically. He arrived on Wednesday afternoons at 3:00 P.M. which was unfortunate on September 21st, 1938. On that date, New England was hit with the most devastating hurricane in its history! Rhode Island lost many lives and the coastal areas were inundated with thousands of homes demolished. "Flying Fingers", who parked in the circular driveway in front of our house, found his car impounded by treelimbs that landed in front and in back of his Dodge. The car, though, was undamaged. He was very proud of his new car that had pushbuttons in the center of the steering column as an innovative idea by Chrysler to make shifting easier. One day, he parked it on a pier overlooking Boston harbor and pushed the wrong button. It plopped into the harbor and "Flying Fingers" became "Flying Feet" as he managed to jump out in time.

I liked Mr. Bray who was a nice guy with a great sense of humor. I found out that his birthday was the same as mine, February 13th, and added his name to my "birthday similarity list" along with, as the years rolled by, Professor Kirtley Mather of Harvard's geology department (a great teacher and scientist), and Harry S. Truman's wife, Bess, (an unforgettable first lady). There was something strange about the aftermath of the deadly hurricane, though. The Unitarian edifice was the only church in Needham that didn't require repair work on its steeple when the winds subsided.

The piano technique I learned from "Flying Fingers" stood me in good stead. Late in the summer of 1939, I spent a week with the Daniels' family at their lovely summer home on Martha's Vineyard. Their only son, Red, was a year older than I, a grade ahead of me in high school and a member of the same Congregational church. He enjoyed similar interests to mine: the sea, yachts, and parties. I took the train from Boston to Woods Hole in Falmouth and boarded the ferry Naushon, which sailed five miles to Vineyard Haven, and twenty-five more to Nantucket. The Daniels' family picked me up at Vineyard Haven and drove a mile to their place on the edge of a salt water inlet known as Lagoon Pond. After unpacking my satchel, I had a sandwich with the Daniels and their housekeeper, Miss Mergendahl, who was a college student and the daughter of a mathematics teacher at Newton High who had taught Bette Davis. Red's parents, John and Margaret (Peg) were of Scottish ancestry and in their sixties. I got the impression that each had been on the stage and consorted with the likes of Sir Harry Lauder, for their anecdotes about him and his contemporaries reflected a close personal association. Mrs Daniels was a reincarnationist who believed that, in an earlier life, she had been Marie Antoinette. This presumption was based on a visit to the Palace of Versailles years ago when each room she entered seemed amazingly familiar. I enjoyed staying with these interesting people and meeting the friends of their son, Red.

The day after my arrival, Red and I climbed into a dinghy and rowed a few yards to his anchored motorboat, a wooden craft twenty feet long, with an open cockpit and single gasoline-powered engine. Several minutes of priming brought the engine to life and we putted up the lagoon to a girl scout camp on the port side opposite the boy's camp on the starboard bank. There, in a wooden recreation hall, I sat down at an old upright piano and, before several dozen girls and a couple of matronly chaperons, played the first song I had ever written:

"I know an island, far out in the sea,
That's always meant, paradise to me, Ohhhh,
Its name is Borneoooo.

And on this little island, way out in the sea,
I met a gal who's gonna marry me, Ohhh,
Its name is Borneoooo.

Chorus:

I was born in Borneooo

Where there is no ice or snow

Where the sun shines all the time in

Borneoooo

And each night when it gets dark

It's time to romance in the park

I'm the wild man of Borneooo."

The girls loved it and Red was amazed by this display of amateur talent. But Helen Moulton would have been deeply chagrined.

I should mention, though, that I had only written the music and at the time did not have lyrics to go with the tune. The audience was mercifully deprived from listening to an untrained voice that could have summoned hogs from Arkansas.

My week spent on the Vineyard was memorable: sliding down the brilliantly-hued clay cliffs of Gay Head (illegal to do now); floating in Vineyard Haven harbor when our sailboat was overturned by 12-foot waves; climbing in the rigging of coastal schooners anchored because of storm conditions outside the snug harbor; sailing among the gorgeous yachts of the New York Yacht Club during their annual August visit to the Vineyard, and watching the girl scouts race in their 18-foot wooden sloops built by the nearby Colby shipyard for a thousand dollars. At Edgartown, I tasted for the first time a new drink called Pepsi-Cola that cost 5 cents for 12 ounces! I watched the lobsterboats unload their pots in Menemsha from which, decades later, the Orca of JAWS would depart with Robert Shaw, Roy Scheider, and Richard Dreyfuss to search for the great white shark. And speaking of films, I saw The Wizard of Oz, starring Mom's boyfriend Frank Morgan, at the theatre in Oak Bluffs. In early September, just prior to leaving the Vineyard, the radio announced a declaration of war between Great Britain and Germany. I departed from the island contemplating what the future would hold for me and my country.

When I arrived at Woods Hole that afternoon and stepped off the ferry, I was surprised to see Dad waving from the cockpit of a cabin cruiser tied up at a nearby pier. The yacht belonged to the Currans, who were his patients and neighbors of the wealthy young widow who lived near the Charles River with her children and exotic animals. Arthur Curran was in the import-export business with an office in Boston, a large ranch-style home on the outskirts of Needham, and a summer home on a cliff in Osterville, Cape Cod, overlooking Nantucket Sound. The cliff house was our destination as we chatted about my week on the Vineyard and the disturbing radio announcement that, among other things, was certain to have devastating repercussions on Mr. Curran's business. No one in the United States could have envisioned what havoc that announcement would ultimately exert on the world at large. Little was said over dinner that evening about international events, but as we bade Arthur and Ethel Curran farewell and headed back to Needham, Dad seemed reticent, as if he anticipated something ominous was about to occur.

The fall term at high school was largely spent preparing for the college entrance examinations in mathematics, English, and general knowledge (equivalent to today's SATS); the day of reckoning to be held at Newton High School in November. Time was also devoted to playing the clarinet in a newly-formed band, as well as in the long-established orchestra. Angus Anderson, the shortest member of the band, umpahhed on the largest instrument, the Sousaphone which, whenever the band took to the field, looked like he was wrestling with a chromium-plated python. In the orchestra, a doctor's daughter beat the skins like Gene Krupa. The staccato rhythm of Gilbert and

Sullivan operettas (The Pirates of Penzance and The Mikado) jogged the brain as rehearsals required weeks of preparation before the musicians and actors melded their respective skills into a cohesive entity that was ultimately appreciated by the audience.

During the spring, I was busy as the manager of the school's baseball team that experienced a perfect season. My duties were unlike those of a professional manager, for I was responsible for managing the equipment as well as keeping up the spirits of the players. The most mind-boggling event of the year happened one Wednesday morning during an assembly when the guest speaker, Mr. John MacGregor, spoke about what was happening to the Jewish people in Nazi Germany. Mr. MacGregor was the owner of the instrument company of the same name, which manufactured stainless steel scalpels and surgical needles in a factory in Needham. He was also the district governor of Rotary for Massachusetts, well-traveled, and a person of considerable eminence.

"I have just returned from Germany and must report to you that what the Nazis are doing to the Jewish citizenry is absolutely revolting! I have personally witnessed Jews having high pressure hoses inserted into their rectums and seen these poor souls literally blown apart!"

The proclamation of Nazi deviant behavior by the esteemed gentleman fell on deaf ears as if no one understood what he was talking about. By contrast, an earlier speaker that year, Paul Siple, who was renowned as the "boy scout with Byrd" and a veteran of several polar expeditions to Antarctica, gave a lecture that was memorable and well received. A quarter century later, I had the pleasure of meeting Mr. Siple and telling him how much the class liked his talk. After citing a few memorable incidents he had related, such as Admiral Byrd's eyeballs freezing from the extreme cold, I asked him how he happened to be available to talk at Needham High.

"I was taking a graduate course at Clark University in Worcester at the time and welcomed the opportunity to make a few dollars," he answered. Sadly, Mr. Siple died of a heart attack shortly after he had lectured on Science Day, an annual event sponsored by the Temple University Chapter of the Society of the Sigma Xi to interest several hundred high school students from the Delaware Valley in considering scientific careers. Appropriately, although the occasion was scheduled the day spring arrived, the polar explorer was greeted with a snowstorm that delayed the arrival of the other participants.

As for Mr. MacGregor's comments, few people knew what was happening in the political climate of Germany other than that their rampant inflation was under control, they were beginning to gobble up land that was once theirs, and were repeating their obnoxious ways of torpedoing non-military passenger ships in the waters off Britain, against whom they had declared war less than a year earlier. They had lost their spycraft, the Hindenburg, which photographed military targets in the Boston area as it headed for finality at Lakehurst, New Jersey. As for the Jews, Needham had only one family, Bill Perlin and his wife, who owned a clothing store a block from where officer McLeod was gunned down on the railroad tracks adjacent to the Needham Trust Company bank. They were highly respected in a town comprised mainly of Roman Catholic parishioners represented by Italians and Irish, and Protestants of Anglo-Saxon origin who attended Congregational, Baptist, Unitarian, and Christian Science churches. The absence of a sizable Jewish population required the Perlins to attend a synagogue in nearby Newton. It would have made more sense if Mr. Perlin had shared the high school podium with Mr. MacGregor to provide further insight into the scourge of Nazism and attempt to explain the bizarre actions of these fanatics.

Before I realized it, the school year had terminated and graduation day, June 13th, was fast approaching. Before that milestone occurred, however, there was the class picnic, which involved taking a bus trip to Rowe's wharf in Boston harbor and boarding the Steel Pier, a ship that used to ply the waters between Havana and Miami, for a cruise to Provincetown. Those who suffered from *mal de mer* were ignominiously photographed by those who had sea legs.

CHAPTER EIGHT

The Harvard Years

(9,999+ 1 or 10,000 Men of Harvard)

The letter of acceptance arrived in the spring, April to be precise, the traditional month for exuberance or despondency, since the oldest college in the United States was founded in 1636. Living in Massachusetts gave me an edge over those who resided elsewhere, for the charter mandated male citizens of the Commonwealth be given preference if their scholastic and extra-curricular accomplishments were equivalent, but not inferior. Harvard was the only college I had applied to for a number of reasons, the least of which was its close proximity to home where the laundry facilities were located. Mother had suggested I apply to Dartmouth, the northernmost Ivy League school, which was not only highly regarded as an intellectual center, but also touted as a healthy environment due to the presence of fresh air swirling around the white mountains of New Hampshire. Ninety pound students, such as myself, might even become better athletes! Dad won out and, happily, Mom and I packed the car with sheets, clothing, and a typewriter and drove down Massachusetts Avenue to the gate that proclaims: *"Enter to grow in wisdom, Depart to serve thy country."* Parking adjacent to the sidewalk, we found the caretaker, signed in for keys, and proceeded to climb the stairs of Wigglesworth with a suitcase and several cardboard boxes. My room was located on the third floor and comprised a bathroom with white tile walls, a large room with fireplace, and three alcoves, each containing a wooden bureau with a mirror and a bedframe with a gray mattress folded like a tortilla. The other residents had not yet claimed an alcove, so I dumped my belongings into the one nearest the fireplace, contemplating that the furnace three flights below might experience a failure during the winter months. According to the caretaker, my roommates were Bill Snow of Abilene, Texas, and a fellow from Shaker Heights, Ohio, named Ed McLeod. They arrived hours later, long after mother had left, and I had eaten in the magnificent Student Union whose high ceilings and walnut-paneled

97

walls lent a library-like atmosphere to a delicious meal served by middle-aged waitresses overseen by a gorgeous hostess.

That evening, while sitting near the ash-discolored fireplace in the rug-free living-room reading a freebie Harvard Crimson distributed at the Union, I heard someone fumbling at the door.

"I'll be right there," I yelled. But the fumbler succeeded in opening the door before I could rise. "Hi, I'm Bill Snow from Abilene," said the tall, bespectacled young man, well-dressed in a gray suit. "Am I too late for supper?"

"Yes, the Union is closed now, but I'll join you for a snack." I introduced myself and, after he had unpacked his belongings from a bulging brown suitcase that resembled an overfed steer, we walked to Harvard Square and entered the Georgian restaurant facing Lehman Hall in Harvard Yard where, in the years ahead, I would consume a ton of apple pie with my parents as they returned me late each Sunday evening with freshly laundered clothes and a few textbooks. After Bill and I returned from the Georgian with full stomachs and enough caffeine-induced stimulation to guarantee a sleepless night, the third roommate had arrived looking tired and disheveled from a long and boring trip from Shaker Heights, Ohio, to South Station. Mercifully, he had eaten on the train, so we helped him unpack and put his alcove in order. The three of us then sat down on what seemed to be left-over furniture from the previous occupants and chatted about our aspirations. Bill and I were pre-med, while Ed was pre-law. Thus, Ed would not be enrolled in the hard core courses of chemistry, physics, and biology, nor would we find ourselves in political science and the famous History I course flamboyantly presented by one of Harvard's most renowned professors in an ancient brick building known as the New Lecture Hall. (Harvard has a unique tradition of naming their buildings in a strange way. For example, I took gym once a week during the Freshman Year in a large brick facility called the Indoor Athletic Building. If it were a building, which it was, and you were inside it, obviously you were indoors unless the architect forgot to cap it with a roof, which he didn't!) Regardless, it was obvious that Ed would not be studying with us unless he took sociology, psychology, anthropology, or geology to broaden his knowledge. Our discussion soon led to interests and hobbies. Ed loved classical music. He didn't say so, but we suspected he played the piano. Bill and I were John

Philip Sousa enthusiasts, loved march music (even circus music), and auditioned for a place in the famous Harvard band. Bill was successful and joined the band as a tuba player. I failed miserably on the clarinet and sulked back to Wigglesworth, realizing how lousy my contribution to the Needham Band must have been! The terrible irony of this was the tragic news I heard the following fall, that Bill had been killed by a truck while crossing a Baltimore street in heavy rain with a tuba in his arms! Texas lost a brilliant future physician and a wonderful young man.

A week after my arrival, a multi-hour session waiting in line at Memorial Hall (another unique structure known affectionately as the Gothic Cow barn), resulted in signing registration cards for English, German, Inorganic Chemistry, and Physics, and procuring subscriptions to the Harvard Crimson, Harvard Lampoon, and discount coupons at a variety of services. I quickly disposed of the laundry services. The following week, courses began and I found myself sitting in a classroom in an old, yellow clapboard house adjacent to the Student Union peering over a notebook at Professor Baker, a remarkable look-alike of the famous movie actor, Walter Abel.

Glancing at the 45 students, he proceeded to introduce himself and describe the course content:

"I'm Dr. Baker from California and this is English A, a composition course to hone writing skills for those who intend to make a living from journalism, and also to improve communicative abilities of future captains of industry, medicine, the military, and politics. You cannot estimate the millions of dollars that misunderstanding of the English language costs the citizens of this country. For example, some of the legal manifestos foisted on Americans by ignoramus politicians remind me of the many interpretations of prophesies by the oracle of Delphi to cover all possible situations.

As for those who plan to write short stories and novels, a word of advice: a few years ago, in the thirties to be precise, I wrote a novel entitled, Orange Valley, which sold 8,000 copies. John Steinbeck wrote The Grapes of Wrath, about the dustbowl "okies" of Oklahoma who fled their parched lands to the fertile fields of California. His book sold hundreds of thousands of copies and, with his other works, earned him many literary* prizes. My novel had a similar premise but was regarded as a copycat work. In short, be original and write about something of interest to the masses. My wife, Dorothy Baker wrote the biographical book, Young Man with a Horn, the life of the famous trumpeter, Bix Beiderbecke, which was so well received that Hollywood purchased the rights. (*In 1950, the movie starring Kirk Douglas and Lauren (Betty) Bacall was likewise well received.*)

Now the assignments for the month of September: 200-word essays on subjects of your own choosing will be due each week with the exception of the final week when a 2,000-word essay will be due. Each month will be the same with respect to the number of pages of my assignments except prior to a national holiday, such as Christmas, when the 2,000-word essay will not be assigned. Are there any questions?"

He peered at the student roster and then looked at me. "Are you related to Thomas Mann (pronounced Mahnn)?"

"No, not that I'm aware of." I replied. "My relatives are of British origin, not German. And I pronounce my name man."

He asked a few other questions of students and then continued:

"Now, about picking up your graded papers, across the hall from this room is a small room with a wall of pigeon holes with your names arranged alphabetically. Three days after handing in an assignment, your graded paper will appear in the appropriate slot. Use thin typing paper and a typewriter with black ribbon. Regardless of the clarity of your handwriting, papers written in ink will not be accepted."

---

*In 1962 he became the sixth American to win the Nobel prize in literature.

The hour passed quickly. Dr. Baker had assigned a 200-word essay on any subject from personal experience, and several hours remained before my next course in introductory German, was presented on the first floor of Sever Hall. I decided to see the buildings where my science courses in chemistry and physics would be given the following day. The chemistry building, known as the Mallinckrodt Chemical Laboratory, was a red brick, reinforced concrete structure which, upon entry, reeked of odors from organic and inorganic chemistry laboratories. In the midst of glass beakers, rubber tubing, and Bunsen burners, geniuses such as Louis Fieser advanced their areas of expertise by synthesizing vitamin K for medicine and napalm for the military. A large auditorium on the first floor seated three hundred students who faced a series of blackboards that spanned the front of the room. (Even that vast area was insufficient for Dr. Fieser who continued to scrawl chemical formulas on the walls above, below, and beyond when he ran out of blackboard space).

The ceiling of the auditorium depicted in exquisite detail the arrangement of elements in the sequence of atomic numbers of the periodic table, a situation often overlooked by professors whose examinations sometimes contained questions the answers to which were readily accessible by looking skyward! The atmosphere of the building, however, was as austere and cold as the masonry from which it was constructed. Yet, in the months to follow, those who taught in this morgue-like environment succeeded in infusing an enthusiastic approach to chemistry that dispelled boredom.

Impressed with the stateliness and aura of Mallickrodt, I walked to the Jefferson Physics Laboratory, a multi-story brick building with a facade that closely resembled the knitting mills of nearby Lowell and Lawrence. Unlike these now-vacant structures, however, the Physics Laboratory was bustling with activity, doors and windows opened to provide natural air-conditioning. I decided not to attract attention and disturb some future Nobel laureate's experiment. Tomorrow would be soon enough to investigate this architectural monstrosity.

Lunch was now being served at the Student Union, so I headed to Harvard Square, walked down Massachusetts Avenue, and turned left on Quincy Street. My German class was scheduled for 2:00 p.m., which allowed me about an hour to eat, pick up some things at Wigglesworth, and walk to Sever. My roommates were not at the Union, so I sat down at one of the tables with strangers and introduced myself. After a sandwich and coffee, I returned to Wigglesworth, grabbed a notebook and pen and headed for Sever.

The class met in a musty room on the first floor with thirty or forty wooden chairs, half of them now filled with young men. I sat near the front of the class and anxiously awaited the arrival of the professor. Within a minute or two after I was comfortably seated, the door at the front of the room opened and in stepped a middle-aged, slightly pot-bellied, red-faced man who resembled a burghermeister. He greeted the class in English, whose delivery and enunciation verified it was his second language.

"Velcome." He removed his jacket, draped it over his desk chair, and sat down.

"This course is for those who enjoy learning a new language and vish a foundation in German maybe to pursue a doctorate degree in the distant future. Class participation is my method of teaching. To avoid the embarrassment of being unprepared, I find this is the best way to learn. I'm not going to demand that only German be spoken in this class. That's childish! Next time we meet, you vill all have your books, so I vill now assign the first chapter, which contains a number of idioms that should be committed to memory. You should also purchase a German-English dictionary as suggested by the course description brochure. Now are there any questions? If not, class dismissed."

There were several comments out of ear-shot of the professor that were whispered as the students walked out of Sever.

"This guy looks like a taskmaster. He's a real German."

"Wonder if the FBI is spying on him? Maybe they're pretending to be students in his class! What happens if they flunk?"

Personally, I felt that he knew his subject and would be a good teacher. I had no idea, then, that one day he would miss his class, and I would volunteer to take over for the hour, giving birth to my career as a teacher!

That evening, having joined my roommates at the Union, we discussed the events that had unfolded since the alarm clock rang at 7:00 A.M. Bill had enjoyed his English A professor whose educational background included a stint at the University of Texas in Austin, and Ed had met a classical music enthusiast, Peter Gram Swing, son of the famous radio commentator, Raymond Gram Swing. Peter became Ed's roommate the following year, and, after graduate school, enjoyed a remarkable career as a professor of music at Swarthmore College.

The next morning I was seated under the periodic table ceiling at Mallinckrodt listening to Professor Forbes, a gentleman in his sixties with a Teddy Roosevelt mustache, describe a chemical reaction in inorganic chemistry:

"I take this beaker filled with blah-blah, and add it to another beaker containing blah-blah, and PRESTO, I get an orange-colored precipitate of blah-blah!"

At the appearance of the colorful chemical reaction, applause resounded from the audience. At each lecture, the same thing happened: new chemicals added to other chemicals created reactions that were colorful. Following the predictable applause, Professor Forbes neatly inscribed the chemical reactions involved on the blackboard. His course will long be remembered for several interesting things: LIFE magazine sent a photographer to Harvard to document activities there of interest to the general public. One morning, the photographer stood in front of my row and took a picture of Professor Forbes in action, What a splendid memento to show our offspring and grandchildren! Daddy or grandpa was sitting there in that class when the picture was taken! The other interesting thing that took place in his course was the sole experiment that I can recall which involved taking a dime (with 90% silver) and reducing it to a pellet of pure silver sealed in a nitrogen-filled glass tube! With today's coinage representing a complicated amalgam of zinc, copper, and what-not, this experiment must have gone the way of the do-do! On the other hand, the prescient nature of Harvard's officials, among other things responsible for the burgeoning endowment during the Reagan years, probably stockpiled pre-1964 dimes by the ton in anticipation of Professor Forbes' famous experiment!

One student became the envy of the entire freshmen class. He was the fortunate chap who happened to be standing next to the statue of *JOHN HARVARD* when the LIFE photographer passed by. He ended up appearing on the cover of the magazine **(See that boy standing next to the statue. That's your daddy!)**

The physics class in the afternoon was introduced by a man who was a Purdue graduate destined to receive a Nobel prize for developing (along with colleagues) the Nuclear Magnetic Resonance technique of ascertaining the structure of matter. He demonstrated how a television camera could pick up an image and project it on a glass screen. Harvard had procured the RCA camera from the New York World's Fair, which closed in the fall of 1940. It was positioned outside the building on a concrete ledge and, as the students in the auditorium peered at the cathode tube, the operator followed a young girl walking innocently on the sidewalk. Again, applause from the students.

After the physics lecture by the eminent Ed Purcell (eminent is redundant because all associate and full professors at Harvard are eminent), the class was divided into sections A and B for laboratory assignments. The laboratory session, involving experiments in mechanics, heat, light, and magnetism, was given in a 3-hour period on Thursday afternoons for my section (B). The laboratory was a large, high-ceiling facility smelling of varnished wood and filled with slate-topped desks on which were displayed the tools of the physicist—various kinds of scales, tuning forks, metric weights, and alnico magnets of different dimensions. A partner was assigned, the son of Dr. Parran, the surgeon-general of the United States. I recall that Richard was a slender chap who was proud of his father's accomplishments and competent as a co-worker.

Late that afternoon I joined my roommates to attend a tea given at the Phillips Brooks' House in Harvard Yard by President James Bryant Conant. It was the first time I had seen the person whose reputation as an educator and scientist had grown rapidly since the days when he started his academic career as an instructor in chemistry at Harvard. A number of rumors, truths, and half-truths had circulated among the newly-arrived student body concerning this man whose accomplishments were of gargantuan proportions. In short order, he had advanced to professor of organic chemistry and had elucidated the structure of chlorophyll, a plant chemical involved in photosynthesis that, among other things, transforms formaldehyde into sugar while removing carbon dioxide from the air and releasing oxygen into the atmosphere. In May, 1933, he had succeeded President A. Lawrence Lowell at Harvard and instituted a number of curricula and social changes that occasionally enraged those who were affected. (One of these was the elimination of geography as a college course, which Conant believed would be more appropriately served as a high school subject). On the lighter side, there was the rumor that, on weekends, Dr. Conant played with his Lionel trains in the basement of the beautiful brick mansion, that resembled a Nantucket sea captain's domicile, where Harvard housed its presidents close to the freshmen Wigglesworth dormitories. In the flesh, this awesome personage appeared as a thin, bespectacled man with a pleasant voice. He shook hands with each student, smiling graciously and, after this task was over, gave a brief address in which he discussed the social responsibilities of a Harvard man as an undergraduate student and as a citizen of the world in the future. Tragically, several hundred young men, who now listened attentively to these words, would soon lose their lives in the oncoming World War II, their names to be inscribed in gold on the wall of Memorial Church; their sacrifices never to be forgotten by future Harvard men and women, (In the early 70's, I saw Dr. Conant in a pharmacy at Hanover, N.H. He had recently purchased a home near Dartmouth College).

The following week I pulled my 200-word essay out of the pigeon hole and noted below the C+ grade a comment: *I enjoyed your descriptive reference to Medfield as a "sleepy town."* I had written about visiting the Cedars on weekends in the thirties when we left Needham at mid-day on a Saturday and stopped at Clement's drugstore in Medfield where the three of us had sundaes. Mine was chocolate ice cream plopped on a white paper insert that was held in place by a slot stamped out of the metal goblet-shaped container. Chocolate sauce was dripped over the ice cream a la Jackson Pollock (then was the right time to start acquiring his works), powdered walnuts were added, then fluffy cream capped with a maraschino cherry completed the concoction. According to Dr. Baker at the next class meeting, the grade distribution was: a couple of A's, a few B's, and a plethora of C's and D's. I was suspicious that Dr. Baker was on a strict diet and I had lost a few points for my description of the ice cream stopover in Medfield.

Classes were moving along smoothly as the days approached the end of the month. I had started my 2,000-word theme, the content of which dealt with John Holland, the inventor of the modern submarine (along with Simon Lake) and his attempts to interest the U.S. military in its potential as an offensive weapon. They were disinterested, so he turned to Russia. As a result, the submarine became the Soviet's most effective offensive weapon in its Navy. What prompted me to investigate this recurring situation (remember Billy Mitchell also had his problems trying to convince the military that an airplane could sink a battleship with a bomb, until he accomplished the feat before their startled eyes), was my interest in submarines in the late 1920's. I was returning from Provincetown when Dad turned on the radio and we heard that a submarine had sunk off the coast of Cape Cod. Divers were sent to examine the hull and over the radio one could hear the tapping of the crew inside trying to send messages to loved ones in Morse code. They soon ran out of air and the tapping stopped.

The Squalus out of Portsmouth, N.H., went down just before WWII and 33 crew members were saved by a rescue chamber that fastened to the hull. The submarine was eventually refloated, refurbished, and re-christened the Sailfish. It sent Japanese ships to the bottom of the Pacific in the months ahead and survived the war.

These were the reasons I had chosen Mr. Holland's Jules Verne-like creation to write about. My experience with the "sleepy town of Medfield" piece would not permit me to reveal that the crew of these submarines enjoyed chocolate sundaes with hot fudge topping, sprinkled with powdered walnut, capped with whipped cream and a red cherry.

The semester passed quickly. Thanksgiving came and went. The football season had ended with the annual Harvard-Yale skirmish, and students happily departed for the Christmas holidays. Bill, with a different time slot, finished his courses several days before Ed and I, and left for Abilene for what was to be his last Christmas. I invited Ed to attend a dance in Boston with me before leaving for Shaker Heights. He consented and we took public transportation to Needham for dinner with my parents before meeting my date, Phyllis, the daughter of Les and Maisie Cook. Nearby, we picked up his date, Margaret, daughter of a Protestant minister (father) and school teacher (mother). The car was my black Plymouth convertible which was quite comfortable provided you wore warm clothing and huddled in the front seat. Ed and Margaret, on the other hand, were shivering in the rumble seat exposed to frigid temperatures. As we headed down Highland Avenue excited about dancing at the Park Plaza hotel to the strains of a new orchestra, a car backed out of a sloping driveway directly into our path. Fortunately, there was ice on the road and we were crawling, but the jolt was enough to cause Phyllis to hit her head on the dashboard. Luckily, no one was seriously injured and the cars were undamaged (they built them in those days). Dad and Les arrived at the scene within a few minutes and Phyllis was treated for a swelling on her forehead the size of a hen's egg. We continued on to the dance where Claude Thornhill, a young orchestra leader, was debuting his swing band. Within a year, his band would be as well known as those of Shep Fields, Freddie Martin, Horace Heidt, and Stan Kenton.

Shortly after New Year's Day, it was time to return to Wigglesworth with a brain bursting with information soon to be evaluated for proper recall by examinations in the Gothic cowbarn, Memorial Hall. Built as a cathedral-like structure to honor those Harvard men who had made the supreme sacrifice in the Civil War in the Union Army, the creepiness of the interior, under normal conditions, would be sufficient to intimidate a visitor who had just admired the glass flowers of the University Museum. The nervous student about to undergo an academic inquisition in this atmosphere was further petrified to be handed examinations that resembled the page proofs of a Knopf novel by a stern-faced proctor.

There were only three examinations: German, Physics, and Chemistry. English A did not require a final because grades were calculated from the 200-and 2000-word themes required throughout the semester. The examination period was given over a period of two weeks, which meant that if you finished after a couple of days, you had plenty of time for fun. I found the tests to be what I had expected: Introductory German was difficult; Physics and Chemistry were fair with no off-the-wall questions. I finished them during the first week and had another week to relax. I treated myself by seeing John Ford's movie, Stagecoach, starring John Wayne, Donald Meek, Thomas Mitchell, John Carradine, and Claire Trevor, filmed in Monument Valley.

The second semester began with the same course titles. English A now discussed the works of American authors, instead of using the period to critique students' prose. During the first semester, Dr. Baker selected a paper (usually a 200-word essay) and read it before the class, inviting criticism while its author remained anonymous. No holds were barred. If the work stunk, the author could be easily identified cringing in his seat. If the essay were of the 2,000-word variety, the professor would read only excerpts of exemplary prose, or examples of garbled verbiage. One student in the class stood head and shoulders above the rest in writing skills; he was a native of Mexico, whose papers were always selected, anonymously of course, for class scrutiny. The sentence structure and concepts of his works were as easily identified by the class as those of Joyce or Hemingway. Strange that pen scratchings can identify an author's works as readily as brush strokes can unmask the artist who painted a masterpiece!

Introductory German had advanced beyond the memorization of idioms and simple phrases. Now the class was translating the poetry and prose of renowned writers. Mein *Kampf* by A. Hitler, was not recommended reading and I'm not sure that the quality of the writing was in the same league as that of even the minor German authors, let alone the magnificent literary contributions of Thomas Mann! I still had considerable difficulty recognizing German s's and f's. (Sever Hall could use stronger lighting.)

Physics class had a new lecturer, a foreign professor, whose accent made it difficult to fathom his enthusiasm for fundamentals of light, refraction, color and optics. However, the laboratory experiments were far more fascinating than those of the first semester, and demonstrated many aspects of the science that were applicable to life beyond the university sphere. In other words, practical knowledge.

Inorganic chemistry had become more complicated for Professor Forbes had either departed for the semester to do research or gone on vacation. His replacement was aware of the periodic table embossed on the auditorium ceiling because his examinations never referred to the atomic numbers of the elements. Imagine my astonishment when I saw Dr. Forbes, hale and hearty, at my 25th reunion in 1969! He must have been in his late eighties. Working with all those chemicals, which would have caused cancer in lesser men, had prolonged his life! The second semester progressed more rapidly than the first; perhaps this had something to do with the coming of spring. Soon it was time to await the letter announcing where I would live the next three years. It turned out to be Kirkland House—Room H-32 to be precise.

There was a minor problem associated with my new living-quarters at Kirkland House—no roommates. Ed, the classical music enthusiast, had decided to room with Peter Gram Swing, a future music historian of Swarthmore College, while Bill Snow, whose younger brother was now a member of the freshman class, had decided to room with him. Because freshmen are housed in dormitories in the Yard and upperclassmen reside in houses near or bordering the Charles River (i.e., Lowell, Winthrop, Kirkland, Eliot, and Dunster), Bill got special permission to choose a university approved private home that was reserved for those who had applied late for freshmen or upperclass housing. The result was, come September, I awaited the arrival of my new roommate, as I had done the previous year at Wigglesworth. When he finally arrived, he looked at me and asked:

"Are you Jewish?"

"No," I replied, somewhat taken aback by the abrupt manner he had posed the question. "I'm a Congregationalist."

I'm reminded of an incident that occurred during my 50th reunion when the alumni and their wives were seated in Symphony Hall enjoying a Boston Pops concert. The conductor was a young man who had come from classes in Tanglewood to entertain the surviving members of the war class of 1944. At the conclusion of stirring music by Wagner, someone in the first row stood up and shouted "Heil Hitler!" The class and their wives gasped at this rude indiscretion and my mind wandered back to the fall of 1941. Was this the same person who had asked me the question: "Are you Jewish?"

Several days later, I had a new roommate from Needham's neighboring town of Natick. His stay was short-lived for, one Sunday morning at 8:00, Ralph Pfeiffer turned on the radio and, moments later, shouted that the Japanese had attacked and sunk our battleships at Pearl Harbor! It was one of the few weekends when I had chosen not to go home. Within a few weeks, Ralph had signed up with the Coast Guard and departed. After the war, he became an executive with Pan-American airways.

The second semester I moved to another room at Kirkland House where my new roommates, Richard Robinson Wood and Albert Gjerding Olsen, resided. I was destined to spend many unforgettable moments with these compatible people, who not only endured my unique sense of humor, but also expanded my social horizons by introducing me to their friends. Unfortunately, Al, who held a doctorate degree in physiology (endocrinology) from Harvard (1951) and taught vertebrate physiology as a professor of biology at Brandeis

University, died of a heart attack in June, 1978. He was unmarried, leaving as sole survivor—Mr. Google, his beloved basset hound. Dick Wood, father of three children, enjoyed considerable success in the real estate business. At our 50th reunion, Dick, still resembling a class J-boat yachtsman, and I, the professorial landlubber, toasted one another as surviving roommates.

The courses for the first semester of the sophomore year represented a vast departure in subject matter from those of the freshman year. Inorganic chemistry had become organic chemistry based upon the presence of the carbon atom, the cornerstone element of all living things. Its foremost protagonist at Harvard was German-born Louis Fieser whose lecture presentations were vigorous, stimulating, and frequently sprinkled with interesting anecdotes. In his department, a young doctoral student named Woodward was synthesizing complicated organic structures and would win a Nobel prize for recreating strychnine and quinine from simple chemicals. Dr. Fieser, at the time I was enrolled in his course, was creating napalm or jellied gasoline as a new type of weapon. An acquaintance of mine (Dr. Stern of Temple) provided the name for this unique creation from its key ingredients: sodium, Na and palmitic acid, palm (+ gasoline).

Introductory psychology was presented by a Robert Benchley look-alike, the famed experimental psychologist, Dr. Edward Boring. To demonstrate reflex activity, he fired a starter's pistol during a lecture and the entire class rose as one (about 12 inches). He had received his doctorate degree from Cornell University after studying nerve regeneration by cutting small nerves in his arms.

Economics A was taught by Professor Triffin who became a professor at Yale during the war years. His graphs of supply and demand curves were accompanied by vivid descriptions of the business of business. His frequent reference to the Cadillac as a kad-E-Lac drove home the fact that possession of this vehicle was the reward for financial success.

Biology, my first introduction to a subject that would be my major, was taught by a thin, articulate professor who had achieved fame as a paleobotanist. Professor William C. Darrah was an outstanding lecturer whose task was to inculcate beginning biologists with the characteristics of living creatures without eyes and feet, the world of plants. Botany can be tremendously boring, learning about gymnosperms, fungi, xylem, phloem, and the unappreciated features of leaves and seeds. Later, when I became aware of the medicinal properties of plants incorporated in the science of pharmacognosy as a mainstay course in the pharmacy curriculum and now referred to as medicinal chemistry, I silently thanked Dr. Darrah for providing the botanical basis for understanding the dynamics of this fascinating medical subject.

I had not blindly entered college as a dilettante, endeavoring to discover an area of academia that might entice me to continue further, even culminating as a career choice. My parents would not have allowed me the luxury of dawdling in courses that were pre-law, pre-theological, pre-business, or pre-med, as those with unlimited funds had done in expensive Ivy-league schools. Like the children of circus performers indoctrinated in the sawdust ring almost at birth, a single child of a physician and nurse is destined to become a member of the medical profession, or a plumber, electrician, or roofer. To reduce my chances of becoming a skilled, blue collar worker, I was exposed to several operations, scrubbed and covered in oversized, freshly-laundered, green pajama-like clothing, and allowed to stand inches away from the surgeon.

The first operation at Needham's Glover Memorial Hospital was a Caesarian. The surgeon was one of Boston's best: middle-aged, red-haired, and hot-tempered; a person I shall call Dr. X. When Dad and I arrived and entered the doctor's dressing room, Dr. X was already there, carefully removing his pants and starting to take off his shirt and undershirt. Suddenly, a chubby nurse burst into the room and stared at the three of us.

"If you burst into this room again without knocking, I'm going to rape you!" snorted Dr. X.

She turned red-faced and fled.

Dr. X uttered a few oaths and continued undressing. In a few moments, he went to his bag and pulled out a notebook.

"Just reviewing the procedure," he said, briefly removing his glasses to glance at me. "There are two kinds of Caesarians—high and low. I must decide which is more suitable for this patient."

Within a few minutes, we were in the operating room staring at the huge belly of a young woman. A nurse coated the abdominal surface with an iodine solution which stained the skin a peculiar shade of orange-brown. The anesthetist administered a spinal anesthetic between the second and third lumbar vertebrae at a point called the cauda equina (horse's tail) where the spinal nerves float in cerebrospinal fluid without the danger of being pricked by the needle. Before we entered the room, the anesthetist had withdrawn a small volume of fluid from the spinal canal to which the local anesthetic was added. Thus, no additional pressure was exerted on the brain and spinal cord unlike what would have occurred if the tap had not been performed. *(Headache was the most frequent side effect when this procedure was overlooked).*

An incision was made in the skin with a gushing of bright red blood as the blood vessels beneath the surface were severed. Gauze bandages quickly sopped up the mess, whereupon a deeper incision was made, this time in the thick musculature of the uterine wall. More blood gushed out and suddenly the baby appeared as a slimy, pink-colored creature. Working feverishly, the surgeon stitched the uterine walls and overlying tissue while a nurse gently washed the baby, having severed the umbilical cord a few inches from its entry into the stomach wall. Deftly, Dr. X sewed the skin together and daubed the surface with additional iodine solution. All this had taken place within several minutes.

However, for a reason not made clear to me, the mother died the next day. I was shocked upon hearing this. Like most laymen, I thought that this was a safe procedure. Dad wasted no time in setting me straight.

"Son, there's no such thing as a hundred per cent safe operation, You would think that a tonsillectomy was completely safe. Yet, there are hundreds who die each year from excessive bleeding, traumatic stress, or too much anesthetic during or after the removal of tonsils."

A month later, I was invited by Dr. X to see another operation, this time in Braintree where an elderly man was about to have his stomach and part of his small intestine removed because of cancer. Mother and I drove to the hospital which was as modern a facility as the one in Needham. After donning the appropriate gown, I joined Dr. X in the operating room and carefully observed the procedure. When the stomach and intestines were exposed, Dr. X showed me the sections he was about to remove—ugly looking tissues adjacent to normal cells. It seemed as if the contact of the intestine with the stomach wall had somehow caused the latter organ to become infected with the disease. While the surgeon carefully excised the cancerous tissue, the patient suddenly coughed from reflex activity and all of his intestines popped out of the abdominal cavity. It took several nurses a few minutes to bathe the intestines in dilute alcohol because they had come in contact with cotton balls placed next to the patient. However, after a lengthy washing, the intestines were plopped back into the abdominal cavity where peristaltic movements would eventually place them in the correct position. The patient recovered and lived several years, according to Dr. X.

The final operation I observed at the Glover Memorial Hospital in the presence of the illustrious Dr. X. This was an appendectomy in a young man who had received a spinal anesthetic (probably procaine in those days) injected between the second and third lumbar vertebrae (a technique pioneered by Temple University's Dr. Babcock). The surgeon deftly poised his scalpel over McBurney's point below which the infected appendix was located, and then made an incision. The youth let out a yell and sat up! Apparently, the anesthetic had not taken effect in numbing spinal nerves (some drugs have a latent period before an effect occurs; others may act at once). The anesthetist should test the sensory response of the patient with an external stimulus such as a pin prick before the scalpel is used. A moment later, the operation continued and soon the appendix, looking like a swollen red worm, was quickly removed and the incision sutured with catgut. The sutures used to seal the portion of the large intestine (cecum) from which the appendix was excised were absorbable; those sealing the skin and underlying tissues were not and thus required removal a few days later by the surgeon.

One of Dad's medical friends was an eye, ear, nose, and throat specialist, a graduate of Harvard Medical School, named Chester Mills who was also a Needham resident. Before his marriage in the thirties, we were invited to spend a weekend with him and his sister Nettie at Hough's Neck on the Massachusetts coast. The morning after our arrival, while eating breakfast he asked me:

"How are the fried eggs?"

"Where's the bacon?" I responded. This was one of many embarrassing moments I subjected my dear parents to at the age of 7.

Dr. Mills married a pretty nurse and had two children—a boy and girl. (The boy became an automobile mechanic; the girl, a concert pianist).

In the manner of Dr. X, he invited me to witness a tonsillectomy in a young girl who was first anesthetized with ether in a sitting-up position, whereupon a wire-like device was wrapped around her tonsils and tightened, severing the lymphatic blobs of tissue. Then, a stainless steel instrument shaped like the shovel part of a steam shovel was used to remove the adenoids. A lot of bleeding occurred at first, but a coagulant solution quickly arrested the flow.

I asked Dr. Mills if he'd ever made a mistake and removed the uvula (that funny-looking thing hanging down from the roof of the mouth), and he answered: "Once in a while."

Under my parents' ploy that I couldn't become a boy scout until I'd had my tonsils and adenoids removed, Dr. Mills came to my house one day and, using the operating table in Dad's office, successfully removed the offending structures. I'll never forget the experience of drinking orange juice the next day. Even ice cream didn't soothe the irritated area. The outcome was I did join the boy scouts and earned a music merit badge that I proudly wore on a sash in the Memorial Day parade. I soon attained first class status and heard the famous mayor of Boston, James Michael Curley, deliver a speech at my first jamboree held in Boston Garden.

Having seen some of the dynamic career options of the medical profession-an appendectomy, Caesarian, gut re-section in a cancer patient, and a tonsillectomy, and few of the negatives, I had entered college with the goal of becoming a member of the medical profession. I was not knowledgeable about the other subdivisions of medicine such as dermatology, pediatrics, clinical pharmacology, urology, hematology, rheumatology, and parasitology, but I had seen a few of Dad's patients before and after their arrival at the office. A lot of his work was seasonal: July 4th and 5th, for example, required the stitching of fingertips blown off by torpedoes (resembling Hershey kisses but were brown bags of explosive powder), and the treatment of burns from Roman candles, sparklers, and skyrockets that sometimes ignited prematurely or misfired. In those days, hastily assembled wooden booths draped in bunting were set up in Memorial

Park at the foot of the high school where any kid with spare change could purchase firecrackers and punk to light them.

Later, in August, swimming injuries would occur as feet were penetrated by rusty nails or glass, bones were broken from amateur wrestling, and head injuries happened from a foolish disregard for the depth of ponds, lakes, and even quarries (where floating debris caused broken necks). In the fall, football injuries, consisting of broken legs, sprained ankles, and snapped collarbones, predictably occurred. Basketball injuries were rare and did not result from bodily contact but from knees hitting the floor. In those days, bodily contact was penalized, for it represented sloppy playing.

Then, there were those patients who suffered from depression (usually in the winter) and acute or chronic drug intoxication. For those who had taken too much alcohol within a short time the depression was counteracted with a hefty dose of black coffee rectally. Chronic alcohol intoxication required the administration of paraldehyde by the same route and this drug was also used to quell prison riots. (I once had a student ask if the rectal route were used in the latter circumstance!)

Accordingly, I was aware of the entire gamut of human frailties of the mind and body with firsthand information about what a general practitioner was involved with on a day to day basis. On the negative side, I was also cognizant of the lack of appreciation some patients had for Dad's medical skills, for they were negligent in paying bills and probably silently rejoiced when Dad passed away because they were taken off the hook. Yes, these characters exist and are not characteristically represented by the *nouveau rich,* but by the phony, showy types who, on the surface appear to be well to do, but in reality are four-flushers who live from paycheck to paycheck.

Now I found myself listening intently to Dr, Ficser's lecture in organic chemistry, realizing that this was a pre-med core course packed with practical knowledge for future physicians and medical researchers. The brief announcement at the end of the lecture in Professor Fieser's characteristic New York accent sent a chill throughout the audience:

"The final laboratory grade in May will depend upon the student's ability to recall one of the experiments that was performed during the second semester. You will not be allowed to bring any notes to the laboratory, which means that you must commit to memory all of those experiments. Also, when you enter the laboratory, you will select a number that will indicate which experiment you will perform. Not everyone will do the same experiment. And, finally, after receiving your number and identifying the experiment you will duplicate, go to the supply room with a list of the chemicals and glass equipment you will need."

In the middle of the semester, one of Dr. Fieser's graduate students inadvertently added concentrated sulfuric acid to potassium permanganate ($KMn04$) and an explosion produced a purple-faced instructor who was taken meekly to a nearby hospital. It required weeks for the stain to leave the epidermis despite constant washings with soap. Of course, the thought entered every student's mind that this kind of accident would be the ultimate disgrace if it happened during the final laboratory examinations in May. Imagine stepping into the outside world with a purple face. People would think you participated in a blueberry pie-eating contest and forgot to wash your face.

Professor Boring's lectures were replaced with those of Dr. Gordon Allport during the second semester. The most interesting concepts in this course were mentioned in those areas of psychology that touched upon abnormalcy. Lectures on the psychoses of manic-depression, the schizophrenias, melancholy, paresis (caused by syphilis), and drug-induced aberrations in human behavior (such as schizophrenia caused by snuffing cocaine) were fascinating and prompted some pre-meds to explore in greater detail the specialty of psychiatry.

The sequel to Dr. Triffin's *Economics A* was Pitirim Sorokin's *Sociology*. This sad-faced professor, a fugitive of Stalin's Soviet Union, taught a subject that related how greed in the highest level of government led to war and genocide, and that a communist regime did not distribute wealth equally to its citizenry, but allocated it instead to the equivalents in our capitalistic society of CEO's, who called the shots. He told us about his most embarrassing moment when he sat next to Lenin's wife at an Embassy banquet in Washington when the conversation was as cold as the Beluga caviar served at the dinner.

Professor Fred L. Hisaw followed the botany course, so ably presented by Dr. Darrah, with introductory zoology lectures in the second semester supplemented with laboratory sessions of frog dissections and anatomical exhibits of invertebrate specimens preserved in ethyl alcohol, especially members of phylum Mollusca. The student quickly realized that the presence of a notochord qualifies an animal, however insignificant-looking, for membership in the hallowed phylum Chordata, to which he belonged. Furthermore, the class learned the words to that wonderful invocation to our vermiform ancestor, and occasionally sang "It's a long way from Amphioxus, It's a long way to go!"

Professor Hisaw was a heavy-set, middle-aged gentleman whose easy-going attitude and Missouri twang belied a person of considerable intellect and scientific acumen. His specialties were mammalian reproduction and endocrinology. In addition to unlocking the intricacies of the menstrual cycle in Old World monkeys (Macaques) from India and Malaysia, he had, early in his scientific career, isolated a human hormone called relaxin. It was first detected in the tridecemlineatus (13-line) ground squirrel, the function of which is to widen the pubic bones of the birth canal at parturition. The polypeptide hormone, unique among reproductive hormones in not having a steroidal configuration, is secreted by both the human ovary and placenta. Consequently, it was believed to play a comparable physiologic role in human beings. Acting upon the pubic symphysis to cause relaxation, its appearance in the blood just prior to parturition in both humans and ground squirrels established its identity as a newly discovered hormone. Some scientists, however, predicted that the only valid human use for such a hormone was to allow ballet dancers to perform the split with greater ease! Regardless, it was prepared as a pharmaceutical product under the name Releasin (registered), and more widely employed for additional basic research studies than for obstetrical use.

I encountered Dr. Hisaw again in my senior year when I enrolled in his graduate course in endocrinology. I can still visualize to this day his blackboard sketches of what happens when a mature female monkey is given high doses of estrogen and progesterone and then the medication is abruptly stopped: she menstruates. Then, on another blackboard he portrayed a monkey without drug therapy. It was anesthetized and the spinal cord was severed: again menstruation. Thus, various kinds of stimuli or physiologic insults can elicit the same response. The stress from a severe automobile accident can likewise induce menstruation in an otherwise normal, non-pregnant, mature human female. Although the course was interesting, a companion course in graduate mycology (fun with fungi), given by Dr. David Linder, consumed a great deal more of my time and was far more demanding.

At the conclusion of the sophomore year in June, I spent the summer working at Tillotson's factory dipping metal molds into hot liquid rubber to form ink sacs for fountain pens. On weekends, I drove to Mirror Lake for two days of swimming, boating, and fishing. I caught my first pickerel, a 16-inch beauty, but refused to eat it.

In addition to demanding academic courses for the development of the mind, the college provided mandatory enrollment in athletic exercises each week during the freshman year, which were held in the *Indoor Athletic Building,* and supervised by a former first baseman of the Red Sox. The uniform was comparable to the outfits worn by prisoners portrayed in Alcatraz movies—gray, over-sized, woolen material with short-sleeve blouses and baggy pants. Before a student became eligible to participate in the various exercises the Red Sox veteran evoked from a vast repertoire of bone -crushing exercises, a physical examination by a university physician was required. Apparently, everyone passed (even those who suddenly died of heart attacks years later) for, at each session, half the class of nearly one thousand strong, shook the building as they jumped up and down. Without these heart-warming exercises (literally), many of the later mid-life fatalities might have occurred years earlier.

After the freshman year, a student was required to indulge in a sport that required physical activity (chess was exempt). Tennis, squash, and badminton were popular, but, as a true Aquarian, I chose crew. At the start of the sophomore year, I walked down Boylston Street, crossed the Lars Anderson bridge spanning the Charles River, and turned right. A few yards beyond stood the Newell boathouse, a dark green-stained, cedar-shingled bungalow with a sloping black roof, that protected the spartan headquarters of Harvey Love, the popular crew coach, and also served as a storage facility for numerous shells horizontally suspended one atop the other like giant noodles from the ceiling. A gymnasium with rowing machines, lockers, and shower facilities occupied the remainder of the building. Barn-like doors, fronting on the river, opened wide enough to enable crewmen to carry their shells overhead on outstretched arms down a sloping deck to the Charles River where they were launched beside a float to which the Leviathan was tied. This famous (or infamous) craft was actually a training raft equipped with fixed rowing gear on either side and manned by landlubber students who'd never grasped an oar in their callous-free hands. To view this vessel underway with its crew struggling to row in synchronization reminded spectators on the shore of a Laurel and Hardy movie. For most neophytes, it only required a few hours of travail before the student was ready to try his hand (and body) in the real thing.

The shell was a delicately -balanced, thin-walled, needle-shaped boat constructed of cedar planking with oak ribs. Eight sliding wooden seats on individual metal tracks accommodated the rowers, while a single immovable seat at the stern belonged to the coxswain. Wood plates, embedded in the sides of the hull, were there to respond resonantly with a loud clunk when tapped by the coxswain to establish a rowing tempo. This was the seat occupied by my roommate, Albert Olsen, who was steering his crew around a bend in the Charles River in February when he collided with another shell and everyone ended up in the freezing water.

I should mention that the coxswain's role, in addition to tapping the sides of the shell to establish a rowing rhythm, is steering the craft with both hands holding a loop of rope attached to the rudder. Traditionally, the coxswain's weight is under 150 pounds, whereas the rowers are above 180. That totals at least 1,590 pounds of human flesh equal to the gross weight of an Indy race car! The weight of the shell and its accoutrements tipped the scales at less than half the total weight of its eventual occupants. And the cost of the shell in the 1940's was 1500 dollars. Harvard had a generous benefactor who replaced badly damaged boats with spanking new ones each year.

Wearing skin-tight shorts with a leather seat and a skimpy cotton blouse, I quickly learned to grip the oar with both hands and row in synchronous rhythm, turning the oar slightly when the stroke ended in the water to cut air resistance. I also took turns at being coxswain.

Al Olsen, who preferred using his middle name Gjerding instead of Albert, had the physique of a jockey. The son of the treasurer of the Bond Baking Company, his home was in Pleasantville, N.Y., near the *Reader's Digest* headquarters. He loved classical music, especially works by Nicolas Andreievich Rimsky-Korsakov (Scheherazade) and Wilhelm Richard Wagner (Tannhauser), and, therefore, was in an unswerving position to critique the song entitled "Let's Remember Pearl Harbor", as one of the worst musical compositions ever written! If it were played on the radio while he was studying, he flew into a rage. On the other hand, he loved my awful composition, "I was born in Borneo," and the opening sentences to an essay I wrote in high school describing the chateaux of France:

"In the fertile valley of the Loire, the scenery is fresh and delightful. Everywhere one sees beautiful trees, fine flowers, and fields that abound in the products of the land." *If only he had been my freshman English instructor.*

I last saw Gjerding at our 25th reunion when he recited those lines to my astonished wife and children. He also revealed some of my idiosyncrasies he had observed during our three-year stint in Kirkland House. It dawned on me during this reminiscence how many of his shenanigans I had missed by going home on weekends.

Crew coach Harvey Love, a small-boned man with a handsome visage, died following a heart attack in January, 1963. Gjerding Olsen, as a Professor of Biology at Brandeis University, brought Fred Hisaw's enthusiasm and endocrinologic background to his students in Waltham where he died of a heart attack on June 4, 1978. How enraged and disgusted he would have been to hear a classmate yell "Heil Hitler" at the conclusion of a Wagnerian composition at his 50th reunion!

My first association with parasitology occurred when I enrolled at the start of the junior year in Professor (actually Associate Professor) Lemuel Roscoe Cleveland's course. As soon as verification of my desire to major in biology was recorded, the university assigned Dr. Cleveland as my advisor. Instead of introducing myself at his office in the strikingly beautiful, brick Biological Laboratory edifice, guarded by two life-size, bronze rhinoceroses (sculpted by a woman) at its entrance, I decided to look up his course listing and enroll in the one scheduled for the first semester. A week later, when I attended his class, a middle-aged gentleman with a round, pleasant face entered smiling and immediately introduced himself and his background. The class was soon made aware that here was a human laboratory, a person who could produce a vast number of parasites (including Plasmodium vivax-a malarial organism) when the laboratory part of the course required their presence. Furthermore, we became aware that this modest man had solved the mystery of why termites consume large amounts of wood. In their gut, he discovered trypanosomes that converted cellulose into simple sugars upon which the termites fed. Despite his parasitic infestations, he seemed to be in excellent health.

As the course progressed, I soon learned that the African continent teemed with horrible parasites—intestinal tapeworms that grew thirty feet in length and robbed the body of essential nutrients; malarial mosquitoes that infected the blood with spindle-shaped sporozoites destroying red blood corpuscles and causing chills, fever and death; worms as slender as dental-floss curled up in eyeballs and eventually inducing blindness. The laboratory manual was authored by Dr. Raymond Cable, one of the country's outstanding parasitologists, who became a friend when I met him in the late forties at Purdue. Professor Cable was an authority on flatworms inhabiting the gut of seagulls.

Dr. Cleveland was an inspiring counselor and dedicated teacher who conducted an interesting and practical course. The mark of greatness of a teacher and his or her subject is the amount of knowledge one can recall decades later. I think I could still pass some of his old exams.

Kirtley Mather's *"geology for rockhounds"* was unquestionably the most exciting subject given at Harvard, and included laboratory sessions throughout the junior year that required field trips to graveyards to examine the geology of tombstones. This activity was the only link with the medical profession. The composition of the headstones, ranging from Vermont marble to Quincy granite, provided superb laboratory material. During Dr. Mather's course, the class listened to lectures by "Sleepy" Wilson, the world's foremost authority on trilobites, the ancient ancestor of the horseshoe crab.

"Why is he called Sleepy?" someone inquired.

"You'll soon find out!" came the reply.

The first yawns were heard five minutes after Professor Wilson began to describe a horseshoe crab's similarity to a trilobite fossil.

Professor of Physical Anthropology, Oxford-educated Dr. Earnest Hooton, lectured, during the first semester, in a course second only to Dr. Kirtley Mather's "dinosaurs, diamonds, and diatoms" presentation (along with "Sleepy" Wilson's contribution on trilobites, and several graduate students' offerings in petroleum geology) for humor, cerebral enlightenment, and sheer fascination. As the author of many well-received texts, i.e., Up from the Ape and Man's Poor Relations, Professor Hooton was regarded as a premier anthropologist on both sides of the Atlantic, and as a brilliant lecturer with a wonderful sense of humor. Any conversation centered around Dr. Hooton, in public or private, soon mentioned that he had a son named Newton. Yet, no one among the student body had ever seen Newton Hooton or even cast eyes upon his likeness in a photograph.

Physical anthropology usually deals with the morphological features of *man* and the evolutionary progression of his (and her) ape-like ancestors, usually starting with the discovery of a piece of a jawbone and a few broken rib fragments scooped up from an African plain, to the Michelangelo-touted beauty of present-day *homo sapiens.* But Dr. Hooton's course was anything but usual. His subjects were the great apes—the gorillas, orang-utans, gibbons, chimpanzees, the Old World monkeys—hamadryas baboons, mandrills, rhesus monkeys, the New World monkeys of South America—the howlers and spiders, and the primate suborder—the goggle-eyed lemurs. For each animal, he described its morphologic and behavioral characteristics and, sometimes with tongue-in-cheek, showed how similar these were to our own. For example, the agile, rascally gibbon has a skeleton that is a miniature version of that of *Homo sapiens.* And, of course, the rhesus factor was first detected in the monkey with the same name.

Our ancestors comprised a disturbing part of the course, for they seemed like Hollywood monsters with misshapen heads, hairy torsos, and overgrown brows. Neanderthal man, Cro-Magnon man, Piltdown man, Minnesota woman (100,000 years old), and Pithecanthropus erectus (Australasian man) were described in skeletal terms first and then, what they would have looked like, with muscle and skin added. Piltdown man turned out to be as authentic as the Cardiff giant, a gypsum statue found in Ohio, that was believed to be the mummified remains of a gigantic man whose skin had pores because someone had poked it countless times with an ice pick! Piltdown man was a chimp's jaw to which were added bone fragments of other primates to fool the experts. It worked for a few decades. I wonder if anyone in the class requires a grade adjustment based on this late finding?

The most memorable part of the course took place during a slide presentation when Dr. Hooton, projecting a photograph of a mountain gorilla standing on all fours, stepped between the projector and screen and, crouching to pick up something, caused his shadow to block the outline of the great ape with no overlapping.

My final course for the first semester of the junior year was the biology of vertebrates given by the world-renowned Dr. Alfred S. Romer who, as a paleontologist and zoologist, specialized in a field that was the antithesis of Dr. Darrah's expertise in paleobotany. I couldn't have enrolled in his course at a more exciting time! A fish, believed to have been extinct for 75 million years, had been caught in a fisherman's net off the east coast of South Africa in 1938. Known as a coelacanth (Crossopterygian), the prize catch was recognized belatedly as an historic find, but extensive decomposition made precise anatomical study of its internal organs impossible. A black and white photograph of the strange fish, packed in ice like a Pacific salmon from Seattle, hung on the wall of Dr. Romer's laboratory. It should have been accompanied by a photograph that captured his ebullient expression when he learned that other coelacanths had been caught and quickly preserved in ice.

When one realizes that, prior to the coelacanth's capture, the only knowledge about them had come from fossilized remains (comparable to observing a pencil tracing of Da Vinci's Last Supper instead of seeing the painted wall version). The repercussions of this fantastic discovery were multifold: the possibility of other ancient forms of life being found in the ocean depths; the dissection of the internal organs confirming the long-held belief that these creatures were the ancestors of terrestrial vertebrates; the presence of lungs as a prerequisite for a terrestrial life; and, finally, the thickness of the lobe-fins that could have served as appendages to transport them from drought-stricken areas.

Dr. Romer's course included a laboratory that utilized Libby Hyman's laboratory "cookbook" of detailed instructions pertaining to the dissection of the cat. I had the pleasure of listening to Dr. Hyman lecture on invertebrate biology a few years later when I was studying marine biology at the MBL at Woods Hole. She had the largest nose I had ever seen on a human female and, with a cold she was unable to shake, her lecture sounded as though it were accompanied by a French horn! She was probably the most intelligent being I would ever see in person—a premier authority, and author of at least a dozen books about her greatest love—protozoans barely visible to the naked eye—paramecia, amoebae, and other miniscule creatures with and without cilia.

Dr. Romer's lectures, on the other hand, discussed the evolution of reptilian jaw structures in Tyrannosaurus rex and other notorious dinosaurs, and how they differed from the jaws of present-day mammals. How he would have enjoyed Michael Crichton's Jurassic Park, as portrayed in Steven Spielberg's brilliant film. And he might have agreed with several of its scientific assumptions, i.e., birds being the heirs of dinosaurs (but what about the avian evolution of the jaw and what happened to their teeth?).

During Thanksgiving weekend, a seemingly innocuous football game was played in Boston's historic Fenway Park before 40,000 cheering fans. Everyone expected the undefeated (in nine games) All-American Eagles of Boston College to defeat the less talented players from the Worcester-based, liberal arts college of Holy Cross. Attending the game were a number of celebrities: Boston's handsome mayor, Maurice Tobin, who the famous movie director, Cecil B. deMille, had once invited to Hollywood for a screen test and potential role in a forthcoming epic, and the mayor's guest, movie cowboy Buck Jones, who was ending a war bond drive and coast-to-coast tour to revitalize an apathetic public's sagging interest in the western genre. After the game, a Boston College victory celebration and announcement of the team's selection to play in New Orlean's Sugar Bowl on New Year's Day were planned as part of the ceremonies to be held at Boston's Cocoanut Grove nightclub.

The outcome of the famous rivalry between the two Catholic schools, however, was astonishing! Holy Cross scored 55 points to BC's feeble 12! No one realized it at the time, but this outcome saved a number of lives. It also took the lives of a few who, with a reversed score, would have been elsewhere. After the game, a few fans were amazed when they noticed that the the cover photograph on the program showed the two captains wearing numbers that represented the final score.

Within a few hours, Boston's most famous disaster, a fire that killed almost 500 people (including Buck Jones) occurred at the nightclub named after the famed Los Angeles' Cocoanut Grove. When the fire broke out in the basement of the Melody Lounge, approximately 1,000 people had crowded into the former film warehouse whose exits were padlocked to prevent dead-beats from sneaking out. The main entrance at the front of the building on Piedmont street was a revolving glass door.

I was home with my parents on that November 28th Saturday when the telephone rang and Dad was informed that a fire had taken a number of lives at the Grove. All physicians within a radius of 25 miles of the Hub, as Boston is called, were being informed of the tragedy in case some of the victims happened to be their patients. This was not an invitation to come to Boston, the caller made clear, because the traffic in the immediate area of the nightclub was impossible. Furthermore, the shortage was not of physicians or nurses, but rather of plasma and antibiotics. Penicillin was in short supply and had to be recovered from patients' urine to be used again and again. The news that evening revealed the number of deaths at several hundred, but this number was doubled a few hours later as more accurate counts were made. The most revolting thing announced that evening was the theft of jewelry from the bodies of those lying burned beyond recognition in the snow and ice of Piedmont Street outside the Grove. In several cases, fingers were amputated to facilitate the seizure of expensive rings. A Harvard classmate-Herbert Collins Arnold, Jr,—lost his life in the conflagration.

A few weeks after the disaster, I asked my parents if they had ever been to the Cocoanut Grove. It was a stupid question because, residing in a town that was only a dozen miles from Boston, they had sampled the fare and entertainment of all of the outstanding restaurants and nightclubs in the area. The Grove did not provide the exotic entertainment of Havana's Tropicana, or the exquisite cuisine served at the Hotel Tamanaco in Caracas, Venezuela, but it did offer, at a reasonable price, decent entertainment and good food within the restraints of Boston society. It fell short, however, in a consideration for the safety of its occupants.

Bodies, overcome by smoke inhalation, were piled up in front of a jammed revolving door. Other bodies were discovered at exits that were locked and, in several cases, concealed. Fireproof furnishings and decorations, which had been tested when the humidity was high, quickly lost fire-resistance when dryness, caused by the presence of hundreds of warm bodies, prevailed. To satisfy the public's demand for someone to blame, a scapegoat was eventually identified as the bar boy who had lit a match inches away from a pseudopalm tree in a dark corner of the Melody Lounge trying to locate the socket from which an amorous Don Juan, seeking greater privacy, had removed the bulb. This idiot will probably remain forever anonymous.

The tragedy affected Needham in at least one awful instance. When Dr. X retired from general surgical practice, he was replaced by a young surgeon, Dr. Frank Chiampa, whose skills were recognized by Dad as comparable to those of the elder master of the scalpel. Dr. Chiampa resided in nearby Newton with his lovely wife and young children. As his mentor, Dad indoctrinated (excuse the pun) the thirtiesh surgeon in financial affairs (buy Boeing), and the trials and tribulations of practicing medicine and surgery in Needham. One day in casual conversation Dad mentioned the Cocoanut Grove fire to Dr. Chiampa and noticed that his eyes suddenly became moist.

"I never mentioned this to you before. My sister was there that night and lost her life. She was the youngest victim. Only 15."

Dr. Chiampa also told Dad that whenever he went by a store window that displayed mannikins, he immediately thought of those bodies piled up behind the revolving door.

Another far lesser tragedy befell Dr. Chiampa and his family when, in the late summer of 1954, my folks, wife, children and I stood on the sandy shore of Mattapoisett and watched his recently purchased summer home break up at the mercy of violent waves churned up by hurricane Carol. Just before the waves hit, his wife darted into the kitchen and retrieved a pair of scissors and several tomatoes.

With the start of the second semester in January of 1943, several course subjects changed. Professor Cleveland's parasitology was replaced with Dr. Clark's ecology at a time when those who had heard of the word, much less knew its definition, totaled only a few hundred scientists throughout the world. Dr. Clark was a marine biologist who summered and lectured at Woods Hole where the world-famous Marine Biological Laboratory (MBL) and Oceanographic Institution (WHOI) were located near Eel Pond, a large tidal pool, serving as a protective anchorage for a variety of yachts. A man-made canal, linking the Pond to the ocean and spanned by a tiny drawbridge carrying local traffic and tourists from Cape Cod to the marine research facilities, was bordered by several commercial buildings, one of which was the only barbershop in the country boasting that salt water fish could be caught while getting a shave or haircut! This was the domain of Dr. Clark and his mentor Dr. Henry Bigelow, a founder of the Oceanographic Institution, world-famous authority on sharks, and an occasional lecturer in the ecology course that discussed limnologic (fresh water) and oceanographic (salt water) biology.

Within a few weeks, I became familiar with the cetaceans and their migratory routes, diet, and intelligence. The largest mammals in the world (the blue whales) dined on one of the world's smallest creatures—krill. When I told Gjerding Olsen that I was scheduled to give a seminar on whales in Clark's course, he said he would attend and purposely ask the question: "What happens when you stick a banana in the blowhole of a right whale?"

I told him I'd check *Moby Dick* for the answer. Today, we know that far worse things clog the blowholes of whales thanks to the practice of dumping garbage from the decks of cruise ships and naval vessels.

An amusing aspect of this course involved returning an overdue textbook to an annoyed Dr. Clark who, in 1950, recalled the incident when I met him at the MBL at Woods Hole following a stimulating lecture on crustaceans. Whenever I eat crab or lobster, I am reminded of him.

Dr. Bigelow, a thin, elderly gentleman who raised beef cattle on his farm in Concord and played tennis daily, interspersed anecdotes from his remarkable life with observations of shark behavior. He held the great white shark in highest esteem as the world's most dangerous predator which preferred cold waters of the Pacific to the warm waters of the tropics, the milieu of most of the other sharks.

Hooton's physical anthropology, attended by the "ape's rich relations" during the first semester, continued as cultural anthropology with the "culture" referring to the life styles of the diverse Indian tribes of North America. It has been said that there are more Indians (now called native Americans) living today on the North American continent than at the time of Columbus. The reason may be that five hundred years ago these people existed in small groups or tribes with competition for food, and the territory that provided it, being extremely intense. Tribes became suspicious of one another and engaged in fatal skirmishes that reduced their numbers. Although most of the lecture material compared and contrasted the characteristic cultural attributes of the Hopi, Sioux, Apache, Blackfoot, Cherokee, Algonquin, and Iroquois tribes, the professor also mentioned the great Egyptian civilization, which, unlike America's first families, left no modern day relatives. In fact, anthropologists were amazed when they discovered that the blood type detected in Egyptian mummies was neither O, A, B, nor AB, but a fifth classification, now called Egyptian, which exists nowhere else in the world today except in those who have been dead for a few thousand years. What happened to these people? Did a plague, the airborne ancestors of the AID's virus, wipe out their culturally-advanced civilization which spawned many of the embryonic elements of paper-making, medicine, surgery, art, and architecture? The veneration in which these remarkable people were held is noted by the action of the Egyptians' successors, who tossed their mummies into the ovens of steam locomotives *in lieu* of coal!

Speaking of man's inhumanity to others, whether dead or alive, an incident occurred in 1943 that resulted in the most famous unsolved mystery in the history of Hollywood! Once in a while, when classes did not permit me the time to trek down Boylston street to Kirkland House for lunch, I would grab a meal at St. Clair's restaurant where the food was delicious, topped off with a chocolate and fudge sauce sundae. Also, as an added bonus, the waitresses were pretty and at least ten years younger than the more mature ladies at Kirkland. One waitress was exceptionally pretty. Standing five feet five or six inches tall, and with her most characteristic features being jet black hair and an alabaster-white face, she waited on me once or twice. The last time I saw her a waitress came to my table and told me Elizabeth Short was leaving to seek a film career in Hollywood. This would be her last day. So that was her name. When she came by my table, I wished her luck.

Four years later, her ravaged nude body, bisected at the waist, was discovered in a field in southwest L.A. According to Kenneth Anger in his book Hollywood Babylon II, Elizabeth Short, 22 years old, from Hyde Park, Massachusetts, whose mother lived in Cambridge, Mass., had become the leading participant of the Black Dahlia unsolved murder case, the most infamous in the history of Hollywood and the subject of a TV presentation with Lucille Ball's daughter, and a movie starring Robert Duvall. In the innocent days of 1943, when Elizabeth worked at St. Clair's restaurant at the age of 18 and dreamed of being famous, fate was destined to play a horrible trick on her and make this tragic figure far better known to the world than many of the contemporary "greats" in the film industry.

At my 50th Harvard reunion in June, 1994, I visited St. Clair's restaurant off Harvard Square now juxtapositioned between Settebello's designer clothing store and a bookstore. As is usually the case, the restaurant, presently known as Bailey's and specializing in ice cream and fine candies, seemed smaller than I had remembered. Meals were no longer served at the marble-topped tables with their wicker wire chairs that used to be standard equipment in ice cream parlors throughout the country. I wondered, as I enjoyed a delicious vanilla ice cream sundae topped with fudge sauce and whipped cream, if the young men scooping up gobs of ice cream behind the counter had ever heard of the Black Dahlia murder case and were aware that the victim once worked in the same place where they, like her, stood on the threshold of life at the mercy of lady luck. Would they have given a damn?

A message posted on the bulletin board in the alcove of Kirkland House read as follows:

Men wanted on weekends at Mass General to assist nurses and other health personnel. Pre-meds preferred.

One evening I hopped aboard the subway, crossed the Charles River, and got off at Charles Street where the gloomy, castle-like bastion, the Charles Street prison, looms near the giant health complex known as the Massachusetts General Hospital. Entering the newly-constructed White Building, I inquired of the guard at the desk where Ms. so-and-so, the author of the note, was located.

"She's up on the 6th floor. Just take the elevator to your left."

The 6th floor had a photograph of Buck Jones on the wall near the counter behind which a number of young nurses were busy writing, telephoning, and chatting.

One of them suddenly looked up at me and asked:

"Can I help you?"

"Yes, I'm looking for Ms. so-and-so about the notice posted at Harvard for pre-med students who wish to work at Mass General on weekends."

"You'll find her office down the corridor. If she isn't there, she'll be back, so wait for her—and good luck."

I knocked on the door and entered, with some trepidation, after a voice said, "Come in."

"What can I do for you?" asked the dyed-blond, middle-aged lady, who was obviously the supervisor of nurses for the sixth floor. Before I could answer, she went on, "I know, you've come to help our overburdened girls with their nursing duties. Right?"

"Yes, I read your note posted at Kirkland. I'd like to try my hand at helping out."

"Are you pre-med?" she asked. "We prefer those who can benefit from being here. If you're scheduled to enter med school, you have a draft deferment, so you won't be snatched from us after a brief training period." Again, she confirmed her statement with another: "Right?"

"Yes," I responded. "But I can only work three weekends each month because I have to go home once a month."

"That's acceptable. Your hours will be 5:30 to 11:00 each weekend evening starting next Friday and ending each Sunday. You'll assist the nurses in giving backrubs, enemas, and feeding the patients in the iron lungs. Of course, these activities will be restricted to male patients only! Are there any questions?"

"Yes, will I be able to see any operations?"

"I would like you to come with me right now. There's an operation in progress on this floor. I think it's a pneumonectomy."

She stood up and opened the door while I trotted behind like a puppy dog. Shortly, she came to a door, unlocked it, flicked the lightswitch, and I saw a brightly lit room with a huge glass centerpiece, surrounded by rows of chairs on an incline. She walked to the front row of seats and sat down, beckoning me to sit next to her. Down below were about ten people dressed in green clothing, wearing masks, and encircling a table upon which a young girl was lying prostrate. She was obviously unconscious and her chest cavity was exposed, revealing a lot of bright red tissue.

"If you press the button in front of you, you'll be able to hear what they're saying."

I pressed the tiny button and immediately the voices of two people, the surgeon and a nurse, were heard over the beeping of several pieces of equipment. I smiled as I recognized that someone had misplaced a stainless steel retractor and incurred the surgeon's wrath.

"That's Dr. Goodsill," said Ms. so-and-so. "Watch and you'll see the nurse standing next to him stare at the floor in embarrassment."

Sure enough, she lowered her head for a moment.

"She'll find the instrument and the surgeon will nudge her with his elbow."

Again, her prediction came true.

"Now before I go back to my office I want to emphasize that when you watch this, there is more to it than an exercise in human behavior. Watch the technique of the surgeon, his skill with the scalpel and needles, and overlook the human frailties that are bound to occur with the tensions engendered by awesome responsibilities. Take away the support of the nurses and it's as if you stepped on the oxygen line to the patient. Remember their work complements that of the surgeon."

As we walked toward her office and she was about to say goodbye, I asked her one more question.

"Why is the photograph of Buck Jones hanging on the wall next to the nurse's station?"

"That's from his family who requested that it hang there in perpetuity or until a bronze plaque is sent to replace it. You knew he died from severe burns and smoke inhalation on this floor?"

"No, I thought he died at the Cocoanut Grove. I don't recall ever seeing him in the movies. My favorite western star was Ken Maynard. And I once saw Tom Mix and his famous horse Tony at the Sells Floto circus in the thirties. Well, I'll be seeing you this Friday evening. It's been nice meeting you."

As I headed for the Charles Street station, I thought about the events of the afternoon and couldn't wait to inform the folks that I had a non-paying job at the great Mass General Hospital, one of the best in the United States.

On the ride back to Harvard Square, I recalled the framed photograph of Buck Jones, which I had studied for a moment at the nurse's station before I entered the elevator, after bidding the superintendent goodbye. In format, his pose resembled a Remington painting of a successful cowboy astride a cow pony in the Arizona desert, where most westerns were filmed before frugal producers discovered the tax benefits of making "spaghetti westerns" in Italy. I noticed something interesting about the photograph, for Jones was wearing a white, ten-gallon hat, the traditional indicator of the "good guy" in the movie beneath which a blemish-free, chalk-white face smiled. My attention was focused on this cosmetic detail, for Ms. so-and-so had mentioned that he had died from severe burns and smoke inhalation. I hadn't realized that Hollywood make-up artists applied cosmetics only to the visages of members of the cast who played heroic roles. Conversely, the supporting male cast members and villains were always unshaven, unclean, and chosen by casting directors for their ugliness. It troubled me that I had never seen Buck Jones in Saturday matinee films at the Needham Paramount Theatre, or at the theatres in Burlington, Vermont. How tragic that this talented film star, about to rejoin his family on their California ranch, should lose his life caught up in someone else's whim to celebrate a football victory that had backfired with people he didn't know or care about! The subway cars arrived at Harvard Square and I climbed up the stairs thinking of the vicissitudes, seemingly inconsequential at the time, that can exert profound influences on our lives.

Back at Kirkland House I thought about Mass General and the great opportunity I would soon have to learn more about the medical profession and actually interact with patients. My experiences at Needham's Glover Memorial Hospital and the one in Braintree had been interesting but "hands off" situations, for there was never an opportunity to talk to people who had recovered from operations. Now, this Friday evening, I would meet face to face with patients suffering from different kinds of health problems. Also, I would have a role, though minor, in helping them on the road to recovery. The next morning, sitting in a lecture room at the Biological Laboratory, I listened to Professor Welsh discuss the invertebrates, my companion course to Dr. Clark's ecology. Like the well-known ecologist, Dr. Welsh held a professorship at Harvard and conducted grant-sponsored research at the MBL during the summer months. Also, Welsh's invertebrate course like Clark's ecological offering consisted only of lectures without the tedium of ancillary laboratory exercises. He was explaining how the isolated heart of the hard-shelled clam, *Venus mercenaria,* known in the New England area as a quahog, was used to detect minute amounts of acetylcholine in biological solutions. Acetylcholine, a neurotransmitter found in the mammalian brain, and the central, peripheral, and autonomic nervous systems, conveys impulses at nerve endings to the tissue sites or receptors, which then respond. It is also a neural transmitter in invertebrates. For example, the lowly clam contains acetylcholine in the vicinity of its three primitive ganglia.

Friday evening finally arrived after a week of anticipation interspersed with lectures on plankton (Clark's ecology), economic differences as a cause of war (Sorokin's sociology), glacial deposits (Mather's geology), and the nervous system of the giant squid (Welsh's invertebrates).

I arrived at Mass General at 4:30 p.m. and had a quick snack in one of the small restaurants bordering the antique shops on Charles Street. Half an hour later, I decided to look at the shop windows displaying marine items for a few minutes before heading back to the hospital. One store had several models of old sailing ships whose actual counterparts I had seen in National Geographic magazines participating in the annual grain races from Australia to England in the 1930's. Scrimshaw, marine postcards, and dusty books by Joseph Conrad and Herman Melville surrounded the windjammers, *Parma* and *Pamir,* bedecked in windless, lifeless sails. It was time to settle my nerves and head for the White building. I made a mental note to return to this store earlier next time to see what treasures were on display inside.

On the sixth floor of the White building, Ms. so-and-so was waiting at the nurse's station with several other youths.

"Right on time! I like that!"

"Yes, I could never be a politician with this fault alone," I jokingly remarked, recalling how long I had to wait, along with hundreds of other boy scouts, for Boston's silver-tongued mayor, James Michael Curley, to arrive at our jamboree at the Boston Garden.

Disregarding this comment, she introduced me to the other volunteers who had come from B.U. and Tufts, and were also pre-meds.

"Let's get down to business. First, a tour of this floor. Except for a visit to the cafeteria on the first floor, your work will be exclusively here on the 6th. If you decide to leave for a snack, please inform the nurses at the central station. Emergencies occasionally arise and are usually dealt with by professional personnel. Now we are going to visit the male patients in the non-contagious ward down the hall. They are cardiac patients and should not be subjected to excitement. If you give them backrubs, be sure to check with the attending nurses that they are eligible."

We entered a large, well-lit room with a dozen or so beds in which men, ranging in age from mid-twenties to seventies, were listening to the radio, reading, or sleeping. Ms. so-and-so approached one of the younger patients, an Italian.

"Hello, Mario. I'd like you to meet our new people. They are here to help our overworked personnel and learn a little about the health profession."

She introduced the three young men from B.U. and Tufts and then turned to me. "And this is Mr. Mann. He's a junior at Harvard."

Mario sat up in his bed and shook the hand of each volunteer accompanied by a high-pitched "Hi."

He looked at the superintendent and smiled feebly.

"Mildred, can I have a backrub?"

"I have to examine your chart first. How are you feeling?"

"Weak. At least I can keep a meal down but the taste is gone."

"You don't seem to be gaining any weight, Mario. If we add salt to your diet, you'll put on a pound or two, but it will be retained water and the electrolyte imbalance is bad for your heart. I'll check your records and, if you're strong enough, one of these young gentlemen will oblige."

We left the room with superintendent Mildred who commented on Mario's condition while we walked to the next room where the faint wheezing of iron lung machines was heard.

"Mario is only twenty-eight, but his heart condition is what is known as cardiac myopathy. I want you to look up the definition of this disease and be prepared to answer a few questions concerning what causes it and its prognosis or outlook."

As the five of us entered the next room, the sound filling the air reminded me of the Walker-Gordon rotolactor or bovine merry-go-round that was a hit at the New York World's Fair in 1940. There were only half a dozen large, barrel-shaped machines in the room, each with a head sticking out of one end, lying on a pillow with a mirror overhead. The metal barrel resembled a miniature submarine with hinged portholes of glass that could be opened. Each unit was protected from a power failure with parallel wiring to an auxiliary generator that functioned when the main one was disrupted. The patients were young and victims of the polio virus; a disease known as infantile paralysis that had incapacitated the ambulation of FDR, but fortunately not his breathing.

As we stepped into the hallway, superintendent Mildred announced that she had to return to the nurse's station and would rejoin us there in half an hour. Meanwhile, she suggested that we visit the snack bar in the basement and get acquainted over a cup of coffee and some pretzels. The four of us took the elevator and dropped seven floors where the decor changed dramatically. Cement blocks, painted light blue, gave the walls a combination of solidity and coldness in contrast to the beige-colored plaster walls that exuded comfort and warmth.

"I'll bet the morgue is nearby. All hospitals have a morgue in the basement where an elevator takes the deceased discreetly to the fridge," said Fred, the B.U. student.

"It's also where it's found in the movies, especially the spooky ones," added Harry, the scholar from Tufts.

135

"Are you forgetting we only have a few minutes for a cup of coffee and some pretzels? We can see the morgue some other time," I cautioned. "Furthermore, do you realize if we get our hands on a second-hand stethoscope and shove it in a pocket of our white jackets, we'll have a passport to anywhere in the hospital?"

"Any place except the sixth floor and this place, or the cafeteria where Mildred and her cohorts can spot us," mentioned Walter, another student who was taking pre-med courses at Tufts College in Medford.

Agreeably, they heeded my suggestion and together we headed for the snack bar where we discussed our courses, professors, aspirations, and, of course, girls over a strong cup of coffee and salty pretzels.

We met Mildred at the nurse's station where she had good news: Mario could have a backrub this evening.

"Do we have a volunteer?"

"I'll be glad to give him a great rub," I said. "I once had to give my father a backrub when he was recovering from an operation. He said I did a good job. All I need is isopropyl alcohol and a towel."

"I know what you'll need," said Mildred. "I can assure you, among other things, our nurses have done backrubs. And don't try to pull the stethoscope trick on me!"

That evening, I entered the ward where Mario was ensconced and greeted him with a spirited "Hello!."

"Hi, Dr. Mann. Am I glad to see you. My back is sore as hell. *(groan)* These people have no idea what torture trying to sleep in these beds is."

"Mario, call me Dave. I've got a long way to go before I become a doctor. Where are you from?"

"Rhode Island. A place called Bristol. Ever been there?"

"No, but I've heard of it. Quite a yachting center. A lot of shipyards there that refurbish yachts. That's the origin of Bristol condition, isn't it? Getting them seaworthy?"

"Yea, that's what I used to do, scrape and paint hulls. Scraping and burning paint off with a blowtorch was o.k. But the docs tell me I got a bad heart from absorbing too much copper from the anti-barnacle paint we used. They call it cardiomyopathy. I shoulda been more careful and used gloves and a mask, but in the summertime it gets too damned hot for those gimmicks!"

"Mario, first the alcohol and then the rub."

Mario removed his Johnny, the jacket-like covering with the slit in the rear that was obviously designed by some weirdo to encourage exhibitionists, and rested on his stomach awaiting the stinging alcohol on his aching back. After sloshing a few ounces on the shoulder blades, I began the massage by putting my palms together and rotating them in a circular motion gently hitting his back. Then, a slow, progressive kneading of the muscles, starting with the back of the neck and progressing down the spinal column to the lumbar region, completed the procedure.

"That was terrific," said Mario when I'd finished. "I could use a backrub like that every day. But these nurses don't have enough power in their hands to give a decent one. Did you ever milk a cow?"

"Yes, on my grandfather's farm in Massachusetts and on my relatives' farm in Canada."

"I knew it! Milking makes strong wrists and steely fingers."

I said goodbye to Mario, satisfied that he would enjoy a good night's rest, and headed back to Harvard.

When I returned the following evening, I went to the ward to see how Mario was doing. A new patient was lying in his bed.

"Mario died last night," said one of the patients.

I was shocked. Had Mildred made a mistake by allowing him to have a backrub? Had my vigorous massage caused his demise? The rest of the evening I went about my duties in a robot-like state. Now I had firsthand experience about what it was like to have someone you knew, although briefly, to die. Doctors experience this dark side of medicine routinely and the younger the patient, the more difficult it is to handle emotionally. The only time I recall tears in Dad's eyes was when children died. And their demise was far too common in the days before anesthetics, vaccines, insulin, sulfa drugs and antibiotics.

I was in no mood to see Mildred but I ran into her in the hallway as I came out of the iron lung room.

"I'm sorry about what happened to Mario," she said in a monotone. Then, as if changing the subject, she asked: "Did you look up cardiomyopathy?"

"Yes. It's heart muscle damaged by a variety of factors such as genetics, nutrition, allergies, drugs, and even metallic compounds such as bismuth."

"Very good. You've done your homework. You probably think that your backrub was a factor in his demise. That's nonsense. Some patients with comparable heart conditions, and of the same age and sex, had weekly massages, yet expired without a reasonable physiologic explanation."

I felt better after Mildred's talk, but I thought about Mario frequently and his occupational disease and the irony that he had put yachts in "Bristol condition" at the expense of his own well-being.

Just before the final examination period began in May, I left Mass General to prepare for exams. I had enjoyed functioning in a superb health-care environment with dedicated nurses who were professional in every sense of the word. They saw to it that the collegiate volunteers only did the heavy work beyond their own physical capabilities and the routine procedures that were learning experiences *by rote.* And my association with the B.U. and Tufts students, although thwarted somewhat by Mildred's assertion that the stethoscope trick wouldn't work, was memorable. And we did visit the morque which prepared us for the anatomy laboratory we were soon to experience during the first year of medical school.

By tradition, the final examinations were held in the *Gothic Cowbarn,* Memorial Hall, built to memorialize 136 Harvard men of the Union Forces who had died in the Civil War, and whose inner tomb-like ambiance was as imposing as ever. Across from this huge structure crowned with a massive clock tower, the Cambridge fire-station stood, a red-brick and concrete chameleon mimicking the Georgian architecture of the freshmen dormitories in the nearby Harvard Yard. Few fans, attending football games at the oldest stadium in the United States, were aware that some musicians in the Harvard band were also members of the Cambridge Fire Department. When they played: "There'll be a hot time in the old town tonight," it was from personal experience.

On September 6th, 1956, during a restoration program with the staging still in place, fire broke out and quickly consumed the entire tower despite the valiant efforts of fire-fighters whose equipment came from across the street. The cause of the conflagration was presumed to be an unattended blowtorch, that unassuming, unpretentious incendiary tool that reduced the great ship Normandie to scrap during WWII, caused fatal shipboard fires in the Navy since its invention, and destroyed numerous buildings throughout the world because of human carelessness.

After the examinations were over, I spent the summer working in the Dewey and Almy chemical factory in Cambridge where some of my duties were to load railway freight cars with barrels of synthetic rubber. These had to be rolled up a wooden ramp and, once inside the freight car, turned right side up so that they occupied as little room as possible. While rolling one such barrel, my gloves caught in a wire used to seal the top, and I did a complete somersault landing on my back. I felt no discomfort whatsoever because my rowing days on the Charles had strengthened my back muscles.

I spent the first week living in a rented room near the Harvard Law school on Massachusetts Avenue, awaiting a larger facility on Sacramento Street, for Gjerding, Richard, and I had to vacate Kirkland House to make room for the V-12 Naval Program that trained "ninety-day wonders" who became 2nd lieutenants, and the bane of enlisted seamen. As a seaman *first class* in the waning days of WWII, while I was in boot camp at Bainbridge, Maryland, I learned that CPO's (chief petty officers) commanded the most respect, while the "ninety-day wonders" were most likely to be thrown overboard when their ships were at sea. I was pleased to leave my musty, drab room with the portrait of Sir Galahad and his horse on the wall for a much larger room in a Victorian mansion whose occupants were young people, pretty waitresses from St. Clair's and the Georgian on Harvard Square, and young men who worked at Sears-Roebuck several blocks away. I noticed a bit of envy when Dad dropped me off at my new location. The girls and I quickly became acquainted. I had many lively discussions with them about the singing talents of Frank Sinatra, Dinah Shore, and some of the problems they had with their boyfriends.

With the arrival of September, I left the Dewey and Almy Chemical Company experienced in moving drums filled with synthetic rubber from one room to another for labeling. Then, rolling them to the red-colored freight cars which conveyed the cargo to Michigan where the white gooey stuff was mixed with chemicals and Amazonian rubber latex and vulcanized to become semi-synthetic rubber for the defense industry. My friends at Sacramento Street appeared to be well on the road to matrimony when I wished them good health and happiness and returned to Kirkland House. One of the waitresses, however, had so infuriated her boyfriend by doing something stupid that he threatened to shoot her. The evening of the anticipated meeting, a *mumps* sign was affixed to her bedroom doorknob, effectively thwarting an immediate confrontation and allowing the situation to cool.

The courses I selected for the fall semester of the senior year were: Professor Hisaw's graduate course in endocrinology, Dr. David Linder's graduate course in mycology, a fine arts course, and a course in American history.

As mentioned previously, Dr. Hisaw's lectures were based on his studies on macaques which were injected with various kinds of hormones to alter the reproductive organs. The textbook was entitled, Sex and Internal Secretions, by several famous endocrinologists, which contained such gems of information as: *stallion urine is rich in estrogens; the pussy willow and the lobster contain estrogenic substances; remove a gonad of a hen turkey and the other gonad becomes an ovotestis.* Procuring the text from a female librarian involved much embarrassment until it was made clear that such a book existed.

Dr. Linder's mycology described in detail the life cycles of fungi and was given at an important moment in medical history when antibiotics were being discovered, cultured, and evaluated for effectiveness. Alexander Fleming's discovery of penicillin in 1929 was superseded by the sulfonamides as medicinal agents until Chain, Florey and associates at Oxford purified the antibiotic and focused attention on its antibacterial effectiveness against syphilitic, gonorrheal, and other bacterial organisms. At the time I was enrolled in the course, the pharmaceutical industry was searching for ways to increase production. There were only three students taking the course (one was female) and at the first meeting of the class, Dr. Linder emphasized that if one of us couldn't be in attendance because of illness, kindly call him at such and such a number so that he could cancel the class. There was a laboratory session of several hours each week where the same edict was in effect. During one such session, Professor Linder inoculated our Petri dishes containing potato agar with Penicillium notatum, the mold which exudes penicillin. Several days after the inoculation, an examination of the Petri dish revealed what appeared to be a miniature, almost microscopic depiction of a Tahitian island with a tiny volcano surrounded by a white (coral) beach. Several days later, a yellow exudate appeared beyond the white area—penicillin! A few days later, it was the weekend and I was home peering at Dad's garden filled with tomato plants, lettuce, radishes, carrots, and cucumbers. Suddenly, a thought entered my mind—*what would happen if the mold, Penicillium notatum, were inoculated in a Petri dish containing carrot agar instead* of potato agar? What effect would the change have on the production of penicillin? I knew how long it took for penicillin to appear after the initial inoculation of the mold in a dish of potato agar. It took three-five days. In the laboratory, I could repeat the process enough times to establish a standard in hours after inoculation. Then, substituting carrot agar for the standard potato medium, I would determine how long it took for the penicillin to appear. The following week, I explained the experiment to Dr. Linder who was excited about the project. Together, we prepared the agar from carrots pulled from Dad's garden and proceeded with the experiment.

To summarize the results: the *Penicillium notatum* inoculated on carrot agar yielded penicillin in far fewer hours than was observed for the mold growing on potato agar. Furthermore, the yellow ring of penicillin in the carrot agar plate grew wider in a shorter period of time than was noted for the potato-nourished mold. Dr. Linder was ecstatic! Could this technique be the answer to the stepping-up of production commercially? Also, if there were indeed a nutritional substance present in carrots that stimulated antibiotic production (i.e., beta-carotene), could it be possible that the hitherto unknown precise chemical structure of penicillin be based on the structure of that precursor of vitamin A? Having obtained the necessary data showing the impetus provided by carrot agar in hastening the mold's output of penicillin and increasing its amount, it was time to act.

Dr. Linder invited me to have lunch with him at the prestigious Faculty Club where I rubbed elbows with Harvard's famous including astronomer, Harlow Shapley, who indeed resembled Napoleon Bonaparte as observed in a student handbook. There, I was introduced formally to Dr. Louis Fieser who invited me to his office to discuss the research as a representative of Merck, the pharmaceutical firm of German origin. Without divulging any secrets, I met with him for half

David E. Mann, Jr.

an hour to discuss superficially the concept of the study. Several other pharmaceutical manufacturers became involved and I eventually received an invitation from Parke-Davis in Detroit to spend several days with them. I arranged a flight on an American Airlines' plane from Boston to Detroit that included a brief stopover in Windsor, Canada. Taking off from Canadian soil, it took exactly 5 minutes to fly from Windsor to Detroit. At the airport, my name was announced over the loudspeaker and shortly I was riding in a limousine to the Book-Cadillac Hotel with a Mr. Fritsch, head of the R & D division of Michigan's foremost pharmaceutical firm, which was also one of the largest in the world.

After accompanying me to the registration desk, Mr. Fritsch summoned a bellhop. Before registering, I was informed that he'd meet in the lobby in half an hour after I freshened up, whereupon we'd go to his office at Parke-Davis and chat with some of the scientists interested in my project. My room was on a floor that overlooked the roofs of neighboring buildings, high enough in the air to muffle the distant cacophonous smorgasbord of screaming sirens, honking horns, and screeching tires emitted by afternoon traffic. The flight from Boston had been uneventful except for the continuing hum in the ears that I hadn't experienced since the seaplane flight in the thirties over the Panama Canal. Someday they'll have insulation that'll completely block out the noise and annoying vibration. I took pride in the fact I'd been able to fly to Detroit. It was almost impossible to procure a ticket during wartime unless the trip were top priority. When I made the airline reservation over the telephone, the clerk had asked: "Why don't you take the train?"

"Impossible," I'd replied. "My schedule at Harvard is especially tight. It really boils down to arriving quickly, getting the business accomplished, and departing the same way."

I'd left the impression that I was a professor or scientist who was aiding the war effort. Mentioning the word HARVARD had helped to consolidate the concept. It wasn't the first time that the name of the prestigious university had tipped the scales in favor of positive accomplishment. And it certainly wouldn't be the last.

I met Mr. Fritsch in the lobby half an hour later. He was a distinguished-looking man in his mid-fifties, balding, and

142

bespectacled with a spring in his step that came from fitness derived from weekly tennis games.

"How are your facilities?" he inquired. "Will you be comfortable?"

"Oh, yes," I replied. "The room is delightful. The bed looks comfortable and the bathroom has a fantastic tub and shower. Reminds me of one we engaged at the Hotel Tamanaco in Caracas during a Caribbean cruise stopover."

"You've been to South America?"

"Several times. My family loves to travel."

"You're fortunate. The only traveling I've done is between the East and West coast on business. Strangely, the higher up I go in this company, the less traveling I do! I envy you already."

The drive from the Book-Cadillac Hotel to the imposing Parke-Davis facility took several minutes during which Mr. Fritsch mentioned some of the scientists I would meet. The only one I recognized was Dr. O. Kamm, the endocrinologist, whose significant contribution to science had been the demonstration of the active principles of the posterior lobe of the pituitary gland: *pituitrin* which contains vasopressin (a pressor agent) and an anti-diuretic (urine sparing) component, and *pitocin* which is oxytocic (capable of stimulating the smooth muscle of the uterus). I only knew him by his last name, but seeing the O of his first name in the literature I assumed it stood for Otto. Later, when I met the elderly gentleman, Mr. Fritsch introduced him as though he too were a visitor:

"Dr. Kamm, I'd like you to meet Mr. Mann. He's the young man with some new ideas about penicillin production. He just arrived from Cambridge."

"England?" asked Dr. Kamm as he shook my hand coincident with the raising of extremely bushy eyebrows.

"No," I answered. "Massachusetts. The locale of Harvard."

"I've been there. Given a talk or two at the medical school. Which reminds me, would you like to hear my lecture on steroids tonight at Wayne State? It's at the medical school."

"I'd be delighted," I answered. "If Mr. Fritsch hasn't made other plans."

"You'll be ending the tour of our facilities before Dr. Kamm gives his lecture. I regret that I'll be unable to attend. Unfortunately, I have some personal responsibilities this evening," said Mr. Fritsch.

I visited with Mr. Fritsch for several minutes in his elaborate, walnut-paneled office before joining several young scientists for a room- by- room tour. In some respects, I was reminded of the sixth floor of Mass General: the clean, odorless atmosphere and the neatly dressed personnel who went about their business without gawking. When we came to a large auditorium, one of the scientists said:

"This is where you'll be giving your own lecture if everything works out."

That evening I waited for Dr. Kamm at the rear of the building. A large black LaSalle drove up and the driver, Dr. Kamm, beckoned me to step in. When I closed the front door, he zoomed the engine, we took off, then he turned the engine off and we coasted a block or two to the entrance of Wayne State medical school. The lecture lasted only an hour, but the words rang in my ears long after I had climbed into bed at my hotel: cortisone, hydrocortisone, testosterone, progesterone, estrogen, *blah, blah, blah!*

The next day I conferred with both biologists who were specialists in the fields of microbiology (then called bacteriology) and mycology. I was pleased that the latter group was familiar with the contributions of Dr. Linder in his speciality, the basidiomycetes. They inquired of his graduate course and smiled when I mentioned the number of students in the class. Of all the disciplines of biology, endocrinology attracts the greatest enrollment and mycology, traditionally, the least. I told them about the time when Dr. Linder told his triumvirate of scholars at the conclusion of a lecture: "My suspenders gave way at the start of my talk and my pants fell down. Thank God no one peeked in the side-door window or I would have gained a reputation for being a pervert!"

That evening with nothing on my schedule, I went to the movies and saw Gary Cooper and Ingrid Bergman in Ernest Hemingway's *For Whom The Bell Tolls.* It was a magnificent film with beautiful scenery and outstanding acting by both the principals and supporting players (one of whom, Katina Paxinou, received an academy award for her portrayal of the counter-revolutionary Pilar).

The following day I bid Mr. Fritsch goodbye and headed for the airport with a contract bound between blue cardboard covers that stated in legal terms "that if a penicillin -type antibiotic is derived from the basic beta carotene structure, it would be known as the antibiotic *caricillin.* Mr. David E. Mann, Jr. of Needham, Massachusetts, would be paid royalties according to the percentage of net sales (10%) with no funds accruing from the sale of antibiotics to the armed forces."

When I returned to Harvard and met with Dr. Linder, his first question was:

"Why isn't my name on the contract, too?"

"Because, for a legal reason, only one name is acceptable and that name had to be signed in the presence of witnesses."

This explanation proved to be entirely satisfactory to Dr. Linder when I said: "Of course, the royalties will be divided if anything comes of this."

The conclusion, which occurred many weeks later, was the discovery by someone in the beer industry, that bubbling oxygen through huge vats of Penicillium notatum greatly increased penicillin production. When I read this in the newspaper, I took the blue bound contract and shoved it in a waste-basket.

"No one will ever get more out of my garden than you did!" commented Dad.

Having devoted considerable time and space in discussing my adventures in Dr. Linder's *mycology* course (regrettably, he died of liver cancer several years later), there were two other courses, in addition to Dr. Hisaw's *endocrinology* (of female macaques), *fine arts* and *American history.* The former surveyed artists from Giotto and his masterwork, Flight into Egypt (13th century), through outstanding artists representative of each century and their paintings (in oil usually), to Gluyas Williams, the famous New Yorker cartoonist, whose prolific works equalled those of Peter Arno in the 1930's and 40's. It was one of the few art courses in the United States where the actual canvases being discussed were taken off the nearby walls (of the Fogg Art Museum where the course was held) and placed on an easel in front of the class. Also, it was offered at a time when the 20th century's foremost artist—Pablo Picasso—was still alive and productive. During the course, it was very difficult for me to acknowledge the greatness of the Spanish master on the basis of what I had viewed. The teacher pointed out that Picasso himself had once proclaimed: "When I was young, I painted like an adult; when I was old, I painted like a child."

In the seventies, my wife and I were fortunate to have dined with a former student, author, art connoisseur, assistant dean, and close friend—Fred B. Gable—and his guests, the famous Philadelphia artist and teacher, Morris Blackburn, and his wife. I couldn't resist asking him the question: "Was Picasso really deserving of the accolades that touted him as the greatest artist of this century?"

Mr. Blackburn, who was often mistaken for the Missouri artist, Thomas Hart Benton, replied: "Without a doubt the critics and his peers are correct! The man was a genius creating art concepts that were far ahead of his time."

Regardless of this astute assessment by a man whose own works were rapidly increasing in value, I still had my own favorite-Jan Vermeer—whose oil paintings always seemed to be slight variations of an open window, an oriental rug on a table, an antique map on the far wall, and a girl alone or with a suitor standing in a shower of glistening golden light. I viewed the largest Vermeer, *The Concert, 72.5 x 64.7 cm.,* at the Isabella Stewart Gardner Museum in Boston a month or so before it was stolen along with Rembrandt's only marine painting.

The remaining course, *American history,* covered the same wars I had studied in high school. The only difference between that course and this one was the reality of the defeats presented at Harvard. For example, Braddock's defeat was attributed to the general being in bed with his mistress instead of performing his duties on the battlefield.

It has been said by gerontologists that an early indication of the ravages of the aging process is the loss of memory of recent events and the vivid recollection of those in the distant past. This is indeed true, for I cannot recall the courses I took in the second semester of the senior year other than one was *geography* and another was titled *The Old Testament.* These were courses I decided to take after getting advice from Leonard Packard, Dad's patient and Jimmy Davidson's neighbor on Warren Street, who was the author of the most widely used geographic textbook in the country, and Uncle Harold, rector of the Episcopal Church in Central Square, Cambridge, whose most famous parishioner was Harold Russell, the young sailor who lost both hands during WWII and won an academy award for his supporting role in *Best Years of Our Lives.* At the time, I thought these would be entertaining subjects that required little effort and yet would yield decent grades.

Geography fulfilled these expectations but at the same time seemed to be a little childish because the professor assigned homework that involved putting on paper a drawing of a hypothetical seaport complete with natural resources responsible for its existence with accompanying industries supporting its economy, and the effect of the port's prosperity on the surrounding territory.

On the other hand, the *Old Testament* was a nightmare out of a Cecil B. DeMille religious movie. The cast of characters rivaled that of Tolstoy's *War and Peace!* Uncle Harold had frequently remarked that I should follow his vocation and enroll in divinity school upon graduation. The thought of speaking before a large group of people each week gave me goose pimples. Incidentally, when Harold was given a larger parish in Lawrence, Massachusetts, at Grace Episcopal Church several years later, he related a weird incident that had occurred before he arrived. During a Christmas Eve service, the janitor of the church, a Roman Catholic, was stoking the furnace when he lost his balance and fell into the red-hot ashes. Not until many hours later did the parishioners realize that his cremation had helped to heat the chancel.

During my years at Harvard, Uncle Harold lived on Clinton Street in Central Square, a subway stop from Harvard Square. I enjoyed visiting him, Aunt Lillian, and their three pre-teen boys. On one occasion, Uncle Harold was counseling a mixed racial couple in his office when I was upstairs molding lead soldiers in a kit that required electricity. Running out of lead, I substituted raw sulfur crystals which immediately fumigated the premises. To this day, my cousins have vivid recollections of this incident, but no memory as to the outcome of the counseling.

The second semester came to a close and graduation day approached, traditionally on a Thursday in June, when the chances for rain in the forecast were as likely as snow. Out of a class of almost 1000 men, only 11 were present for the ceremony beneath the stately elms of the Tercentenary theatre, a vast open space in the Harvard

Yard extending from Memorial Church to the steps of Widener Library.

I was enjoying a long weekend at Harold and Lillian's newly purchased summer home at Hampton, New Hampshire, where just weeks before, celebrating the settlement, a gaggle of relatives were standing in the kitchen when the floor collapsed, sending them all in a rumpled mess to the ground below. Although the stove was boiling water for tea, the section of the floor on which it stood remained intact, so no one was injured beyond a few splinters and bruises.

A few weeks later, a cardboard tube arrived at 863 Great Plain Avenue (the number had been changed from 921 for some unclear reason) and, upon opening it, a Harvard diploma appeared printed in dark India ink. There is a common complaint among educators that today's graduates can't read their diplomas. Well, I have never read mine entirely which, in contrast to modern Harvard sheepskins, was in Latin. It states at the top: Universitas Harvardiana Cantabrigiae; the recipient's *name-Davidem Edvinum Mann,* appears midway; the date awarded is Junii XXVIIII in the year MDCCCCXXXXIIII, and bears the signatures of Jacobus Bryant Conant (Praeses) and Alfredus Chester Hanford (Decanus). The preponderance of science courses with only one language, German, required that the degree be a *Baccalaurei in Scientia.* Now I had to set my sights on additional summer courses in biology (bacteriology and embryology) at Tufts College in nearby Medford, where Dad had started out his practice, before entering the medical school on Huntington Avenue to initiate my medical career.

CHAPTER NINE

TUFTS' Decisions

During the first week of July, I enrolled in two biology courses, *bacteriology* and *embryology,* at the Tufts University campus in Medford, several miles from Harvard Square. I found a room a short distance from the campus on the first floor of a two-story dwelling. Five other male members of the freshman class of Tufts Medical School were quartered in single rooms at my place, also taking scientific subjects that they had avoided as college undergraduates. We introduced ourselves the morning of our arrival at breakfast in the student dining-hall, then, fortified with fried eggs, bacon, and toast washed down with coffee, we registered for biology and/or chemistry, according to scholastic need, and then toured the beautiful campus areas before entering the biology and chemistry facilities. One of the students, my old friend Harry from the Mass General experience, seemed pleased to renew my acquaintance. When I asked about the other Tufts pre-med student named Walter, whom I had also met at Mass General, he reminded me that Walt had a year to go before being admitted to medical school.

"I don't think his heart is in medicine. He had some lousy experiences with patients at Mass General", added Harry.

We entered the biology building where graduate assistants were busily assembling laboratory apparatus, winding Harvard kymographs, and weighing chemicals in spacious rooms in which long rows of oak desks, capped with dark slabs of slate, were surrounded on both sides by wooden bar stools. Stepping into the lobby, we were confronted by a huge, stuffed, African elephant standing a few feet from the doorway, its trunk lifted as if reaching for peanuts. An inscription on the metal plaque set atop a mahogany stand near the trunk identified the behemoth as *Jumbo,* or what remained of the world's most famous pachyderm. The great circus showman, P.T. Barnum, had donated the famous beast's remains, along with a briefly inscribed history relating how Jumbo had been purchased by Barnum from a London zoo in 1882 for a Rajah's ransom of $10,000. After crossing the Atlantic by freighter, he became the foremost attraction of Barnum's show. On September 15th, 1885, while crossing a railroad track in Canada with a smaller elephant named Tom Thumb, both were killed by a freight train. Newspaper reporters, embellishing the incident to increase circulation, wrote that Jumbo had sacrificed his life to save his companion. The bones were sent to the Smithsonian for anatomic study and the skin was mounted by a taxidermist on a steel frame and donated to the University. Harry told us later that he had purposely entered the biology building through the rear entrance so we could experience the full impact of suddenly being confronted by the huge pachyderm. As the weeks passed, I discovered that Barnum had donated dozens of other exotic animals that had died at his circus and were sent to taxidermists before ending up in the attics of the science buildings. When the University put on its annual summer circus show with participants from the student body, I saw firsthand the lions, tigers, zebras, leopards, and jaguars that stared at me with amber-colored glass eyes as my friends and I searched these attics for suitable props.

Dr. Sweet's *bacteriology* was a fascinating course given by a pleasant-looking man in his forties. The laboratory was an added joy for the person in charge of microscopic work (absolutely essential in a subject soon to be called *microbiology*) was a pretty, natural blond, blue-eyed, young woman who spent far more time crouching to help the boys identify elusive organisms than standing. In some respects the course reminded me of Dr. Cleveland's *parasitology,* for a small part of his subject involved bacteria (no pun intended). During a lecture on *Treponema pallidum,* the causative organism responsible for syphilis, the scourge of human sexual behavior, Dr. Sweet remarked that "because these bacteria exist in the human body within a narrow temperature range, the inoculation of patients with *Plasmodium vivax* to afflict them with malaria represented one of the earliest therapeutic measures. The resulting *chills and fever* eliminated the bacterial infection completely at the expense of replacement of syphilis with malaria, a disease that could be controlled but not cured with quinine."

The main objective in treating syphilis is to eliminate the spirochetes from skin and mucous membrane lesions to prevent spread of the disease. Mercury, introduced in 1495, was the first antisyphilitic agent. In 1910 Paul Ehrlich introduced 606, his 606th experiment to find a suitable agent. It was known as salvarsan (arsphenamine) and contained arsenic. Penicillin, the first antibiotic-antisyphilitic agent, had a great advantage over preceding drugs with its low toxicity, although some recipients died from anaphylactic shock. I was intrigued with the specialized uptake of certain dyes or stains by bacteria that resulted in their differentiation into *gram positive* and *gram negative,* which also *offered a clue centuries later to their susceptibility to antibiotics.* When the bacteriology course was over, I found out that I had creamed it for a straight A, something that I hadn't succeeded in doing with Dr. Cleveland's exciting and wonderful course.

Dr. Carpenter's *embryology,* a lecture and laboratory offering of avian (the chick), and amphibian (frog) development, provided a good foundation for a somewhat comparable course given at the medical school during the first year as a component of gross anatomy. Dr. Carpenter, like Dr. Sweet, was in his forties, a dynamic teacher with a wonderful sense of humor. Although these people were not as eminent as their peers at Harvard, I felt that their teaching skills and attentive behavior towards their students were comparable and, in some cases, superior. By a remarkable coincidence, cousin Marion, the offspring of Uncle Harry and Aunt Anna of Hampton, New Hampshire, was an undergraduate student at Tufts while I was there, but the demands of our studies were such that we rarely saw each other except at meals. There are more members of my family who have attended Tufts than any other school. There are even some who, having been accepted at Harvard, have decided instead to attend the University in Medford.

When September arrived, I moved from my off-campus room at Medford to a second-floor apartment near Huntington Avenue that consisted of a large living-room and three small bedrooms. The medical school was a two-minute walk (traffic permitting) across the street, an inconspicuous facility that had once been a garage and dealership, for the former presence of large plate-glass windows, which had allowed prospective customers to admire the latest creations from Detroit, was now demarcated by an indentation made by the slightly uneven placement of concrete blocks to provide security. Upon entering the building, the smell of motor oil mixed with the acrid odor of formaldehyde revealed the past and present functions of the facility. Its location, however, was handy for both students and faculty. The Wentworth Institute, where N.C. Wyeth's brother, Stimson, taught French, was adjacent to the medical school as one proceeded toward downtown Boston. Beyond the Institute, the multi-story, ugly brown YMCA provided athletic facilities and, a few blocks beyond, was located one of the best seafood restaurants in the city, the famous *Lobster Claw,* where at least once a month my parents would visit for baked lobster stuffed with lobster claws (hence the name) following an appetizer of 6 baked oysters. It was owned by a Greek restaurateur who always looked impeccable in a tuxedo and employed a half dozen waitresses, one of whom had been there over thirty-five years! Few students from the nearby Wentworth Institute, the Y, Northeastern University farther up the street where three white brick buildings, in various stages of construction, revealed its expansion program, or the Tufts medical school, dined at the *Lobster Claw,* the major reason being its high prices. A lobster dinner at 5.00 dollars with oyster appetizer at 2.50 was one-fifth of the average person's weekly pay! Therefore, the only time I dined there was to celebrate a special occasion, which was usually accompanied by a glass of fine wine.

My two roommates were also from Massachusetts, unlike the foreign diversity that characterized the student body at undergraduate Harvard where one's roommate was as likely to come from Europe or Australia as from New England. Aristotle from Peabody and Robert from Malden were dental students who took the same courses as medical students during the first year, a practice common to many other universities who have consolidated such programs for greater utilization of faculty while reducing instructional costs. Consequently, the three of us were able to study together evenings while attending lectures and laboratory sessions in the daytime, the latter periods as partners.

Our first lecture was in anatomy, known as gross anatomy, where the instructors, Dr. Sam Magruder, a thin gentleman in his mid-forties with a southern accent, and Dr. Spector, a slightly overweight, middle-aged man, who resembled Dickens' Mr. Micawber, were introduced by a person who taught histology and was known to students as the Amoeba, a much older woman than the young blond scientist who had so conscientiously assisted students in bacteriology laboratory in Medford. Despite her unicellular monicker, the Amoeba proved to be a multifaceted educator.

The anatomy lecture was held in a large room filled with metal tables on which our cadavers were soon to be placed. At one end of the room, a wooden platform constructed five inches or so above the surrounding linoleum-covered floor represented the lecture area where a large blackboard bedecked with rolled-up anatomical charts awaited the instructor's use. Beneath the blackboard was a large wooden box filled with arms and legs bathed in a sweet-smelling solution which prompted the class to mumble "Tootsie Rolls" every time the lid was opened. Along the other walls, shelves containing sagittal sections of various portions of the human anatomy in strips a centimeter or so wide bathed in formaldehyde were available for individual study. As the lecture began following the introductions by the female histology professor, Dr. Magruder, the youngest member of the trio, ascended the platform, smiled, and welcomed the class, 90% of whom were men.

"The seats you are sitting on are not the most comfortable in the world, so I'll make this discussion brief. First, please purchase two books that you'll use throughout the year: Morris' tenth edition of *Human Anatomy,* edited by J. Parsons Schaeffer of Jefferson Medical College and *An Atlas of Anatomy* by J. C. Boileau Grant of the University of Toronto (1943). Most schools of medicine use *Gray's Anatomy,* but we prefer Morris for its clarity and scope. And you'll find the Grant's Atlas a valuable aid in locating anatomical structures that are difficult to discern in the cadaver. Remember that no two cadavers are identical with respect to where the nerves, blood vessels, and lymphatics lie. You'll learn their general locations and relationships to other structures, but sometimes you'll be amazed at the variance in both of these parameters. Now are there any questions?"

A student raised his hand and was acknowledged by a nod.

"How much should I charge for an appendectomy?"

Dr. Magruder stared at him coldly for a moment and then replied:

"This kind of frivolity will not be tolerated here. And that reminds me, we'll be starting the dissection of cadavers shortly and anyone who engages in frivolous behavior will be subject to dismissal! For

153

example, students who swipe wrist bones to fashion into charm bracelets for their girl friends, or anyone caught engaging in childish behavior with parts of these cadavers, such as throwing organs around the room, will be shown the exit—permanently!"

Reaching for a chart of the abdominal cavity, he pulled it down vigorously like a stuck shade and began discussing the dissections planned for the next two days.

After a few hours of discussion during which Dr. Magruder described various components of the abdominal cavity from a colored chart, occasionally emphasizing salient features with a metal pointer, he announced it was lunchtime and that we had an hour to eat and relax. As we filed out of the laboratory, the whispered conversation centered on the demeanor of the instructor who had read the riot act to the class in response to a thoughtless remark.

"It's embarrassing that someone of that ilk should be admitted to med school," remarked one student, cautiously looking around to see if the offender were in hearing range.

"Yes," agreed another. "Thanks to him the class got off on the wrong foot. You can bet our reputation will spread to the rest of the faculty that we're a bunch of money-grubbing medicos!"

At one o'clock we returned to the laboratory and were surprised to see each table occupied by a cadaver, loosely wrapped in moistened layers of cloth that reeked of ethyl alcohol and formaldehyde. While the class was peering at the mummy-like shapes, Dr. Magruder entered the room and announced that five students would be allocated per table to comprise a dissecting team.

"Any less than this number, and you'll have a lot of work to do to catch up with the rest of the group," he added. "Now remove the cloth over the mid-section of the body and examine the skin. You'll notice that the cloth is folded in such a way as to be easily removed over specific areas without uncovering other sections. When you have finished with your dissection this afternoon, kindly rewrap the abdominal area in the same way that you found it. This prevents undue drying of other parts of the body not being dissected. Now we're ready to make our initial incision."

He gave instructions as to how the scalpel should be held and where to cut so that with a few additional incisions the entire abdominal area was exposed.

A female student raised her hand and, upon acknowledgment, asked: "I just peeked at the head of my cadaver and was surprised to find it tightly wrapped in cloth. Why?"

"We won't be working on the triangles of the neck or the musculature of the face for several weeks, so these structures must be kept moist to facilitate their dissection. If you allow desiccation to occur, which happens rather quickly in this room, your work area will become leathery and careful dissection will be extremely difficult."

"Thank you," said the young lady and proceeded to work.

Within a week the class had become proficient in the anatomy of the abdominal cavity, and everyone knew the boundaries of the epiploic foramen, knowledge of which each person kept in abeyance, in the back of the mind, like the combination to a safe, awaiting the first test: "also known as the foramen of Winslow, it is located just below the liver and bounded superiorly by the caudate lobe of the liver, inferiorly by the duodenum, posteriorly by the inferior vena cava, and anteriorly by the right margin of the lesser omentum." The Amoeba had given several lectures in histology lab and spent a few hours hovering over our microscopes as we attempted to identify cross-sections of the linings of the stomach, small and large intestine. By studying the cellular make-up of the intestinal wall, it was actually possible to determine from which part of the intestine the slice came.

The next sequence of lectures was in neuroanatomy and involved the brain, that miraculous organ that is the basis of organized existence, of the precise control of physiologic functions, and of the preservation of being and well-being. Again, Dr. Magruder reached for a rolled-up chart and gave it a yank.

"You're perhaps wondering if it's now time to unravel the cloth from the head of your cadaver and saw the top off the skull to remove the brain. If you do, you'll get a surprise. The brain is no longer intact but a mushy, indefinable mass of gray gunk. As soon as death occurs, nervous tissue deteriorates immediately. The brains you see preserved in formaldehyde in the jars lining the walls have been removed within a few minutes after death. Now, the cranial nerves are the following: olfactory, optic, oculomotor, trochlear, trigeminal, abducens, facial, acoustic, glossopharyngeal, vagus, accessory, and hypoglossal. When requested on an examination you silently recite this: On Old Olympus Towering Top A Fat-Assed German Vainly Ate Hops. As one of the bonuses of this course, you'll learn poetry that can only be recited among friends in bars!"

Dr. Magruder picked up his painter and proceeded to show where the cranial nerves originated in the brain. By the time the lecture ended, he had pulled down a half dozen charts depicting various views of this remarkable organ, including several that revealed how soft tissues, fluid, and bone provided a protective cushion to absorb otherwise traumatic head blows. The students came away from the afternoon session in neuroanatomy with a newfound respect for the mysteries of biological creation. In the days to come, this respect would be further deepened when the architecture of the human body would reveal a structural scheme hitherto unappreciated for its ingenuity.

While anatomy and histology lectures and laboratories were underway, another subject, biochemistry, representing the final curricular commitment of the first semester was also given in a lecture and laboratory format. These were the days prior to the discovery of DNA and other remarkable strides in a specialty whose content borrowed from a large number of scientific specialties, including the principal meld of biology and chemistry. If one were to examine the content of 10 biochemistry courses given at 10 colleges throughout the country, the following assumption would be made: *all covered essentially the same material, but an emphasis on certain areas, which happened to be the field of greatest interest to the instructor and in which he or she researched for advanced degrees, comprised a significant portion of the course.* Thus, carbohydrates, fats, and proteins would be discussed in every course, but the significance of protein and sugar in the urine, and the production of cataracts in the eyes of rats after excessive intake of galactose would be taught only to medical and dental students. After the biochemistry course at Tufts, I enrolled in a graduate course in the same subject at Purdue a few years later and was astonished to see how different the course contents were! I don't even recall the name of the Tufts instructor, although I do remember a dour person whose gruff manner was better suited for that of a football coach. In short, I gleaned very little educational gems from his course.

The examinations were far more challenging than the delightful brain exercises I had encountered at Medford. I succeeded in passing the two major courses, anatomy (which included histology, embryology, and neuroanatomy), and biochemistry (a boring course and a *bona fide* time-waster). In the brief period of several months, I had absorbed the pages of Morris' Human Anatomy, knew the origins and insertions of countless muscles, followed the pathways of spinal nerves, the sympathetic and parasympathetic nervous systems, examined the cardiovascular and lymphatic systems and the location of their nodes which served as filters, and memorized the endless terminology of the parts of the skeletal system. What remained during the second semester were the various systems: endocrine, urogenital, respiratory, digestive, and the special sense-organs which would be correlated with one another to emerge as an organismic whole.

My roommates, Aristotle and Robert, also survived the vicissitudes of the first semester of the freshman year and were anxiously awaiting courses that would be more applicable to the dental profession. Basic courses are a necessary evil, but they provide future health care professionals with a biological blueprint without which the practical aspects of medicine and dentistry cannot be performed.

When the second semester began in January, Dr. Spector took over the lectures from Dr. Magruder. His area of expertise was the respiratory system, but as the semester progressed, the class became aware that his knowledge of other systems was just as awesome. As mentioned previously in discussing a teacher's research interests comprising much of the course content, the man with the Micawber-like physique seemed to be an exception to this presumption, for with the introduction of each system, no redundancy of information that could have been interpreted as earlier research efforts was noted. Dr. Spector went far beyond the anatomy of each system. For example, along with the anatomy of the respiratory system, he talked about collapsed lungs, pulmonary edema (fluid in the lungs) caused by drugs and disease, and how a stethoscope can be used as a diagnostic device for lung as well as cardiac disease. As a master teacher, Dr. Spector utilized with brilliance the principle of association to instruct young minds in the relationship of structure to function and malfunction.

Biochemistry was replaced by physiology; a dour teacher was replaced by an even more dourly one. This subject deals with the functions of organs, including those such as the blood and the skin without definable boundaries, which, together with organs of distinct form, perform important duties in maintaining the internal environment, or what Claude Bernard called the *"milieu interior."* I don't recall the professor's name, but his unique style of lecturing was often peppered with the word *dogmatic,* so frequently in fact that before long the entire class responded on cue upon hearing the word with a scarcely audible "bow-wow."

The laboratory began as a traditional experiment that was a "kick-off" exercise found in the laboratory manuals of most medical, dental, and pharmacy schools before the days when animal rights groups were organized. The little frog *Rana pipiens,* that slimy amphibian participant of first-year biology courses throughout the world as a miniature demonstrator of chordate anatomy, revealed physiologically how skeletal muscle tone increased when an electrical stimulus was applied to a sciatic nerve that terminated in gastrocnemius muscle. In a similar preparation, South American arrow poison (curare) was added to the muscle with a medicine dropper and the nerve was stimulated with an electric current. The result?—no muscle response. The muscle was then stimulated with a weak current and it responded with a twitch. Conclusion: curare acts between the nerve ending and the muscle or myoneural junction to cause a loss of tone (tone or tonicity of skeletal muscle defined as *a partial state of contraction of muscle in the resting state).* As the course in physiology progressed, I learned to respect the little amphibian whose heart furnished the means by which Otto Loewi received a Nobel Prize in Medicine by proving that vagal stimulation slows the heart beat through the release of a chemical, acetylcholine, the first neurohormone and recognized chemical mediator of the brain, spinal nerves, and autonomic nervous system.

Toward the end of the second semester, on April 12, 1945, President Franklin Delano Roosevelt succumbed to a stroke in Warm Springs, Georgia, where he spent a few weeks each winter to rest and strengthen his polio-weakened lower torso and leg muscles. At the time of his death, he was sitting for his portrait in the presence of the famous watercolorist, Elizabeth "Mopsy" Shoumatoff, cousins Margaret Suckley and Laura Delano, and longtime companion Lucy Rutherfurd. Every biographical account of the tragic incident relates that he suffered a severe headache and subsequently a massive cerebral hemorrhage that his physician, Dr. Bruenn, treated with amyl nitrate to relieve the intense vasoconstriction. His heart stopped beating at 3:35 P.M. The biographers are inaccurate for, in reality, Dr. Bruenn administered amyl nitrite by inhalation. Amyl nitrate is not employed therapeutically and would be ineffective under these circumstances.

Millions of grateful citizens mourned the heroic father figure who had used drastic measures, sometimes subverting the Constitution, to restore confidence in the economy. During the dark days of World War II, he mobilized the seemingly inexhaustible natural and human resources of the nation to support the war efforts of Great Britain and Russia against the Nazi threat in the European theatre and the Japanese encroachment in the Pacific.

To this day he is criticized for placing U.S. citizens of Japanese descent on the West coast in "concentration camps", penalizing them for being *potential saboteurs.* No one realizes that the temper of the times would have resulted in the deaths of many innocent persons if these unfortunate citizens of the United States had not been protected from the erratic behavior of patriotic zealots! (Did you know that, during the bombing of London by Hitler's bombers and V-2 rockets, some of the British people stoned dachshunds to death as they wandered along rubble-strewn streets?) "Why," some people ask, "weren't the U.S. citizens of German ancestry similarly corralled?" The answer is obvious; the physiognomy of a person of British, French, German, and even Russian descent isn't distinctive enough to reveal their country of ancestral origin. Yet, Roosevelt will be blamed for this constitutional violation of citizen's rights for decades to come as will Harry Truman's decision to nuke Hiroshima and Nagasaki, which prevented thousands of children from growing up in single parent households. Why is it that those with the greatest degree of ignorance become the most authoritative on subjects they know the least about?

As I mentioned previously, my parents and relatives disliked FDR as a president, but they all shed a few tears on April 12, 1945, along with millions of others, who realized that a statesman had left his mark of greatness on the world's stage.

On September 14th, 1945 I entered active service in the U.S. Navy (file number 805 13 76), having decided to postpone my medical career in a *better late than never* situation. Because of my proneness to several allergies, one of which was hay fever, to enter the Armed Services in the early stages of the war would have resulted in refusal and a 4-F classification. Years before that fatal day of December 7th, 1941 at Pearl Harbor, I had planned to follow in Dad's footsteps and, following graduation and completion of my internship, to work at his side in general practice. However, the year at medical school was disillusioning, unpleasant, and unrewarding. The instructors may have had children in the Services and regarded the student body as a bunch of draft-dodgers. A year of this nonsense was enough! On a crisp September morning I boarded a train at the Needham station and went to Boston where, after a physical examination, I left for Bainbridge, Maryland, for 12 weeks of naval training on the banks of the Susquehanna River. I didn't do this all on my own. I persuaded Dad and Mom I was adamant about entering the service despite the allergic situation (hay fever a detriment in the Navy?). One of his patients, a Mrs. Kennedy whose husband was a distant relative of JFK, set the stage for my induction.

The twelve weeks I spent at Bainbridge were educational. I lived in a single-story barracks with a central corridor lined on each side with neatly arranged double-deck bunks beside which small cabinets containing handerchiefs, underwear, and stockings stood. Neatness, tidiness, and cleanliness were the hygienic watchwords of the U.S. Navy and, unlike the conditions on some foreign ships during World War II, tuberculosis was unknown among our naval personnel.

A chief petty officer (CPO) was in charge of each barracks of a hundred men and was responsible for allocating appropriate punishment for misconduct. One evening someone refused to shut up. A whistle blew and everyone, some awakened from a sound sleep, rose and ran a mile or so on a dirt track in the cool October weather in their white scivvies. When the CPO discovered I had been a medical student, he asked me to discuss venereal diseases before the men of his barracks. Standing on a wooden table surrounded by gaping neophyte sailors, I discussed the biology of *Treponema pallidum (syphilis),* the ravages of *gonorrhea,* and why consorting with prostitutes could lead to serious trouble. I finished my lecture with a reference to my first year at Tufts medical where my apartment shared with dental students Aristotle and Robert was occupied by prostitutes on the floor above. The only contact we had was the day when Robert knocked on the door to ask if they'd look after his goldfish during the Thanksgiving holiday. They said "yes."

A young man with nine children, who reminded me of W.C. Fields and worked as a trolley- car operator in Boston, was sent home after only a week at Bainbridge. He and I had chatted together on the train trip to Maryland and he had prayed that the Navy would keep him for at least 90 days so he'd be eligible for benefits of the G-I Bill.

David E. Mann, Jr.

CHAPTER TEN

I'm in the Navy now

Boot camp at the U.S. Navy's Bainbridge, Maryland, facility was a gigantic bustling complex of hastily constructed, metal-framed, buildings of cream-colored aluminum siding. They were spread like giant rectangular mushrooms over several hundred acres of gradually sloping countryside that bordered the wide, slowly flowing Susquehanna River beside the Havre de Grace bridge. Narrow asphalt roads intersected and connected the various structures that collectively were designated as regiments. The third regiment comprised buildings that housed the newly arriving raw recruits and their training areas. The sailors of this regiment were assigned funny-looking canvas puttees, more for recognition of lowly status than for protection of the legs. A dining-facility, infirmary, and Quonset-hut structure were located within walking distance of the single story barracks that served solely as sleeping quarters. Although the entire regimental facility appeared to have sprouted quickly from the resources of the immediate area in response to the relative unpreparedness the country found itself in on December 7th, 1941, the sylvan beauty of the landscape enhanced by the majestic presence of the broad-flowing Susquehanna imparted a serenity that belied the frantic activity that took place each morning at daybreak.

For some reason, known only to those who commanded or designed the facility, there was no first or second regiment. There was, however, a fourth regiment, entry to which occurred only for those who had completed many weeks of training that included loading 5-inch shells in a destroyer's cannon by hand, tying knots, studying naval lore (John Paul Jones especially), and taking turns preparing food in the mess hall. When the sailors moved a few hundred yards to the fourth regiment, it meant a promotion to *seaman second class* and, after taking intelligence tests, being assigned to posts anywhere in the world.

I remember the IQ test. It was simple, but after completing it, I realized that I'd missed a question about a tugboat docking at a pier. I

162

never did get the results, but someone in my barracks learned that the Pennsylvania Railroad detective, who accompanied us to Bainbridge as an inductee, had received the lowest grade ever recorded on a Naval intelligence test. I found this scuttlebutt amusing but doubted its authenticity, although I'd chatted with the man on the train trip to Maryland and not been particularly impressed with his repartee. He was no Fred Allen!

Before joining the fourth regiment, I fell victim to what is called "cat fever," a syndrome characterized by irritated eyes, nausea, dizziness, fever, and G-I upsets. The bug infected many of my buddies and required an ambulance to transport the groaning bodies to a large infirmary several miles away. It was there that I saw my first Red Sox game (against the Washington Senators) with Dave "Boo" Ferris the losing Red Sox pitcher.

The infirmary stay lasted a week and, upon my return to the third regiment barracks, I learned that the CPO had assigned me the title of *Athletic Petty Officer,* a task that required settling arguments among my colleagues by arranging boxing bouts. The ring, part of a health-recreation facility, was the first I'd ever seen up close. I was given a key to a locker containing boxing gloves, towels, liniment, bandages, and sneakers. After each skirmish ended, the items had to be checked, which theoretically saved the taxpayers money. If only those in the rest of the military establishment had been as conscientious.

There were only two "bad eggs" in the entire barracks and both were from West Virginia. One was a Bible reader, a kid with a bunk at a remote corner of the barracks, whose high-pitched voice reciting Biblical passages could be heard long after the signal for "lights out" resounded. The religious mumblings so enraged a few sailors one evening that he was attacked and had to defend himself with a knife he stashed under the mattress for just this kind of confrontation. He was quickly apprehended by the CPO and, after visiting the resident psychiatrist, was discharged and sent home to Wheeling.

Oddly, the other "bad egg" was a sailor I thought was a decent chap. He seemed to be older than the rest of us, a young version of an "old timer" who aroused the ire of some who didn't understand his rantings about the military. I had to arrange more than one fight for him, which he managed to win consistently by beating his opponents to a bloody pulp.

When it was time to join the fourth regiment after the twelfth week of boot camp, everyone in the unit was required to visit the infirmary for "shots", assuming that duty for some might be in parts of the world rampant with disease. The brave men of the barracks turned into a bunch of sissies at the thought of getting jabbed in the arms or buttocks with needles which their minds had magnified to the size of ice picks! Some became nauseous; others suffered from Montezuma's revenge the day of inoculation. The day we were due to leave for duty in the outside world was also the day we were scheduled to step into a chamber and inhale various poisonous gases in subtoxic doses. The week before, we had entered a steel chamber containing a vat in which several hundred gallons of oil flowed. The oil was suddenly ignited and each sailor was required to extinguish the conflagration with a steel wand that emitted a fog spray. It was truly amazing how quickly the mist suppressed the flames. At that demonstration we saw the *Handy-Billy* in action, a gadget resembling an outboard motor (costing $75) and just as noisy. It came with a thirty-foot length of hose that was tossed over the side of a naval vessel to suck up sea water. For civilian use, it was an ideal fire extinguisher for those with homes near a swimming pool, pond, or river. These were exciting events, but departing for my immediate destination, Washington, D.C., and missing the poisonous gas exercise was a real coup.

The OGU or out-going unit left for Washington in a bus filled with sailors anxiously awaiting the next military assignment, having survived the rigors of boot camp with its sociologic challenges. The trip ended near a military installation on the banks of the Potomac where a number of small craft were dragging nets like the sea-going scallopers off the New England coast. As I stepped out of the bus lugging a large duffel bag containing a ditty bag (with a toothbrush and Dr. West's toothpaste and Gillette shaving equipment), summer whites (I was wearing the warmer blue bell bottoms), underwear and socks, and a hammock, the realization came to me that the boats were dragging for something other than fish. A civilian standing near the water's edge informed me that a seaplane had crashed and already the river police had picked up a few bodies. Coming to Washington for the first time under these conditions set an eerie tone to my arrival which manifested itself by a peculiar sensation in the stomach and a feeling of sadness for those who had lost their lives. Perhaps this feeling had something to do with a subconscious thought of my own flight across the Panama Canal in a seaplane a decade ago that ended without incident, and anger that these people had been unfortunate enough to have lost their lives because of incompetence or some other inane reason.

I was assigned a roommate, a humorous chap from Ohio, who was a few years younger than I, and given temporary shelter in an apartment complex rented by the Navy near George Washington University about a half mile from the Main Navy Building on Constitution Avenue. Early each morning reveille sounded and, after showering and breakfast, a group of us marched down the street to the Main Navy building, walked up the stairs to the second floor and crossed over the reflecting pool on one of two bridges spanning the water to the four temporary buildings (I, J, K and L) housing Busanda (the Bureau of Supplies and Accounts). Before entering the two-story structure, I had to assemble my group of thirty or so sailors by yelling: *"Attention!"* Then the command *"Dress right!"* And lastly, *"Dismissed,"* whereupon the sailors dispersed to their work areas.

Inside my building (J), a cluster of secretaries sat at typewriters busily sending out orders for all kinds of naval supplies. The person in charge of this operation was Commander Story who sat behind a huge mahogany desk from the Normandie. This and other pieces of furniture had been removed by the Navy from the magnificent ship several months prior to its loss from fire. The first day on the job I met my boss, a young widow named Rosalie, Charlie, a fellow sailor from Michigan, and two young single women who at first believed that Charlie and I were spies sent to monitor their work. Their suspicion was warranted because once a week Charlie and I met with Commander Story to discuss our work which involved activities that were similar to those of our female associates, typing. The only difference among us was that Charlie and I used only two fingers.

The temporary stay in the shadow of the Main Navy building lasted only a few weeks before several hundred sailors were transferred to the recently vacated Wave's barracks located across the street from American University and at the end of Embassy Row. The facilities were much nicer than those I had just left. Instead of sleeping quarters for two in a small room, the new barracks had cubicles consisting of two double-bunk beds in a much larger area with cabinets for clothing, chairs, and a couch. At least a hundred sailors were housed in one building in relative comfort. Among the four occupants of each cubicle, each person was responsible, on a rotation basis, for dusting the furniture, sweeping the floor, and arranging the bed covers neatly in preparation for the daily inspection. Responsibility for the daily task was known in advance and inscribed in black ink (from the newfangled ball-point pen manufactured by Reynolds and available for $12.50) on a Betty Grable calendar. Messing up meant a penalty would be imposed. I missed my day only once and was handed a pair of scissors to trim the grass of the front lawn. It wasn't much of a penalty work-wise, but the glances from passing personnel made one feel like a clown. Of course that was the idea and it worked.

One of the occupants of my cubicle was a sailor named John whose home was in Santa Monica. He was in naval intelligence. Although reticent about his military activities, one evening, after a beer or two, he mentioned that President Roosevelt had committed suicide by shooting himself. I told him I thought that was preposterous.

"Well," he remarked, somewhat perturbed that I didn't believe him, "he slept with a pistol under his pillow ever since that assassination attempt in Miami that killed Mayor Cermak of Chicago."

"So what!"

"He had access to a gun. Furthermore, his casket was closed in the rotunda of the Capitol."

"You're forgetting he wasn't alone in Warm Springs. And why on earth would he want to shoot himself! He wasn't in any pain until the hemorrhage. And as for the closed casket, isn't it the custom nowadays for the president's body to lie in state in a closed casket?"

John was a nice guy and became a good friend, partly because he knew I had visited California and seen the pier stretching out to sea from the beach at Santa Monica, a clamshell's throw from his house. As a member of naval intelligence, however, I wasn't particularly impressed with the grasp he had of his occupation.

Weekends were free to have fun in the Washington area. In center city Milton Berle was starring in a play called "Springtime in Brazil" that was riotous, but short-lived. It would have been more successful if it had starred "Mr. Television," a monicker he was shortly to earn, for his verbal and physical approaches to humor were not well known during the war years, but were immensely popular at the dawn of television in the early fifties.

An amusement park in Chevy Chase known as Glen Echo was a favorite recreation spot for servicemen and women. The most memorable feature of the park was the minute size of the popcorn. In some respects Glen Echo reminded me of Paragon Park in Massachusetts opposite Nantasket Beach where a ferris wheel, carousel, and other accoutrements of a typical carnival comprised the attractions along with booths dispensing goofy-looking dolls for those skilled in darts and pellet-gun shooting. Like Glen Echo, Paragon Park closed during the winter months. Unlike Glen Echo, the popcorn at Paragon was of normal size.

Hailing a taxi in downtown Washington often meant sitting on the rear seat between a general and an admiral, or between a senator and a House member. The conversation was nil under these circumstances unless I happened to step on the polished shoes of my companions while entering or leaving the vehicle. A grunt or other barely discernible verbal acknowledgment would be uttered as I left the cab and it speeded away with its unique cargo of earth-shakers. Once, after I had visited my parents in Needham and they had graciously lent me Mom's blue Dodge coupe to take back to Washington, I parked it near the Capitol and was soon confronted by a policeman.

"Do you know you've parked in the space of a U.S. Senator?" he asked.

"No," I replied.

Instead of handing me a ticket, he smiled and said: "Watch it in the future. These guys get cantankerous if their givens are taken!"

I never found out whose space I had taken, but there were a few senators who would have been gracious and understanding. Senator Aiken of Vermont, for example, one of the finest persons who ever represented a state would have demonstrated these attributes.

The greatest feature of my new domicile, the vacated Wave's barracks, was an Olympic-size, heated swimming pool. Now that the Waves had departed, skinny dipping was encouraged. After a long day of arising before daybreak (boarding a bus for a half hour ride to

167

Constitution Avenue, working from 8:00-12:00, an hour for lunch, back to work for four more hours, bus ride back to barracks by 5:30, a meal in the mess hall), a swim at 7:00 p.m. was relaxing. Some sailors played cards in the barracks while others visited the recreation room to play board games, billiards, ping-pong, and badminton. The music of current Broadway shows (Oklahoma and Carousel) was heard on the radio or record player more frequently than the National Anthem.

A U.S.O. Stage-door Canteen was located in downtown Washington (there was one in every large city) where servicemen and women could socialize with civilians while dining and dancing and being entertained by semi-professional and professional musicians and relatively unknown stand-up comedians. On one occasion, the band suddenly launched into *Stars and Stripes Forever,* there was a drum roll a few bars of *Anchors Aweigh,* and suddenly Chief of Staff U.S. Army, Dwight D. Eisenhower and Fleet Admiral Chester W. Nimitz, World War II commander in the Pacific walked on the stage.

Everyone in the auditorium stood up and applauded wildly while the two uniformed men smiled and bowed. Then, each spoke for several minutes, thanking the members of the armed forces for their contributions toward the attainment of unconditional surrender in the European and Pacific theatres. It was akin to Jesus thanking his followers for believing in God. Before they left the stage, General Eisenhower nodded at Admiral Nimitz and said: "If you're interested in obtaining our autographs, leave your name and address on a postcard that will be handed out and drop it in the box at the door as you leave the Canteen. You should receive them in a few weeks."

I followed the directions and sure enough I received a telephone call from Mom about a month later informing me that I had received a letter from Washington. When I got home for Christmas vacation and opened the envelope, inside were a small card signed C. *W. Nimitz* and a slightly larger card with the words War Department at the top, and beneath *Dwight D. Eisenhower* signed in ink, and below that the words *Chief of Staff,* U.S. Army. In *The Evening Bulletin of Philadelphia, October 3, 1961,* I found a photograph of former president Dwight D. Eisenhower and Fleet Admiral Chester W. Nimitz arriving at the Newport, Rhode Island, Naval Base, where Eisenhower delivered a classified address to students of the Naval War College. This clipping provided an appropriate illustration for my two autographs, deftly mounted against a black background and framed in black and gold. What a wonderful memento of World War II!

Maine's contribution to the sailor population at the former Wave's barracks included many lanky lads whose fathers were loggers, lobstermen, and game wardens. They were nice kids with distinctive accents who answered affirmatively with an "ayah" when they meant "yes." One sailor looked like Buddy Ebsen, the dancer, movie star and television detective. I don't recall his name, but his boorishness was unmatched among those who had come to Bainbridge from rural parts of Maine. Imagine how astonished his acquaintances were when he introduced his hometown girl friend and she turned out to be gorgeous! Whoever said opposites attract was right on target.

In Mom's blue Dodge, I returned to Washington after a spring break to find the fringes of the tidal basin bursting with the ephemeral beauty of blossoming Japanese cherry trees. The pink cotton candy of Glen Echo was called to mind along with the irony that these botanical gifts to the government had come from former friends who were now our enemies. I went on two dates with a secretary from Tennessee (who worked in building "J") on the suggestion of my widowed boss, Rosalie. On the first date, the Dodge got stuck in mud as I was attempting to turn around on what appeared to be a cowpath of a large dairy farm. I left my date sitting in the car and walked up the road to the farmhouse. The owner, who was black, told me to wait by the car and he'd be along with his tractor in a few minutes. Sure enough a silver-colored Fordson tractor with small cement-filled front wheels and huge steel, lug-studded, bright red rear wheels pulled the car back onto the highway. He accepted my thanks but refused to take any money. On the second date, I was enjoying a meal in an outdoor cafe where metal tables and chairs were set on a rock and grass surface. Pushing the chair back to stand up, the metal leg caught on a rock and she fell backwards with both feet pointing toward the sky.

Before leaving the Navy, Rosalie, Charlie, two of her female typists and I visited Rosalie's sister in Maryland who had a home on Chesapeake Bay. From her sister, I learned that Rosalie's husband had been killed in the early months of the war while serving in the Army in the European theatre.

David E. Mann, Jr.

On my last leave from Washington, I had taken the train home and left the Dodge parked outside the Naval barracks. At the dinner table that evening, Dad asked where the car was. When I told him, he got up from the table and angrily informed me that I depart at once by plane and drive it back. In an hour I was flying out of Logan airport headed for Washington. The reputation for criminal activity in the Nation's Capitol had apparently been well publicized in the Boston papers.

In May, having received a promotion to Seaman, First Class, and assigned guard duty in the basement of a naval building located behind the imposing Tennessee marble Grecian edifice of the Supreme Court, I spent 12 long hours walking the dimly lit corridors without seeing another human being. Finally, the ordeal ended and I was replaced by another sailor of the same rank whose home was in Lexington, Massachusetts. When I asked him to identify himself, he replied: "I'm David Mann."

"And yours?" "David Mann," I answered.

We had a good laugh over the coincidence. A few months later, when I was discharged (honorably) from the U.S.N. Personnel Separation Center in Boston on July 5th, 1946, his personal files were in my folder and, I presume, my records ended up in his. Those responsible for the mistake should have filed the information according to the file or serial number instead of by the sailor's name. I discovered the error when I saw my hometown listed as Lexington and his dental records, which showed fewer cavities than I had. One of the most memorable experiences of my boot camp days at Bainbridge took place in the dentist's chair. When I was a teenager, an orthodontist had wired both jaws. After their removal, the only time I visited a dentist was when a toothache became unbearable (the predicted behavior for most of the adult population).

Now, sitting comfortably in a padded chair, the dentist carefully examined the oral cavity and reached for a drill.

"You have a number of gum cavities," he muttered, and began drilling.

More than a dozen cavities were filled without the injection of Novocaine. Navy dentists were unsurpassed in their skills as were Navy surgeons. My fillings were retained for more than three decades. My recall of the experience lasted far longer.

CHAPTER ELEVEN

Deer John!

"Now that you have your ruptured duck*, a 100-dollar bonus, and the G-I Bill for education, what are your plans?" asked Dad over the dinner table.

"I haven't decided yet, Dad," I replied. "I want to continue my medical studies, but not in a medical school. I'm going for a Ph.D. in a medical science, either physiology or endocrinology. These were the subjects I enjoyed most at Tufts."

"So you're going to be a professor and teach," said Dad. "Well, you've got to find a school that has a graduate program suited to your needs. Have you given it any thought as to where?"

"Yes, I met some sailors from the mid-West and they said the schools there are fantastic. Michigan, University of Chicago, and Purdue in Indiana are all outstanding as graduate schools for the medical sciences."

"In the meantime, instead of sitting around doing nothing until you have decided which school you want to attend, I have some work for you. There are a lot of dead trees lying around on my land in Norfolk that I want sawed in four foot lengths and brought home for the fireplace. After a cord or so of wood has been cut, I'll rent a truck to bring it home. And the exercise will be healthy for you."

Thus, each morning I drove the blue Dodge to Norfolk and parked it on a triangular section of land called the flatiron. Armed with a handsaw, I moved slowly on the moss-covered soil to prevent stepping on a cute little garter snake that seemed to be at the same place each morning. Finding an appropriate tree trunk, I began sawing the trunk into sections the desired length, carefully removing the limbs as the diameter of the trunk became smaller and smaller. Within a few weeks, I had accumulated enough firewood to heat the living-room for several years, some of it resembling batons.

Dad was thrilled with my accomplishment and hired a truck to bring the booty back where it was piled alongside the carriage-house. But I needed another job.

"I was talking to Dick Salamone, the Fire Chief, and he has a job for you on the Fire Department. It's only for a couple of months replacing a fireman who is presently on vacation. The salary, incidentally, is 32 dollars per week."

"I'll take it," I cried. "I took firefighting 101 in the Navy one afternoon!"

---

*Ruptured duck, vernacular for pin given to honorably discharged WWII veterans.

The next morning I appeared at Needham's Fire Station on Chestnut Street awaiting instructions as to how to proceed as a firefighter. John Cotter, one of the older members of the department, showed me the beautifully-maintained, sparkling-red hook-and-ladder truck, the pumper, rescue vehicle, and other pieces of equipment including the large, canvas-wrapped rubber hose that is screwed into a fire hydrant, and the smaller diameter 2 1/2 inch hose that, after use, is hung on hooks at the rear of the station to dry before being wound like an anchor chain on a reel in the back of the pumper truck. I also met John's friend, Clarence Humberstone, who became a fishing companion and, in the fall, accompanied me on several duck hunting trips.

Chief Dick Salamone, who had been busy inspecting several recently-constructed commercial buildings along Route 128, drove up in his big blue Chrysler, lights still flashing on the roof. Boston's newest major highway, which threaded its asphalt band through land that zoomed in value due to its accessibility to Massachusetts' capital city, had created a number of fire inspection problems for the Chief because of the accompanying building boom. He was pleased to see me, asked how Dad was, and spent a few minutes showing me the upstairs facilities where a galley and sleeping quarters were located. Then, he showed me the steel pole with an orange peel-shaped metal closure that opened as soon as one grasped the pole and slid to the floor below. He grabbed the polished pole to demonstrate how the metallic peel opened, like a giant clamshell, and then closed when he disappeared to the floor below. He landed with a thud on a thick fiber mat about two feet from the hook-and-ladder truck. I followed him and slid gracefully to the bottom within three seconds. Next, he showed me the mobile equipment and introduced me to a few younger members who had just arrived in a jeep from extinguishing a rubbish fire that had gotten out of hand. Instead of carrying a hose, they had used Indian pumps, metal containers filled with water that sprayed the fire when a pump was activated by hand.

My schedule called for arriving at the station by 6:00 p.m. and leaving the following morning at 8:00 a.m. I was assigned a cot and a pair of heavy-duty pants with a rubber boot inserted in each leg. This was to be placed next to my cot and, when the alarm sounded, I was to jump into it and pull the pants up by suspenders. A heavy rubberized coat and metal helmet were to be added downstairs just before I hopped on the appropriate vehicle. I would not be driving right away until I had learned the gear shift sequence. Until then, I was relegated to a seat behind the driver, oftentimes far behind if the pumper with its platform at the rear were used.

It must have been the most fire-free period in Needham's history that coincided with my brief sojourn on the department. For one month, I arrived at 6:00 p.m. and enjoyed a nice meal, then listened to the radio or read until 10 p.m. After a nice uninterrupted sleep, I arose at 7:00 and departed for home after breakfast by 8:00 a.m. Also, there was a check for 32 dollars to pick up each Friday.

In August, the person I had replaced temporarily returned from his Cape Cod vacation refreshed and ready to confront the uncertainties associated with fire-fighting. With foresight, he would have picked any month after July to vacation because climatic conditions and human stupidity resulted in a sudden upsurge of fire activity in the late summer and fall. In August, I was assigned to the fire station in Needham Heights to fill in for another vacation-bound fireman. The station was a two-story wooden relic of the 1890's whose white clapboard sides gave the building the appearance of a Baptist or Congregational church so characteristic of New England's small towns and villages. Directly across the street, the multi-story, gray-stucco buildings of the Carter factory spread over several acres alongside a railroad track from which the famous manufacturer of underwear, pajamas, and other undergarments shipped garments to Boston and the rest of the civilized world. The late founder's offspring, William and Horace Carter, were Needham's foremost citizens and philanthropists, having contributed funds toward the construction of the new Glover Memorial Hospital and, ironically, the magnificent brick structure on Chestnut Street that housed both the fire and police departments. Horace Carter, who like his brother received an annual salary of one hundred ten thousand dollars a year, was a patient of Dad. One Christmas he presented him with an autographed book of poems he had composed and privately published. Dad was thrilled (he loved poetry and could recite a diverse repertoire on demand) and much preferred this gift to that of underwear. Carter's factory was the principal employer of the adults of Needham and Needham Heights. Those who worked at the vast manufacturing complex were there for decades eventually to be replaced by their children.

The antiquated fire station was obviously a conversion from the days when horses' harnesses dropped from the ceiling. If one looked closely, attachments of the fastenings could still be seen as rusty but probably still functional units. It didn't require much of an imagination to picture the presence of two white stallions (they were always white) where now a red *Maxim fire truck with silicon-painted tires was poised ready to rumble forth onto the tarmac when the alarm sounded. The fire truck appeared to be of ancient vintage, an evolutionary step beyond the horse-drawn, steam-fired, fire-pump which, in turn, was only a step ahead of Benjamin Franklin's fire brigade vehicle that was pulled to the site of the inferno by several strong men.

What was especially unique about this pumper, aside from age and glistening silicon-coated tires, was its place of manufacture, Massachusetts. Most of the fire trucks used in the small towns of New England were modified from standard truck bodies built in either Pennsylvania or Wisconsin. The configuration of a truck adapted for fire fighting is largely determined by the height and commercial use of the majority of buildings in the town. For this reason, small town fire trucks differ sharply in appearance from those designed for city use. You'll never see an extended hook-and-ladder truck in a small town with a man or woman valiantly struggling with a steering wheel in the rear to maneuver the behemoth around a sharp corner!

Manufactured by the Maxim Motor Company, Middleboro, Mass.

The complement of the Needham Heights fire station was two firemen:—an experienced officer (a lieutenant, rarely a captain) and a rookie (usually a man in his twenties with limited firefighting experience). I met the requirements of the latter position—my firefighting experience was extremely limited. The senior officer was a lanky veteran named Brown, who had reached the status of lieutenant in his early fifties and was officer Francis Haddock's best friend with whom, on Friday morning, February 2nd, 1934, he'd been chatting inside the fire house when the bleating sound of a distant police siren caused the policeman to run outside and stand on the cement incline, gazing in both directions. As soon as the bank robbers in the onrushing car, with Arnold Mackintosh standing precariously on a running board, spotted Officer Haddock, one of the Millen brothers pulled the trigger and machine-gunned him on the spot. Lieutenant Brown never mentioned the tragedy during my month long stay nor did his son, also an officer, who replaced his father on the night shift. They were both aware that Dad and several other physicians had tried desperately to save the lives of McLeod and Haddock, and that I knew more about the medical aspects of the tragedy than they. It was not a topic for idle conversation. Furthermore, at the time a guy named Ted Williams was putting the Red Sox into serious contention for the pennant despite rabid competition from their perpetual nemesis—the New York Yankees and their star outfielder, Joe DiMaggio.

Within a day or two, I had learned the gear-shift sequence of the Maxim pumper, its hose layout (3 sizes, the largest for fastening to a hydrant), and how to use the First Aid kit that contained an oxygen tank, medicine, and a tourniquet. Several weeks passed before the first and only alarm sounded. It was a brush fire off of Route 128, several miles away. I climbed aboard the ancient Maxim, heart pounding, while Lieutenant Brown took his position to my right. I started the engine, the door opened, and we rolled down the concrete ramp. I turned sharp right through the center of the town and came to a stop light. I slowed down and, as it turned green, revved up the engine but couldn't shift into the next gear. By then it was time to turn off Highland Avenue and enter Route 128 where a small fire could be seen in the distance. I drove off the road and stopped the engine. Lieutenant Brown started the pump while I unraveled the small hose from a reel at the rear. A few drops of water emerged from the nozzle and then nothing. The hose had kinked somewhere. Shortly, we found the problem and quickly extinguished the fire which by now was beginning to die on its own. As we were rewinding the hose, officer Ed Wainwright, destined to be chief of the Needham Police Department, appeared on his Indian motorcycle, a rubber boot under each arm.

Laughing, he said: "When you slowed down at the stop light, they fell off the running-board."

Thus ended my career as a firefighter.

With my firefighting duties over in August (forever), I began preparing for advanced study in the health sciences. I disliked mathematics, especially high school geometry, as taught by Miss Fessenden who had correctly predicted that, if I didn't work harder, I'd be driving a truck (I should have sent her a photograph of me sitting behind the wheel of the Maxim). Fulfilling an educational deficiency, I enrolled in two evening courses at Boston University-trigonometry and introductory physics. To my amazement, I loved both courses and belatedly realized that my problem had been the personality of the teacher, not the nature of the subject. As a result of exposure to these courses, I became enamored with two amazing people who became my lifelong idols: Albert Einstein and Madame Curie. After reading a biography of the former, I was able to reduce his theory of relativity into a single sentence which, years later, I spouted in a lecture: "The phenomenon of nature will be the same to any two observers who move with any uniform velocity whatever relative to one another."

As for Marie (Marja) Sklodowska Curie, the Polish-born physicist who spent most of her life in France and was the first person (and a female at that!) to have received two Nobel prizes, one in chemistry and one in physics, her accomplishments provided me with the appropriate repartee to those fools who cracked Polish jokes suggesting ethnic as well as female inferiority in my presence! In the entertaining movie of her life, I thought Mary Astor would have been a more suitable Marie Curie than Greer Garson. Walter Pidgeon, on the other hand, seemed to be perfect as the somewhat aloof, yet vulnerable Pierre Curie. Regardless, these two giants (one a giantess?) served us with lofty standards of human intelligence and creativity that could only be distantly approached, like a skyrocket launched at the moon whose goal was never actually attainable.

During this time of academic endeavor, I purchased a copy of *Popular Science* monthly for a quarter and, after seeing a column titled: *"I'd like to see them make,"* which offered 5 dollars for a "pet idea of a gadget *he'd like to see in general use* that its editors decided to publish," I sent in mine.

In the February, 1947 issue, page 110, there appeared my idea for a *Back-Seat Ejector.* "The automotive device (above) is the interesting brain child of David E. Mann, Jr., of Needham, Mass.," read the comment beneath the cartoon depicting an irate husband pressing a dashboard button and saying" Stop nagging me!" as his wife was ejected from the back seat!

After the same magazine published an article about James Bond: Thud and Blunder in the sixties, I sent the editors a letter, which was subsequently published in their March, 1966 issue, reminding them "they were unaware of the ejector seat for cars used in *Goldfinger* had first appeared in PS in Feb., 1947, as my creation for getting rid of nagging occupants." The original cartoon was also reproduced. Almost two decades of environmental exposure had not faded its ink.

The aftermath of all this was a telephone call from Firechief Dick Salamone, who had just returned from a vacation in Miami. One of the guests at his hotel, knowing he was from Needham, asked him if he knew Dave Mann.

"Of course," Chief Salamone replied. "I've known him and his family for years. His father is a well-known physician in the area."

"Well," the man continued. "Could you contact him and find out if he has a patent on the ejector seat invention? I want to buy it."

I was slightly bemused when the chief related this incident. Then, a few months later, when the James Bond movie *Goldfinger* was released starring an Aston-Martin with an ejector seat, I realized that the request made sense.

A few years later, while watching the Merv Griffin show on television, the actor who portrayed the inventor of exotic gadgets was asked by Merv where the ideas came from that were such an integral part of the plots.

He replied: "Our writers come up with them out of the blue."

Let's get it right! Your writers read magazines like everyone else and got the idea for the car ejector seat from looking at a cartoon in Popular Science.

Toward the end of November, my parents decided to pack the Buick Roadmaster with rifles and ammunition and head for the coastal woods of New Brunswick, Canada, for a week of deer hunting. I was invited to accompany them because my ten-week courses had ended successfully and it was time for a change of scenery. Our destination was a newly-constructed one and a half story bungalow perched on a rock outcropping that overlooked the Bay of Fundy at Pocologan, N.B. The owners were Dick and Marge Foley, formerly Needham residents (and patients of Dad), who had moved to Canada when their daughter married. Dick had visited Canada as a young man and fallen in love with the beautiful scenic area of seaside Pocologan and dreamed of settling there in retirement to hunt and fish and build a hunting lodge large enough to accommodate eight or ten paying guests. When he met Marge, he solidified his dream by marrying a woman with a sense of humor, business perspicacity, and, best of all, was a great cook.

Rising before daybreak, we drove along the coastal highway through Portsmouth, N.H. and spent most of the day driving through Maine, finally reaching the village of Calais at sundown. There, at the U.S.-Canadian border, we stayed overnight at a frontier hotel. Inside, a brisk, crackling fire in a huge stone fireplace, over which the head of a bull moose stared vacantly, warmed us before we entered the dining-room for a main course of clam chowder and boiled lobster followed by a slice of pumpkin pie.

The next morning we stopped at Canadian Customs where we met a friend of Dick Foley, a jovial, heavy-set man named Fred Dory.

"You're going hunting with Dick?," he confirmed after checking our automobile licenses. "Well, you couldn't find a better guide and hunting companion. He's a scrawny-looking little guy, but he knows the woods better than the back of his hand."

After recording the serial numbers of our rifles and quantity of ammunition on an impressive-looking master sheet, he asked how long we were going to stay in Canada.

"Regrettably, only a week. I'm a physician and have to get back to my patients. Is there any chance of snow while we're here?"

"No, not until December, It's unusual to get snow along the coast even late in November."

Dad then asked if many cars had gone through the U.S. Customs with deer. Fred smiled and answered:

"Last week my counterparts across the way investigated a car headed for Boston with several deer tied to the roof. Unfortunately, they had driven over potholes that caused a beaver pelt tail to pop out of one of the carcasses. They'll get a stiff sentence for that indiscretion…maybe twenty years!"

Having cleared the Canadian Customs, we headed for St. Stephens for lunch, and then on to Pocologan where the gray-shingled bungalow at the end of a gravel-filled driveway stood alone on the edge of a rocky cliff on the Bay of Fundy. During the war, Lockheed Hudson bombers from the nearby Royal Canadian airfield at Pennfield had flown over the bungalow as they headed for long flights across the Atlantic to England. A few of them, leaving in heavy fog, experienced engine failure and crashed in Nova Scotia or disappeared in the frigid waters of the Bay of Fundy.

At the sound of the approaching car, Dick turned on an outside light and greeted us with a hearty "welcome to Canada!." Marge, in the midst of preparing dinner, burst into the living-room and gave her favorite physician a big hug, then kissed Mom and me. I had never met either one before, but was immediately impressed with their friendly behavior. The two rooms we were assigned behind the stone fireplace were small, but furnished with comfortable-looking beds covered with thick blankets.

"You'll sleep better with the windows open an inch or so," suggested Marge. "The air is cold at night but you'll find it is conducive to sleep."

Within half an hour of our arrival, with luggage unpacked and clothes stashed in the drawers of the early American maple bureaus and hung in the cedar-lined closets, we were seated at the dining-room table chatting with our hosts about their friends back in Needham.

"How's Herbie Mackintosh?" inquired Dick. "Does he still take the train to Boston with those brown-colored bananas for lunch?"

179

Herbie was a graduate of Harvard Law who practiced his profession in the city and always ate a ripe banana for lunch. Oddly, two housekeepers had died at his home (in succession) and were belatedly found by Herbie when he wondered why dinner was delayed. Of course, he called Dad each time.

"Herbie's fine," responded Dad, "and, to the best of my knowledge, his dinners have recently been on time!"

The meal of roast beef, mashed Maine potatoes, carrots, beets, and celery slices stuffed with blue cheese was sumptuous, but Marge's *nothings* were manna from the culinary gods. They were so delicious I had to ask how she made them.

"It's very easy. Pillsbury's biscuits that are packaged unbaked in a cardboard container are twisted into small segments, rolled into a ball and then the ball is filled with half of an anchovy. Be sure to remove the excess oil from the fish before putting it into the roll. Next, seal the roll and bake like a regular biscuit."

In 1952, Dick visited my wife and me (Marge had died the year before) in our tiny apartment in Philadelphia where he stayed overnight on the way to see friends in the South. As a late repast, I offered him anchovies on Ritz crackers that had become my favorite after-dinner snack.

"No thanks," said Dick. "I don't care for your Aunt Chovies."

He was right. With a slightly elevated blood pressure, their salt content could cause trouble. Because we had only a single bedroom, I asked Dick:

"Will you be comfortable sleeping on the couch in the living-room?"

"Dave," he replied with a smile, "I've slept on the floors of logging cabins when they were covered with porcupine droppings! This will be quite comfortable."

Early the next morning, awakened by frigid air sweeping across the Bay of Fundy into my room through the slightly-raised window, I quickly dressed and walked into the living-room where Dick had started a fire of dried hemlock logs in the stone fireplace. A commotion emanating from the kitchen indicated that he and Marge were preparing breakfast. Not wanting to interfere with their activities, I opened the door to the porch, a glassed-in area extending the length of the lodge, whose greenhouse effect from the sun's rays provided natural warmth for the guests who desired to sit in rockers and contemplate the ocean. Seaward, the distant horizon was obscured by a large wooded island. In the summer, the quietude of the bay was transformed into a cacophony as the local fishermen pounded freshly-hewn trees stripped of branches into the ocean floor to which nets were attached forming weirs, which rampaging whales delighted in crashing into to free thousands of young herring destined for the nearby cannery as sardines. As I was watching a seagull about to land a few feet away from the porch near the edge of the cliff, I heard Marge say:

"That's Sammy. He drops by for a snack every morning You can feed him by hand in a few days when he's used to you. Did you sleep well?"

"Like a rock," I replied. "The air is like an anesthetic. Are my folks up yet? At home they never rise before ten."

"Yes, I heard them talking. We'll be eating breakfast in a few minutes. Do you prefer cereal or eggs and Canadian bacon?"

"Fried eggs and bacon sounds great! I haven't tasted Canadian bacon."

Shortly, my folks emerged from their bedroom and headed for the dining-room. After greeting Dick, Marge and me, they sat down and hungrily devoured the food.

"Those who prefer eggs and bacon say *aye,*" requested Dick from the kitchen. Then he added, "I got a letter from Les and Ruth several days ago. They'll be dropping in tomorrow. And Lovell and Maude Richardson will be here for a couple of weeks of deer hunting. This place will soon be humming."

"I knew the Cooks were coming, but I didn't know when," said Dad. "We'll be able to hunt together."

Although Dad referred to them as "the Cooks," Les and Ruth, both of whom had lost mates to cancer, were unmarried, yet presumably on the road to holy matrimony And the Richardsons, who spent two weeks each fall deer hunting in Pocologan, were also Needham residents who had known the Foleys long before they had dreamed of retiring to Canada. Their arrival made a tremendous visual impression on me, for Lovell was shaped like Humpty-Dumpty and weighed more than 300 pounds, while Maude was tall and thin. If Dickens had seen him, he would have been inspired to rewrite his description of Mr. Micawber. The contrast in their physical appearances quickly dissipated upon engaging them in conversation, for their erudition, displayed before a roaring fire each evening, was fascinating, worldly, and obviously gleaned from personal experiences. I learned from Dad that Lovell had been of normal height and weight (whatever that meant in the 1930's) when the car he was driving was hit by a steam locomotive. The accident occurred at a grade crossing in Needham that was about a hundred yards or so from the Chrysler dealership where Les Cook worked and approximately the same distance from the highly respected Eaton funeral home on Highland Avenue. Fortunately, Lovell did not require the latter's services, but he was badly injured and hospitalized for months at the Glover Memorial. As a result of this horrible accident, he suffered a glandular problem that caused a tremendous weight gain. This physical aberration, however, was hardly a handicap, for he was an excellent hunter, gliding through the forest with the fleetness of an Iroquois. Maude was also a competent Nimrod and the two of them, according to Dick, were the most likely candidates to bring back a buck apiece *(does* were shunned; hunt for antlers and meat not just meat)

When Les and Ruth arrived late in the afternoon after the Richardsons, Dick celebrated the occasion by serving Canadian Black Horse ale in the evening after dinner. This liquid nemesis, I encountered in Chateauguay and tried to improve its taste by adding sugar, had ended up as a stain on the ceiling of Alfred's farmhouse. Now, as Dick poured ale into my glass, I avoided the sugar. Above me was a wooden catwalk of slats that permitted guests to enter a bedroom at either end which, because of the absence of a full second floor, was actually an expanded crawl space. And a few feet above the walkway was a cathedral ceiling. Slowly, I sipped the ale after searching the room unsuccessfully for a plant that might be suffering from desiccation rather than risk wetting the slats on the walkway above and causing a guest to slip over the side and become impaled on the antlers of the moosehead guarding the fireplace.

Before long, Dick's resonant voice sounded.

"Folks, it's time to retire. We'll be up and at it early in the morning."

"How early?" inquired Dad.

"Five o'clock," said a muffled voice from the crawl space above.

The next morning promptly at 5 a.m., the stentorian tones of Dick's wake-up call resounded throughout the lodge:

"Rise and shine, folks!"

It was still dark when I looked out the window and saw, beyond the edge of the cliff, waves speckled silver by moonlight. After a quick breakfast, Dick packed a few cans of sardines and bottles of soft drinks in an old suitcase and then checked his WWI Springfield Army rifle, making sure it wasn't loaded. (In Canada, as in the States, it is illegal to transport a loaded gun in a vehicle.) Because he was a licensed Canadian guide, it was illegal for him to shoot deer when performing professional duties. Carrying a rifle, however, was a precaution in case a client was attacked by a bear or moose. In the late summer, he was also armed with the old Springfield in case he or Marge encountered a bear while they were picking blueberries His intent was not to harm the animal but frighten it away with a skyward shot, the sound of which invariably made the animal run away except when it was a mother protecting her young. Dick, the experienced woodsman, knew better than to interfere with bear cubs whose presence alone in the woods suggested that they were orphans. For a person to interfere with their playful demeanor would oftentimes put the "orphan" situation, so to speak, on the other foot!

Now Les, Dad, and I jumped into the pick-up truck with Dick at the wheel and headed south along the the shoreline. Passing through what appeared to be a grassy meadow with a few small boats lying on their sides, we crossed a creaky covered bridge and soon were traveling along a dirt road lined with tall trees on either side. Presently, we arrived at a clearing whereupon Dick stepped on the brakes and let out a *shoosh!* With the motor still running, he called softly for me to get out and load while pointing at a deer far back at the edge of the meadow. As the animal continued to munch grass, I aimed my Winchester, bolt-action (.30-30) rifle at the animal's chest and pulled the trigger. A loud *bang* resonated in the damp air as the buck leapt and disappeared into the woods. Where the deer had stood moments before, Dick pointed to blood-stained grass that trailed into the woods.

"You got him! He's close by. Be careful! If he's only wounded, he can be dangerous!"

Several minutes later, we came across a sprawled 12-point buck. The bullet had pierced the heart killing instantly.

"Good shooting, Dave! This is your first, right?"

"Yes," I replied, still too stunned by the incident to realize its significance.

Returning to the lodge, we again passed the spot where boats had been lying in a vast grassy meadow, seemingly discarded as derelicts of no further use to their owners. Only now the boats appeared revitalized, jauntily bobbing in deep water with anchor lines stretched taut by the incoming tide. When Dick heard me laughing from the rear of the truck, he knew what prompted the outburst.

"You've never seen our tides in action, have you? When the moon, sun, and earth are lined up in the correct position, the tidal influx exceeds a height of 25 feet above normal. If you tie your boat to a dock when the tide has peaked and don't leave enough leeway in your lines, a few hours later, when the water level drops, the lines will snap and your boat will float out to sea."

As I was pondering this remarkable marine phenomenon, we reached the lodge and were greeted by Lovell and Maude who, as Dick astutely predicted, had also been successful in bagging a fine buck the first time out. They were suspending it from a 2 X 4 placed between the crotch of two birch trees. Dick had told us that this was where he had hung their Thanksgiving turkey by the neck for a few weeks until the body fell to the ground signifying the meat was properly cured.

The problem with eating meat is that hardly anyone takes the proper time to cure it. Yet, no one eats unripened vegetables and fruit. Dick had given us this gastronomic advice during the dinner meal the day of our arrival. Visualizing a turkey losing its head while I was eating was an appetite turnoff like what viewing Saint Pedro Claver's skeletal remains in Colombia had done for my love of chicken soup. Then, Dick had asked me:

"Have you ever tasted moose meat?"

"Yes," I answered, nodding toward Dad. "When I was in the second grade, he and a friend hunted moose in Quebec. He brought home the hoofs, a moosehorn made from the birch bark of an old canoe, some photographs of the landscape with Dad standing next to a pair of oxen, and meat which a butcher had prepared. It was delicious and tasted just like roast beef."

"I'll vouch for that!" interjected Dad. "It certainly didn't have the taste of venison."

"The key to the flavor was having a butcher prepare it. A young moose isn't particularly tender, but the meat from a bull moose 10 to 12 years old requires very little curing and practically no help from a butcher," said Dick. Then he added, "Young Dave just shot his first buck. Let's toast him with a glass of ale."

For some peculiar reason, this time the ale had a delicious taste.

The next morning the men were up at 5 a.m. again while the women, with the exception of Maude, remained in bed exhausted, unaccustomed to the long hours and hearty meals. In contrast to the men, they didn't burn up excess calories by creeping through the woods like Groucho, an especial challenge to the thigh muscles. Mom's idea of enjoying nature was watching flying fish gliding aimlessly out of foam-capped, azure waves of the Gulf Stream in response to a ship's bow slicing into their domain. Also, Mom had Meniere's disease and partial deafness, a sensory handicap aggravated by loud noises, that resulted in ringing in the ears (tinnitus) and severe nausea. Ruth, on the other hand, loved to walk in the woods and preferred to see deer and other forest inhabitants succumb to old age rather than become a target for a swiftly moving chunk of metal.

Despite differences in hunting philosophy, Mom and Ruth got along nicely with Maude, who, like them, was an expert contract bridge player. Having been brought up in a house of card-playing parents each Saturday evening (and having had my sleep disturbed by their rantings), it was only natural that I would detest cards. I even hated the feel of cards, which, years later, was exactly duplicated when I touched the horny scales of a boa constrictor draped casually around the neck of a zookeeper. The transformation of like to dislike of cards must have occurred during my teenage years: my weekends at "The Cedars" during the thirties often entailed playing cribbage with Grandpa by the light of a flickering kerosene lamp. These were the weekends when his daughters didn't arrive to discuss FDR's politics. Come to think of it, my general dislike of politics was probably seeded on those weekends when they did arrive.

David E. Mann, Jr.

Our deer hunting destination on this bright, sunny day was a few miles inland where the terrain was hilly. Dick's plan was for Les, Dad, and me to position ourselves on the highest point of a hill where, from a treeless, grassy area, one could look down on a spread of forest. Beyond the woods lay a large meadow from which hay had been cut several months before. The stalking strategy involved Dick, John, and Jethroe, our guides, to walk slowly downhill through the woods in different directions towards the meadow to drive unsuspecting deer into the open meadow below. There would be no shooting until the deer had been properly identified. (In the past, Dick's guests from Michigan came bedecked in bright orange clothing that made them look like crossing guards. Hunters from nearby Maine wore olive-colored gear decorated with red patches. Our clothing was also olive-colored, trimmed in red with the complete absence of white, for white to a hunter is frequently mistaken for the upright tail of a retreating *whitetail* deer. Despite these precautions, Jethroe was to lose his life several years later when he sat down to smoke a cigarette and was killed by a hunter who mistook him for a bear exhaling cold vapors).

Within ten exciting minutes, the three game flushers emerged from the woods into the meadow within a dozen yards of each other without arousing deer. They joined forces and continued on through the Guthrie farm where Dick's truck was parked to convey the spoils home. However, as the three of us watched them disappear behind a barn, Les suddenly pointed to a large buck that cautiously entered the meadow almost at the exact spot where Jethroe had emerged from the forest. As it trotted across the broad expanse of meadow, its head moving from side to side, it quickly disappeared into a glen.

"Did you see the size of that buck?" groaned Les. "I thought it was an elk."

"Without a telescopic sight there's no way you could hit a fast-moving target at that distance," added Dad. "Only Dick's Springfield could bring that baby down!"

"Yes," I concurred. "I saw him knock an owl out of a tree half a mile away with that rifle! And he doesn't use a scope either."

"Jethroe must have almost walked up his ass without seeing him!" said Les. "I guess we'll have to try tomorrow. I'm getting tired of snacking on these damned sardines when Marge's culinary delights await us! Let's head for Guthrie's place and go home."

As we started to descend toward the farm at the edge of the meadow, Dad checked the direction on his compass and off we went, moving slowly through the thick underbrush and stately evergreens. When Les, leading the group, passed by the old wood barn, a tall figure approached out of the shadows and suddenly threw his arms around him in a bear hug.

"I wub you," said the deep voice.

As Les gasped for air, Dick came over to allay his fears.

"It's all right. This is Guthrie's boy. He has Down's syndrome. He wont hurt you."

As the grizzly grip was relaxed, Les shook hands with him, smiled, and walked toward the truck, gasping for lost wind.

Later, recalling the incident over dinner, Les looked at Dick and said: "I wish you had warned me about that kid at the farm. He scared the hell out of me!"

"Sorry about that," answered Dick. "I was more concerned about finding a good location where you could bag your quota than thinking of the hazards."

"Speaking of quotas," said Dad. "Les and I have two deer each to go before we leave on Sunday. We only have Friday and Saturday to hunt."

"And young Dave has one to go," chimed in Dick. "Tomorrow we can hunt along the railroad tracks a few miles inland. The train passes by twice a day and you can predict its arrival by putting an ear to the track."

By six-thirty the next day, we were walking slowly through wisps of fog cloaking shrubs beside the track. According to Dick, this was an ideal place to see deer browsing on vegetation until the sun rose over the tops of pine trees, whereupon, at 8:00 a.m., they disappeared in dense thicket. As the sun's rays struck the cold steel of the rails, I heard a loud crack that echoed along the track's corridor like the resounding snap of a gigantic whip cracked in a canyon.

"That noise doesn't seem to bother deer," whispered Dick, correctly interpreting my frown. "They're used to thunder claps, too. Remember when I left the engine running when you shot your buck? If I'd cut the engine, your deer would've skittered away. Discharging a gun has the same effect." He paused for a moment, then said: "Dave, put your ear to the rail…quick!"

I ran to the track, slipping on a couple of icy ties, and leaned down to hear a funny humming sound.

"Dick, you're like the deer. You heard that sound while standing yards away!"

"You didn't see me glance at my watch," replied Dick with a grin. "This train has the best on-time record in Canada!"

Within several minutes, black smoke could be seen billowing from the forest as a small steam locomotive, hauling a mail car, passenger car, and caboose, came chugging by.

"It's after 8," noted Dick. "We're not going to have any luck here, so let's go home and plot tomorrow's strategy."

"Thank God for small blessings," mumbled Les. "I've become allergic to those damned sardines!"

Friday, the women went by car to the seaport city of St. John, approximately fifty miles up the coast, where they spent the day shopping in the granite buildings whose architecture resembled that of their British counterparts across the Atlantic. During their brief spree, they glanced at the harbor and saw several rusty trawlers and draggers unloading cargoes of cod, flounder, and deep sea scallops. In contrast to our Gloucester fishermen, who have to sail long distances to reach teeming grounds, the fishermen of St. John have a rich bounty of piscatorial delights practically at their doorstep. And, like those who inhabit the Maine coast, the people of New Brunswick are practically awash in lobsters. These were the good old days of the late forties when natural marine resources were far from being depleted by electronic devices that transformed rich ocean beds into vacuous deserts. Today, injudicious overfishing of cod, haddock, flounder, and mackerel has resulted in an atrophy of the fishing industry in the United States. Sadly, a comparable situation exists in coastal Canada.

Back in Pocologan, the hunters were walking through woods near the shore under the assumption that deer might cross their sights on the way to salt licks left by the receding tide. The weather was turning cold, and Dad was pondering whether to leave a day earlier because the weather report indicated snow was a possibility. After several hours of fruitless searching, Dad, Les, and I called it quits and piled into Dick's truck for the trek home. Arriving at the lodge and reveling in the warmth of a roaring log fire, we saw through a window Dick's nephew, Ned, drive in the yard in an old, beat-up truck and climb out, casually dressed in camouflaged Army clothing, carrying a gun that had two bores, one for a shotgun shell and the other for a 0.22 calibre bullet. He was in his early twenties, with an old burlap bag slung over a shoulder.

As the young man stepped into the living-room, Dick, who had been in the kitchen with Marge, greeted him with a forceful handshake and then noticed his gun.

"Why didn't you bring a machine gun along with you, Ned? If you don't get a deer with this weapon, you'd better try hand grenades. What's that you have in the bag?"

"Uncle Dick, it's a beaver. I hit it a few minutes ago rounding a curve."

"Christ get rid of it! We've got a Mounty coming for dinner tonight!"

That evening, while we enjoyed a lobster dinner with Dick's Mounted Police friend in full regalia, in the basement a few feet below the dinner table, Ned was busily skinning the beaver while munching on a lobster sandwich.

The early dawn of Saturday, our final day, was, contrary to the weather report, snowless. Instead of packing to leave after breakfast, Dad confronted Les and Dick for a last-gasp attempt to fill our allotted game quota. It was decided that Dad and Les would hunt in the coastal area again near the covered bridge with their respective guides, Dick and Jethroe, while John and I would hunt a few miles inland on the game warden's land known as the *christmas tree sector,* his seasonal source of income. Dick dropped us off at the sector before heading with Dad, Les, and Jethroe to the shore area.

John was a young man in his thirties who owned a small farm several miles north of Pocologan where he lived with his wife and two preteen children. In the summer, he raised vegetables and milked a couple of Jersey cows, while his wife worked in Pocologan's sardine cannery. In the fall, John supplemented his income by guiding deer hunters in the adjacent forests, steering them away from a dangerous area known as *the floating city* that lurked only a few miles away as a place where hunters had been known to disappear after supposedly stepping on solid ground only to plunge dozens of feet through layers of decaying peat. In addition to guiding, John also trapped fur-bearing animals (except beaver) to supply a demand for the relatively unglamorous coats of rabbit, raccoon, and lynx that provided comfort in the absence of class.

We walked several yards into the woods when I looked to my right and was amazed to see a bobcat on its haunches staring at me. I turned to see if John had seen the animal and then back at the cat which had disappeared. Breathing rapidly, we continued to walk until we came to a crater resembling a huge earthen bowl dotted with spruce trees. Suddenly, the entire area came alive with brown forms darting frantically in all directions. I aimed at a large brown form that had paused about 15 yards away behind a small evergreen tree and pulled the trigger. The deer dropped in its tracks with a .30-30 boat-tail bullet in its neck.

"My God, it looks like an elk!" cried John as he and I ran to the deer. "And look at that rack! The antlers are palmated like those of a moose! How're we gonna get it outta here? We need Dick's truck."

"Dick said he'd be picking us up in an hour or so if Dad and Les end up empty-handed. They're going to get the shock of their lives when they see this baby!" I added.

We headed for our pick-up place, our hearts still beating rapidly, and our emotions bursting with excitement. Suddenly, John turned to me and said:

"Dave, did you realize that bobcat was about to enjoy a venison meal?"

What seemed an interminable period during which I pictured in my mind a bobcat reducing my trophy buck to a skeleton, we heard the engine noise of Dick's approaching truck. It crossed my mind that his nephew's truck made the same rattling noise. Perhaps both vehicles were maintained by the same mechanic, used the same oil, and were of the same vintage. Remarkable that my twelve-point buck wasn't spooked by that sound! Maybe it was deaf. Just then the truck, dark green and dust-covered, drove up with Dad and Dick in the cab and Les and Jethroe sitting on a tarpaulin in the rear, all four men smoking cigarettes.

"How'd you make out?" asked Dick, after looking at our smiling faces. "You'll notice we're empty-handed. Did you guys get yourselves a buck?"

"You'll see," I answered. "There's something lying behind a christmas tree a few yards from here."

The six of us walked to the earthen depression where, an hour or so before, a half dozen frolicking deer (a buck and his doe harem) had been startled by the sudden appearance of two hunters. Dick ran behind the evergreen as soon as he saw the massive brown shape, followed by Dad, Les and Jethroe.

"My oh my, oh my," gasped Dick. "You've hit the jackpot with this one! In all my years of hunting, I've never seen an equal to this! What a trophy rack you're going to have for your den!" Then, sizing up its dimensions he added:

"We'll need help to get him on the truck. Ground's too soft and mossy to drive on. We've got to drag it on the tarpaulin. But first, I'll have to dress it to prevent the meat from spoiling. Then, I'll get a couple of friends to help. It'll take about a half an hour."

He unsheathed his hunting knife and deftly removed the animal's entrails before hopping into the truck and leaving the rest of us to stare at the trophy buck, which John named the Monarch of Pocologan. When he returned with two buddies, after much heaving and yanking, the eight of us were able to lift the monarch and position it in the back of the truck with its massive antlers propped up against the rear window of the cab. Lastly, its huge body was covered with the tarpaulin.

Fortunately, Dick's two friends had followed him in their own trucks and we were able to divide the passengers among three vehicles, with John and me sitting beside the monarch as we bounced along a dirt road that led to the highway and home.

Reflecting on this unforgettable happening that occurred fifty years ago, I don't recall Dad's reaction upon seeing the monarch for the first time, but when Dick suspended the carcass from the 2X4 in his backyard, Dad and Les heartily congratulated me, as did Mom and Ruth, Marge, the Richardsons, Ned, and the game warden, who arrived within the hour with a reporter from St. John, who happened to be in Pocologan working on a moose poaching story. In those days, hunting moose in New Brunswick (and in New England) was illegal, but the game warden didn't press charges when a father with hungry kids was caught. This situation, however, was vastly different for the game warden had recently stumbled upon several headless moose carcasses during routine woodland hikes. Furious at this affront to nature and abuse of the law, he had contacted Mounties and Customs officials to apprehend quickly the culprits. A young man accompanied the reporter to photograph the remains lying in the woods, now partially eaten by scavengers.

"And this beauty was shot on my property?" inquired the warden, a close friend of Dick. "Seems like I'm growing more than christmas trees out there!" he grumbled, scanning the impressive antlers and feeling the thick neck with calloused fingers.

"Oh, oh." He suddenly stopped his palpation to feel a small lump beneath the surface of the neck hair. Grabbing a pocket knife, he made a tiny slit beside the lump and removed a slug of metal, the remains of a.30-.30 boat-tail bullet.

"That's what's left after a bullet hits bone. If you examine the bullet's point of entry and insert a wire along its path to its muscle exit point just beneath the skin on the other side of the neck, you'll discover it isn't a straight line because the bone caused the bullet to deviate slightly. Also, if it had only penetrated soft tissue, the shape of the bullet would have been retained."

He reminded me of the famous movie detective, Charlie Chan, talking to his number one son, Keye Luke.

"This is for you," he said, handing me the metal pellet. "Maybe someday you can have a pendant made from it for your girlfriend."

We photographed and weighed the monarch with a borrowed butcher's scale and, even though degutted, were surprised when the arrow stopped at 340 pounds! The antlers, according to Dick, were atypical and definitely worthy of a Boone and Crockett evaluation. On January 17th, 1997, thanks to Jim Gardner and son Scott, we had the antlers evaluated at the Sportsmen's Show held at Fort Washington, Pa. Net typical =150 3/8; Non-typical=173 1/8 and the Gross Score =186.08. I only wish the witnesses were alive to see the results.

The reporter from St. John had done his job well, for the trip back home, with frequent stops at filling stations and restaurants, occasionally attracted people who recognized the trophy buck from the brief associated press account published in their local newspaper.

"Isn't this the buck shot in Canada?"

One of the first things Dad did, after arriving in Needham, was call Phil Stack, a professional photographer, to record the monarch's bulk, strung head down with antlers touching the floor, in the carriage house while I grasped the tail with my left hand. This, and several other views, resulted in three outstanding black and white photographs that Dad deemed worthy of publication in a sportsmen's magazine. Together, we checked the telephone book and discovered *Hunting & Fishing Magazine,* 275 Newbury Street, Boston, Mass., with 350,000 circulation, The Leader in its Field! I made an appointment with Bernie Roth, the editor, and a week later, took the train to Back Bay and walked to Newbury Street where Mr. Roth, a young scholarly-looking gentleman, welcomed me in his office cluttered with taxidermists' renderings of small game and fish, magazines, and wildlife books.

We spent several minutes discussing my hunting experiences, educational background, and future plans before getting down to business.

"These are fine photographs, but I'd like you to go home and write a short article to accompany this one." He pointed to the full-length shot of me holding the tail.

A few hours at my Underwood portable typewriter, inspired by the hunting principles I had picked up from Dick, resulted in several pages entitled: HOW TO HUNT WHITETAIL DEER. Satisfied with my literary effort, I sent the article and photograph to Mr. Roth.

On December 30, 1946 I received the following letter:

Dear Mr. Mann:

My apologies for being so tardy in dealing with your deer hunting article HOW TO HUNT WHITETAIL DEER which I have recently gotten down to examining with real pleasure.

I will schedule it (plus the photograph) for a suitable fall issue if our arrangement suits you and Mr. Stack-said arrangement being payment of $40.00 for the story and picture upon publication. I hope this is agreeable with you. I think it is a corking item.

David E. Mann, Jr.

Sincerely,
Hunting & Fishing Magazine
BAR/tlb   Bernie Roth, Editor

P.S. -I hope your school problems have become straightened out since I last talked with you.-B.A.R.

I had to wait a year for the article to appear in the December, 1947 issue along with a reprinted wildlife murder mystery by Ernest Thompson Seton, penned a few years prior to his death, and sent to the magazine by his wife. This was especially gratifying to me because it was an honor to be published in the same magazine with my childhood author/hero. His best-selling book, "Two Little Savages," told the story of young boys living on farms surrounded by virgin forest whose fathers taught them how to survive when lost in the woods by eating berries, fungi, even the roots of water-lilies yanked from ponds which were as nutritional as potatoes. The reader learned how to convert animal skins into leather, build dams, and even fashion bows and arrows from certain kinds of wood (usually ash). Most important of all, armed with an axe and hunting knife, one could venture into dense woods and create a shelter that offered protection from the elements. No wonder I almost got into a fight with the owner of a lodge during a fishing vacation in Maine! We were paddling a canoe across a lake when the idiot remarked that I wouldn't last five minutes if dropped off a few yards in the forest! I was furious at this assinine remark, knowing full well that the information provided by Seton's book was sufficient to guarantee a return to civilization. I think the man had had too many beers! The next day he was socially acceptable but still stupid.

Mr. Roth had been kind enough to refer to my "school problems" which involved only the selection of the right graduate school. When I received his letter dated December 30th, 1946, I had found a job with the Boston Consolidated Gas Company, headquartered in Everett, Mass., which was the site of my first day at work, the day of the Christmas party on December 24th! It was a strange feeling to celebrate with people who had worked there for decades while I had signed on only hours before! The most unusual thing about the party was meeting Mr. Sullivan, who was none other than Fred Allen's brother! There was neither a facial resemblance nor a semblance of Fred's humor about Mr. Sullivan, who seemed a bit grouchy, perhaps inwardly resenting his brother's spectacular show business success. During the next few months, I worked in a laboratory outside Boston measuring specific gravity and B.T.U.'s of the gas manufactured by burning coal into coke, the by-product of which was heating gas. The Company also owned the Boston Towboat Company, which allowed me to cruise Boston harbor occasionally in a tugboat and also provided me with authentic red paint for a model I had purchased at an arcade in Providence, R.I., and which took longer to build than an actual tugboat.

CHAPTER TWELVE

Hail Purdue!

In mid-August of 1947, I boarded the New England States, the train whose reputation for quality service was equal to that of the Broadway Limited, and headed for Chicago and its foremost university, the Nobel Laureate-laden University of Chicago. My itinerary, carefully prepared with the help of parents and friends, was pared down to the two great Midwestern Universities of Chicago and Purdue, having eliminated the University of Michigan as being too distant (and expensive) for a commute home I had planned at intervals of two months. Years later, I learned that the first department of pharmacology in the United States, incorporating my eventual area of interest, was established at the University of Michigan Medical School in 1890, and chaired by John Jacob Abel, the dean of American pharmacologists, who later continued his illustrious career at Johns Hopkins.

One word described my initial impression of the University of Chicago-vast! The campus grounds resembled a large commercial airport bordered by huge white buildings that required thoughtful course selection to ensure that a student would be able to get to sequential classes on time when several minutes made the difference between arriving promptly or being tardy.

An interview with one of the professors of physiology revealed that there were no vacancies available at the graduate level, although an application submitted now might be successful in filling a vacancy for the second semester.

"You should have sent for our catalogue describing the correct procedure for admittance to the graduate school," commented the interviewer. "You just can't walk in off the street and expect to enroll in our graduate school without following the proper procedure, which includes submitting your scholastic credentials. Are you a veteran?"

When I answered affirmatively, he nodded his head and said: "That's definitely in your favor. But on the other hand, the availability of the G-I Bill has resulted in a tidal wave of applicants of older students. And these people are far superior to the ones we had applying prior to the war. There is nothing like a war to instill the value of a decent education in a person whose life in the military has been controlled by others."

Happily, a few hours later, I departed from Chicago on the Monon railroad and headed south through the monotonously green cornfields of Indiana on my way to West Lafayette, the home of Purdue University. The refusal probably saved me a small fortune in shoe leather.

I wasn't aware at the time that the railroad I was riding on bore the brunt of countless jokes engendered by Midwesterners, and was being bantered among the officials of the more prestigious railroads such as the Nickel Plate, Atchison, Topeka, and Santa Fe, and Union Pacific, because of several indignities it had endured in recent years. As the first railroad in the United States to convert completely from steam locomotives to the more efficient diesel behemoths (Switzerland was the first country in the world to accomplish this), the Monon proudly exhibited their recent acquisitions, both of them. Then, shortly thereafter, the powerful locomotives collided head-on. With just under 1000 miles of trackage, the Monon gallantly boasted that "we *may not be as long as our competition, but we are just as* wide!"

When I arrived at the station in downtown Lafayette, achy and disheveled with suitcase in hand, I started to cross the street to a white-painted hotel when a police car approached with two officers sitting in the front seat. The driver stuck his head out the window and invited me to sit in the backseat.

"You don't want a room in that place," cautioned the other officer. "We'll drop you off at the best hotel in town. Where are you from?"

When I answered "Boston," they chuckled and asked how the Red Sox were doing.

"They're the greatest team Boston ever had, but the Yankees are better because their pitching is better."

"Did you leave Beantown to study at Purdue?" asked the other officer, turning to glance at my suitcase and acknowledging my affirmative nod. "Sort of like carrying coals to Newcastle in reverse," he philosophized.

The police car pulled up to the elaborate entrance of a stately, dark-brown brick hotel in the center of town before several pedestrians who undoubtedly assumed their police had gone into the taxi business to supplement their incomes.

While I thanked them for being so hospitable, the driver pointed at a bus stop.

"That's where you get the bus. Purdue is only a ten-minute ride over the Wabash and up the hill. Good luck!"

I registered at the counter and followed a bellhop to a room overlooking the stores of Lafayette, where a theatre marquee across the street caught my attention. A western movie starring John Wayne was showing, a reminder that this town resembled Burlington, Vermont, though lacking the neighboring aquatic majesty of Lake Champlain.

I wondered if the restaurants were as great as those in the city by the lake. So, after a hot shower, I dressed casually and ventured forth in search of a place comparable to Vermont's Black Cat of lobster fame. The best eateries were actually in West Lafayette in the basement of the Purdue Memorial Union where, on Sundays, the townsfolk traditionally partook of the wholesome succulent farm-fresh vegetables, delicious beef, lamb, pork and poultry entrees, and exquisitely baked cakes and cookies. Also, there were several fast-food counters with swivel chairs housed in former mom-and-pop stores, whose owners had either died or moved to Florida, where traditional bowls of chili were served to students hard-pressed for time, that kept their tongues afire throughout the rest of the day. This information I would learn tomorrow when the ten-minute bus ride stopped in front of the university library opposite Sheetz Street. Not finding a restaurant to my liking in downtown Lafayette, I settled for Chinese food served in a tiny room with an Eastern culture decor suggested by delicately executed landscape watercolors placed on maroon-colored walls with nary a gilded dragon in sight. The food was delicious! I even enjoyed a Thanksgiving dinner there one November when the only other non-Oriental, Caucasian patron was an Army general who, like me, couldn't make it home for the holidays. I never did find a seafood restaurant comparable to the Black Cat Cafe. One place did serve *scallops,* which were prepared by punching circular disks out of the slender "wings" of skates, the elasmobranch relatives of sharks. The appearance on the plate was the same as the muscle of the tasty deep-sea mollusc, but the taste was terrible. That experience taught me an important lesson: when a Midwesterner recommends a great seafood restaurant to a New Englander, head instead to a good meat-and-potatoes place, for invariably the food of the former will be downright lousy.

I ended the day at a John Wayne movie with a bag of fresh popcorn *(this is popcorn country)*. Back in the hotel, I perused the local papers for university-approved living-quarters, assuming that my interview with the physiology department tomorrow would be more successful than my ill-fated Chicago experience, then turned off the light.

The bus ride the next day took me across the famed Wabash River, a stream about the width of the Charles River at the Needham-Dedham line where portly Mr. Sweeney, the organist at the Needham-Paramount theatre on Great Plain Avenue, gave swimming lessons without charge in the late twenties to those kids who faithfully attended his Saturday matinees. His avoirdupois occasionally got him stuck in the mud of the river bottom, and once he almost lost toes to a snapping turtle he inadvertently used as a stepping-stone. In spite of these minor setbacks, he taught the Australian crawl, breast stroke, and dog paddle to his young charges, skills that probably saved some lives during WWII when naval vessels were torpedoed! Ah, Wabash, eat your wet heart out!

Before the thoughts of Mr. Sweeney and the swimming lessons in the Charles River left my mind, the bus had reached the top of the hill, where bookstores, stationery shops, and a hole-in-the-wall popcorn joint marked the edge of the Purdue campus. Several brick buildings, partially obscured by elm trees planted as shade for the adjacent sidewalks, displayed architecture spanning two centuries. Stepping out of the bus, I inquired of a student the location of the physiology department.

"The brick building on the left is *Home Ec* and the building just beyond is the *Biology* Annex where bio courses are taught. I think a few Psych courses are taught there, too. A *Life Science* building is going to be built across the street in a cornfield near the Vet school, but that's in a few years."

I thanked the student and walked toward the Bio Annex, which appeared as an architecturally-deprived brick structure behind which a tall smokestack loomed, indicating the presence of a coal-burning power station. The wooden floors creaked as I entered the building and walked down the corridor past filled classrooms and a bustling laboratory that reeked of burnt paper, an olfactory sign that students had not yet mastered the technique of smoking paper for the Harvard kymographs. I turned right and saw a door at the end of the corridor with a brass plaque identifying the office's occupant as *Dr. Wm. Hiestand, Professor of Physiology*. I listened for the sound of a meeting or seminar and, hearing nothing, I knocked several times.

"Come in," said a professorial voice within.

Professor Hiestand, a balding, bespectacled gentleman was sitting behind a large desk cluttered with test papers he was in the process of correcting. He smiled and looked up, relieved to be momentarily interrupted from an onerous task.

"If I seem a little upset, it's because Sheila, our Irish setter, has a bladder problem. I gave her an intramuscular shot of ephedrine to increase the tone of the sphincter this morning but, according to my wife, she still urinates on our orientals! How can I help you?"

He offered me a chair when I expressed an interest in pursuing a graduate program in physiology. Yes, there was a vacancy and I would have to take a qualifying examination in English and send for my undergraduate grades. In the meantime, I should find a place to rent on a monthly basis until my status as a full-time graduate student was established. After our brief meeting, I headed for the Purdue Memorial Union for housing information and discovered several basement restaurants and the aroma of delicious food.

The following day I bid farewell to the hotel clerk in Lafayette, boarded a bus with suitcase in hand, and headed for my new home at 142 Sheetz Street where a white, traditional Hoosier-style home owned by the Jackson's stood on a quiet, tree-lined street several minutes walking distance from the campus. The Jacksons were an elderly couple who rented spartan single rooms to three male students on the second floor that were slightly larger than the dollar-a-night facilities of the YMCA. (I stayed in such a facility in Chicago for a night and had to shuffle sideways to reach a bureau next to the bed.) Because one of the students upstairs was just completing a summer school session and was now packing to go home, I had to occupy a partitioned portion of the Jackson's living-room for a week or so before I could settle down in a permanent location. Thus, for one week during which I registered for graduate school, took the graduate record examination in English, and became familiar with the facilities and staff at the Bio Annex, I also experienced a toothache that kept me awake for two nights until, mercifully, Monday finally arrived and the offending molar was removed. When at last I entered through the side door and climbed the stairs leading to my room in the front of the house, I encountered the former occupant on his way home to Indianapolis, which he scurrilously referred to as *Naptown, the Crossbones of the Nation.* After a brief introduction at the head of the stairs, he remarked, as though discussing the local attractions with a tourist.

"We have two attractions in *Naptown*—The *Indy 500,* which is actually located in Speedway, and the *Indiana State Fair* where you can see Sally Rand, the most famous fan dancer in the world, and Purdue's contribution to agriculture from farm animals to hybrid vegetables."

"The Indy 500 is one of the reasons I came to Purdue. I can't wait for May to arrive when, according to the Boston papers, the whole city goes berserk. And the time trials are supposed to be almost as exciting as the actual race. Have you been to many?"

"Yes, several times, but I'll let you in on a little secret. There are dirt tracks near here where midgets race and, believe me, they're just as exciting as the big one. The Indy track is two and a half miles long. When the cars roar around the fourth turn and head down the straightaway for the green flag, everyone stands up and remains standing for an hour or so. The start is the most exciting time, but eventually monotony sets in because the cars are spread out over the track and only a few are visible on the straightaways and four curves unless you happen to be sitting in the gondola of a blimp. The midget races are much more fun and far less expensive."

I thanked him for this information, wished him luck, and proceeded to unpack my suitcase and make a list of necessities. My jaw was still throbbing, so I decided to ask Mrs. Jackson for a glass tumbler and some table salt.

"I wish you'd told me about your toothache," she said, after learning why I wanted to gargle with salt water. "I have medicine that would have relieved the pain. Now, if there's anything else you need, please let me know."

After thanking her, I gave her the glass and started to head for my room. But she handed it back, saying:

"Please keep this in the bathroom along with this smidgen of salt." She handed me a small envelope folded several times upon which she had written the chemical formula for table salt—NaC1.

"I was pleased when you told me you were a graduate student," she added as I started to walk up the stairs. Then, speaking in a whisper, "I dislike fly-by-night summer school kids you don't really get to know personally. And the ones I've boarded weren't worth knowing in the first place. If they'd gotten down to business during the regular school session, they wouldn't need to waste time in the summer taking snap courses."

That was the last time I talked to Mrs. Jackson at any length other than to say "hello." Her husband, a tall gentleman with a full head of gray hair, was even more secretive and discreet than his wife, acting like he feared confrontations. Later, when the Jacksons served turkey dinners to those student boarders who couldn't go home for the Thanksgiving holidays, including me on one occasion, I discovered that Mr. Jackson's reclusive behavior had nothing to do with being henpecked or having an inferiority complex. A resident told me he was suffering from cardiac asthma, a severe respiratory problem due to an enlarged heart. His wife, who did her own housework, was not exactly in good health either, for she had a goitre, an enlarged thyroid gland, that was a distinguishing characteristic of Midwestern adults (and those living in Switzerland), who spent their adolescent years deprived of iodized salt in the early part of the twentieth century. Ascending the stairs, I wondered if Mrs. Jackson realized that the white powder in the little folded envelope not only reduced the bacterial population in my mouth but, if taken as a child, could have prevented the swelling in her neck.

Back in my room, the mail brought good news: I'd passed the English examination for graduate students. Although it had consisted principally of checking which words were spelled correctly and which sentences were devoid of split infinitives and double negatives, a few sections of the test were sprinkled with throwbacks of finer points of the language last encountered in high school. That afternoon, I inquired of Dr. Hiestand whether my status in graduate school had been determined.

"I've received your undergraduate grades from Harvard and Tufts and am pleased to inform you that your status as a student in the graduate school is probational, as it is for all students, other than transfers, enrolling for the first year. When you have successfully completed your graduate courses for the first year with a minimum average of 85 or B, your probationary status is dropped and you then become a full-time graduate student. On the other hand, if you receive C grades (a 75 average or less) in more than two graduate courses, you're automatically removed from the rolls. Now that my summer school duties will be finished next week, I'd like to meet with you about course selections and a research project for your Master of Science degree. Do you have any ideas concerning the latter?"

"Yes, I'm fascinated with tissue regeneration in lower species such as salamanders and planaria worms."

"That's indeed fascinating. But I should have mentioned that the project should be accomplished within the equipment and financial scope of this department. Tissue regeneration studies might be stepping on the feet, so to speak, of our biologists who also occupy this building—the Goodnights and Professor Cable. Have you met them?"

"No, but if the other person is Raymond Cable, I used his laboratory manual at Harvard in my parasitology course."

"Well, that's he. Ray Cable is an authority on flatworms, the Platyhelminthes, especially those that inhabit the gut of seagulls. He usually spends part of each summer teaching and conducting research at the MBL (the Marine Biological Laboratory) at Woods Hole on Cape Cod. One summer he visited our cottage on a lake in Wisconsin to examine fish for tapeworms, such as *Diphyllobothrium latum.* This worm reaches a length of 30 plus feet in the human species when a person is stupid enough to eat raw or poorly cooked, parasite-infected, freshwater fish such as the northern pike and muskellunge."

"I recall studying that at Harvard under Professor Cleveland. The smallest tapeworm is *Hymenolepis nana,* the one found in dogs. By the way, how is Sheila doing?"

"Not too well, She's still wetting our orientals. You just reminded me to get in touch with Dr. Edwards, the professor of pharmacology at the School of Pharmacy. He must know a drug that works better than ephedrine in causing urinary retention. Have you met L.D?"

When I answered negatively, our meeting concluded with the scheduling of another to discuss course selections and research projects within the scope of his department.

During the days before my curriculum conference, I spent the time introducing myself and chatting briefly with Professor Cable, the Goodnights, and Donald Stullken, who was a predoctoral student and physiology instructor in Dr. Hiestand's graduate laboratory. Dr. Cable, a bespectacled man in his early forties, whose wiriness and digital skills were fashioned on a Kentucky farm, smiled when I mentioned I had taken Dr. Cleveland's parasitology course and used his lab manual.

"Lemuel is a prince of a person," he commented. "You were fortunate to have had his expertise as a teacher and advisor."

In retrospect, the most vivid recollection I have of this brief encounter occurred after he mentioned his agricultural heritage, offered undoubtedly as one of the reasons why he was teaching at Purdue.

"Teaching is a far safer occupation than farming. The farm where I was born in Kentucky was also where my father, in his nineties, died when he jumped from a hayloft in his barn onto a three-tined pitchfork that pierced his chest! Farms are dangerous places!"

The Goodnights were a husband and wife team who taught general biology courses at the undergraduate level and ecology in the graduate school. They were well-known for their summer research projects in Mexico and Central America where they sought to confirm the theory of continental drift by collecting arthropods on this side of the Atlantic and comparing their anatomical features and distribution with those collected by entomologists in western Africa. On a map, one can fit both land masses neatly together on the Atlantic side, but a problem arises when the Pacific areas are treated similarly. Obviously, when the earth expanded eons ago, both major oceans were involved. The Goodnight research was directed only toward establishing which portion of the Euro-African coastline broke away from Mexico and the Yucatan peninsula.

The Goodnights, like many married couples, resembled each other: short, slightly overweight, and cherubic. Their lectures were extremely popular and presented before standing-room-only audiences. The graduate students, who accompanied them to Central

America and Mexico during the summer months, invariably suffered from severe bouts of diarrhea while their physiques were reduced to skeletal frames, but the Goodnights seemed to thrive in tropical and semi-tropical environments. During my educational experience at Purdue, I didn't take any of their courses because the contents were similar to those I had taken at Harvard.

Donald Stullken, a balding, heavy-set, personable, young man, whose major research interest was aviation physiology, had come to Purdue, after graduating from DePauw University with a B.A. degree in 1941, to teach physiology while taking courses toward the M.S. (1942), and Ph.D. degrees (1950). He had chosen Dr. Hiestand as his mentor based on the professor's reputation as an excellent teacher and meticulous researcher. Interestingly, Professor Hiestand had begun his professional career as an entomological physiologist, but a debilitating disease had transformed his biological expertise to that of a mammalian physiologist to learn more about his own condition.

Donald, out of a blue sky during the first laboratory session, had asked me a strange question: "Did you ever see Dr. Hiestand walk?"

"Why no. Come to think of it, I've only seen him sitting at his desk. I haven't even seen him come into this laboratory."

"You will a little later when he summarizes today's results."

At the end of the laboratory period, I heard a peculiar scraping sound in the corridor whereupon Dr. Hiestand appeared, cane in hand, moving slowly toward the podium next to the blackboard. Carefully, he propped himself up with both hands resting on the top of the podium after placing the hook of the cane in a ring designed to hold a glass tumbler. Donald walked to my table at the rear of the laboratory, where I had been observing muscle contractions in a frog, and whispered:

"This is the reason he became a mammalian physiologist. He has M.S."

This incident occurred after I had met with a seated Dr. Hiestand to discuss which courses I should take during the first semester *(physiology lecture and laboratory, animal husbandry, biochemistry, and physiology seminar),* and a research project, based on a recent article in the *Chicago Sun.* Russian scientists had administered lethal doses of potassium cyanide, one of the deadliest poisons known, and rendered the toxic agent harmless to the nervous system of animals.

Vsevolod Galkin, professor of pathological physiology at the Leningrad Naval Medical Academy, had neutralized cyanide poisoning in cats by giving them a simple anesthetic. Dr. Hiestand's plan was to inject potassium cyanide in mice and try to antidote its poisonous effects by administering the anesthetic, pentobarbital sodium (Nembutal), beforehand. The Russian scientist, unfortunately, had not revealed which anesthetic he had used. Thus, my research project was going to be an interesting toxicological study. After a few weeks of experimentation, the preliminary results were reported in the Purdue Scientist (Vol. 1, No. 1, November, 1947), founded by Warren R. Young, who eventually became an editor at LIFE magazine.

A few weeks into the project it became evident that the preadministration of pentobarbital sodium (Nembutal), followed by a second injection of a lethal dose of potassium cyanide, both given intraperitoneally, resulted in a few mouse survivals. As the work progressed, however, the outcome was statistically non-significant. Another potential antidotal drug was selected, dinitrobenzene, an agent that transforms the normal red pigment of the blood, hemoglobin, into an oxidized form (a condition known as methemoglobinemia). The oxidized hemoglobin, unlike the normal type which carries oxygen to each cell, combines with cyanide ions and prevents them from disrupting respiratory enzymes inside the cells, whose inhibition results in death. My research goal now firmly established, I turned my attention to the preparation of my first physiology seminar scheduled for the following week to be presented before 8 fellow graduate students, most of whom were total strangers, and my instructors, Donald Stullken and Dr. Hiestand in the latter's Bio Annex office.

The subject I had selected (out of a felt hat) was entitled, *"The Influence of Testosterone Administration on Antler Formation in* the *Whitetail Deer"* by a Harvard Medical School anatomist named *Wislocki and his associates. A casual perusal of the monograph, obtained from Dr. Hiestand's personal library, described how the authors had fashioned bullet-like pellets of testosterone propionate that fitted the bore of a 0.22 calibre rifle. Utilizing captive deer restrained in fenced wooded areas, they shot pellets into their rear ends (intramuscularly) and observed how long it took before their antlers dropped off. (The mature antlers of whitetail deer are composed of bone and start their growth in the spring. By late summer, they are calcified with the velvet shed in September.) In male deer (bucks), testosterone levels rapidly elevated by an "injection" of testosterone and then allowed to fall through excretion caused antler loss (which coincides with what happens in nature when gonadal secretions decline). In whitetail does (usually antleress), a shot of testosterone caused antler nubbins to appear. The gist of the article, therefore, involved the influence of high and low testosterone levels on the appearance and loss of antlers in both sexes. The talk was well received though presented with a dry mouth, and much gulping, dead give-aways that the speaker was experiencing a bout of nervousness. The questions that followed were logical and answerable. How I wished for a copy of Hunting & Fishing magazine to impress my listeners with my deer hunting piece, but its publication date, according to editor Bernie Roth, was two months away in December. When the article finally appeared, Dr. Hiestand was the first to point out a grammatical error that had escaped me and the printers! Don Stullken loved the procedures involved in stalking deer and, now aware of my interest in nature and the outdoors, invited me to accompany him on a drive one weekend to visit his sister who lived in a small town east of Gary, Indiana, near the sand dunes of Lake Michigan.

Wislocki, G.B., Aub, J.C. and Waldo C.M. Encrinology 40:202-224 (1947).

Don and I borrowed a couple of sleeping bags, anticipating that our arrival in early March would be chilly in the evening when we bedded down on the sand along the shore of Lake Michigan. I kept thinking that our location, only a few yards from the lapping water, was precariously close to an eventual inundation, forgetting that the tidal inflow of New Brunswick's Bay of Fundy does not occur on the shores of the Great Lakes, while the phenomenon of sand flea bites, like those experienced by sun-bathers on Old Silver Beach at Falmouth on Cape Cod, does.

In addition to visiting his sister, Don had come to her home to pick up a discarded steel brake drum which, when hit with an iron rod, would emit a noise loud enough to cause seizures in audiogenic mice. Dr. Hiestand had ordered a shipment of DBA (dark brown agglutins) mice, a rare variety of seizure-prone rodents which, when exposed to appropriate auditory stimuli, became convulsive. On the day of their arrival, the strange-looking, peanut-sized animals were placed in cages and moved to the quietest area of the animal facility to enjoy a grace period of relative silence for several weeks before sound experimentation began. In the meantime, Dr. Hiestand and Don had studied the literature in preparation for a series of studies employing different kinds of sounds. It was decided that the sounds of woodwinds (clarinet and oboe) would be employed in one group, while a second group would be exposed to the staccato sounds of percussion instruments. The sound of a drumstick on a stretched animal skin was first suggested by Dr. Hiestand. Don's idea, derived from the thought of his sister's brake drum as the source of a foolproof clanging sound, was finally accepted. As a result of this latter discussion, our trip to visit his sister and the sand dunes of Lake Michigan was initiated in the early spring.

After our return, we had to wait a week or so before the animals were conditioned enough from their raucous journey to Purdue to confront the "music." Unfortunately, the university did not possess an anechoic (sound-free) chamber where they could have resided under near-ideal conditions. Instead, they were confined to small animal quarters located a few yards from the railroad track where once a week a hopper car arrived loaded with coal which it dumped with the noise of a mini-avalanche to provide fuel for the power station in back of the Bio Annex. When the experiments commenced, nothing happened. In desperation, we subjected the animals to music by Beethoven, Brahms, and Bach and then, assuming they were of the swing generation, to music by Glenn Miller, Benny Goodman, and even the jazz of Cab Calloway! Delving further into the literature, we discovered that adult mice, 35 days or older, of either sex, were susceptible, whereas the newborn, the size of peanuts, were resistant to audiogenic seizures and, unlike their elders, refused to do the "twist" even in the presence of a muffled sneeze.

In the fall of 1947, a 17-year-old freshman student walked to his classes at Purdue University on a route that would eventually cross the path of Donald E. Stullken. In two years the youth would leave the university to join the Navy as a combat pilot, flying 78 missions off the aircraft carrier Essex during the Korean war. Upon completion of his Navy service, he returned to Purdue and received a degree (EE'56), whereupon he joined an organization with an obscure name which eventually became known as the National Aeronautics and Space Administration (NASA). After spending almost a decade at Edwards Air Force Base as a highly skilled test pilot, he joined the Dye-Soar project to fly a hydrid airplane/spacecraft vehicle, the limited success of which resulted in the program's cancellation in 1963. Anticipating its demise, he applied for training in the astronaut corps in 1962 and was accepted as its first civilian member. His path crossed that of Dr. Stullken when the former Purdue student, Neil A. Armstrong, accompanied by fellow astronauts Col. Edwin E. (Buzz) Aldrin, Jr. and Lieut. Col. Michael Collins, returned to planet earth after the historic landing of Apollo 11 on the moon (Monday, July 21, 1969 at 4:17:40 P.M. Eastern daylight time) and were greeted by the first person to welcome them back (recorded as an historic photograph published in LIFE magazine), my former instructor in physiology. Don had received his Ph.D. degree in physiology in 1950 and fulfilled a desire to become an aviation physiologist by joining the project Mercury recovery planning board that was deeply involved in the space program. Seeing that photograph in LIFE magazine was one of Dr. Hiestand's proudest moments! The photograph also recalled a few memories for me of instructor Stullken that had occurred outside the classroom. During my second year at Purdue, Don was successful in trapping a male mink in a device he designed with a door that closed when a small lever containing bait was touched. He reset the trap a number of times to catch a female companion to study their breeding habits. Failing to do so, I decided to help by ordering a female mink (25 dollars) from an ad I saw in Outdoor Life. Within a few weeks, a loud outcry from Mrs. Jackson signaled she had encountered the animal in a small crate on her porch. Don was delighted when he learned of the arrival and conscientiously prepared a love nest for the pair. One morning, I got a frantic call at my Jackson residence to come to the animal house at once. When I arrived, I saw Don working desperately to revive two comatose minks by administering injections of caffeine citrate.

"What happened?" I inquired, when I saw the limp bodies that were barely breathing.

"Someone fed them rats left over from a pharmacology experiment. They were anesthetized with a barbiturate," he answered in disgust.

Despite his valiant efforts, the slight caffeine-induced stimulation of the minks' respiratory centers could not overcome the severe depression induced by the barbiturate. Don and I were both upset by the demise of these beautiful animals.

Sitting on a hard chair in Dr. Hiestand's office for an hour and a half each week listening to seven other graduate students, who delivered their respective assignments verbally in the physiology seminar, afforded me an excellent opportunity to examine the various artifacts and doodads that cluttered the professor's desk and walls. A bronze rendition of a grasshopper, positioned as though attempting to escape the sudden onslaught of erudition, stood next to small framed photographs of the Irish setter, Sheila, and Dr. Hiestand's wife, Betty, amid books jammed between metal bookends in the shape of largemouth bass. The latter were obviously purchased in Wisconsin where their summer retreat, a wooden bungalow with a huge stone chimney, stood a few yards from a lake, which was the aquatic home of its living counterparts that managed somehow to evade capture by him each year. I assumed that the presence of the grasshopper had something to do with Dr. Hiestand's earlier interest in insect physiology.

The barrage of words accosting the ears from an overzealous student made close observation of the surroundings difficult, for eye contact with the speaker implied that the learning process was taking place, whereas looking elsewhere meant otherwise. Regardless, I was fascinated with a number of small, ink drawings protected by a sheet of transparent plastic that hung on the walls. When the seminar was over, I lingered long enough to examine several of them closely. They were Jimmy Hatlo reproductions of his famous "They'll Do It Every Time," cartoons that appeared daily in hundreds of newspapers across the country in black and white and on Sundays in color. They were copyrighted by King Features Syndicate, Inc, with a thank you note from Mr. Hatlo in a tiny square at the bottom of each panel. Remarkably, in each the thanks were directed toward Dr. Wm. A. Hiestand, Lafayette, Indiana.

"Yes, I sent them in a few years ago and was surprised when they accepted them," said Dr. Hiestand. "Its become a sort of hobby for me. The first one I sent in was about the guy who bought a tiny camera but needed a couple of bags of equipment to operate it!"

On my way back to Sheetz street, I decided to compete with the professor by sending some of my own amusing situations to Mr. Hatlo. The first one appeared on November 1st and showed a man who attended 2,428 baseball games and never caught a ball, but when he went to a midget auto race, he was hit by a flying wheel. In the corner of the cartoon were the words: *Thanks and a tip of the Hatlo hat to David E. Mann, Jr., 142 Sheetz St., W. Lafayette, Indiana.* In the next couple of years, five more were published. The contributions to Mr. Hatlo ended when, after I requested an original drawing, he informed me that they belonged to the Syndicate. At that time, Mr. Hatlo, living in Carmel, California, was earning $50,000 per year. About ten times Dr. Hiestand's salary.

In addition to physiology and its seminar adjunct, which represented my major and served as an academic backbone for my other subjects, I took the mandatory courses, biochemistry and animal husbandry. Fortunately, because of the time factor they were didactic (only lectures) but, instead of a laboratory period, biochemistry required additional hours prior to mid-term and final examinations to review important material. A doctoral student, Irving Smulevitz, conducted the review sessions with a grasp of biochemical concepts second only to that of his mentor, Dr. R. C. Corley. Professor Corley was a man of exceptional talents. An outstanding biochemist, linguist, fly-fisherman, amateur photographer, and music lover, he taught biochemistry without notes and kept nervous students on the edge of their seats hoping the questions asked spontaneously and sporadically were answerable. Dr. Corley was a stickler for correct pronunciation of words beyond those found in the vocabulary of a biochemical scientist. He frowned on the verbosity of those who lived by the spoken word (lawyers and politicians), and from gushing pens (journalists and novelists). His concept of a good lawyer or novelist was one who spoke like Gary Cooper or wrote like Hemingway. When a bell clanged at the end of a lecture, a highly audible gasp of relief could be heard from those who had successfully avoided his scholarly inquisition. Half way through the first semester, Mr.

Smulevitz informed his review sessions that he had changed his name to Sherman and hoped to live for a hundred years.

When he made this announcement in my session, I raised my hand and congratulated him.

"As for living to be a hundred, it's biochemically possible if you keep away from Georgia, especially Atlanta! If you'd changed it to Fahrenheit, you might have made it to 212!"

Animal husbandry, which was actually the endocrinology of farm animals, was taught by *Professor Frederick N. Andrews, a native of Weymouth, Mass., who was a member of the university's department of veterinary science. It was housed in a building shared with the agricultural school and stuck in the middle of a cornfield near the land set aside for the new Life-Science Building. Dr. Andrews bore a close resemblance to the Scottish actor, Robert Donat, who appeared in Hitchcock's *"Thirty-Nine Steps," "The Count of Monte Cristo,"* and *"Goodbye Mr. Chips."* As he lectured on the estrous cycles of cows, sheep, and pigs, his New England accent often sounded like Donat's Scottish brogue. He was a superb instructor, close friend of Donald Stullken, and a driving force who was mainly responsible for the consolidation of the fragmentary veterinary disciplines into a full-fledged veterinary school. In January, 1951, I received my doctorate degree in physiology and was honored to have Dr. Andrews place the black and gold Ph.D. hood over my shoulders on the stage of the magnificent **Music Hall before my proud wife, Mary, who was sitting in the audience.

* *Now Vice President for Research and Graduate School Dean Emeritus. **Purdue's Music Hall has 7 more seats than are in Radio City in New York!*

Purdue's great Hall of Music, without question its foremost architectural pride and joy, served not only the university family as a commencement hall and entertainment center, but also the townsfolk who, after consuming wholesome food at the Student Union, spent evenings watching professionals from the arts and entertainment worlds perform. In the mid-thirties, Frank Buck, of *Bring'em Back Alive* fame, fresh from seeing his recent film of the same title become a surprising hit at Radio City's Music Hall and now on tour to publicize it nationwide, lectured at the Music Hall about his daring adventures while capturing exotic animals in Africa that were caged and shipped to zoos throughout the United States and Canada. The faculty, sitting in the first row by special invitation, were enthralled with his fascinating presentation. At its conclusion, Mr. Buck met with a few of the professors who remained after the lecture to ask questions. In the middle of the question-answer period, the famed animal procurer suddenly announced he'd like a stiff drink. Like a gathering of monkeys looking skyward and seeing a meteorite coming toward them, the crowd quickly dispersed, leaving Dr. Hiestand and Mr. Buck staring at each other. Breaking the embarrassing silence, Dr. Hiestand said:

"Come to my place. My wife would like to meet you if you promise not to capture my Irish setter, Sheila."

That evening, after downing several drinks, Frank Buck retold the adventures he had just vividly described before 7,000 people. This time, however, what he related, with a marked alcohol-induced speech impediment, was quite different from his earlier professional utterances, leaving the Hiestands with the impression they were hearing what actually happened in the wilds of Africa.

During my first year at Purdue, Bob Hope, his side-kick Jerry Colonna, and Les Brown's Band of Renown appeared onstage one Tuesday evening (for their weekly Pepsodent-sponsored radio show) to entertain a standing-room-only audience at the Music Hall. The only time I'd seen Mr. Hope in person was from my seat inches from the ceiling of Boston's Metropolitan Theatre (the Met) in 1943 when he was touting his film "The Cat and the Canary." From that location he had appeared the size of an ant but now, sitting only a few feet from the massive stage, I saw a man in his forties who was of average height and possessed tremendous energy. In one scene, he played a salesman stuck in a hotel room when Mr. Colonna, the hotel detective, asked, "Do you have a woman in there?"

"No," replied Hope.

"Sissy," commented Colonna.

The arrival of May was accompanied by the onslaught of publicity surrounding the opening of the Indianapolis Motor Speedway's two and a half mile track for time trials, scheduled for the first two weekends of the month. During each week, however, the track was open for tire, motor, and chassis testing prior to the actual qualifications on each weekend, involving the average speed accruing from four laps (ten miles), which determined the poll position (the fastest car during the first day) and the eventual placement of the other drivers in the remaining 32 cars on race day. The poll position, incidentally, is the inside location of the first row, which traditionally affords the driver the best opportunity to charge ahead of the other two cars in the first row, thereby executing the first turn and leaving the others in the dust. This maneuver is also the most dangerous of the entire race and always leaves several hundred thousand people breathless with a deep sigh of relief when all the cars whizz past turn No. 1 without hitting the concrete wall. In the 1947 Indy 500 race, Shorty Cantlon, attempting to avoid two cars swerving recklessly in front of him, smashed almost head-on into the wall of turn No. 1 on the fortieth lap and was killed instantly. In those days, shortly after WWII, hitting the wall at speeds of 110-115 mph always meant serious injury and occasionally death, for drivers didn't have the protection afforded today by enormous tires, carbon fiber chassis, flame-retardant clothing, and aerospace plastics of superb strength. Furthermore, the presence of highly-trained personnel from the racing community, from helicopter pilots to fire-fighters in asbestos suits equipped with the *jaws of life,* dramatically increases the survival statistics of both spectators and drivers.

In May, 1948, I could not attend the time trials for my research demands occupied each weekend, while lectures and labs took care of the rest of each week. I was able to procure two tickets to the big race from Dr. Hiestand's friend, Dr. K. K. Chen, a famous researcher at Eli Lilly, who, along with Dr. Carl Schmidt of the University of Pennsylvania Medical School, had introduced *ephedrine* to medicine as an antiasthma drug several decades before. Dad was excited when I phoned him with the good news and he immediately made plans to take the train to Chicago and then, by Monon, to Lafayette for the Memorial Day weekend. The day before the big race, he arrived at the Lafayette railroad station and together we went by taxi to the Purdue Memorial Union where a comfortable room awaited.

"Wait'til you sample the food here. You'll love it!" I remarked exuberantly.

He loved to eat and that evening he had the dinner that would have made even a Roman emperor's mouth water. The next day we would see the gladiators in action.

In my room that evening, after enjoying a chocolate sundae with Dad, I set the alarm clock approximately ten times, checking the tiny plug in the rear to be sure it was out all the way, and then climbed into bed. But it was impossible to sleep thinking about the excitement soon to unfold. Eating chocolate ice cream containing caffeine, the mild central nervous system stimulant that had failed to revive the barbiturate-depressed minks, was also a bad idea. Lying in bed staring at the blankness of the ceiling, I realized I had something in common with all but the most experienced race car drivers who were about to tempt fate in the most dangerous sport on this side of the Atlantic (bull *fighting probably ranks as high on the other side).* Finally, after many hours of relentless squirming, the alarm sounded. Within 20 minutes I knocked on Dad's door at the Memorial Union and breathed a sigh of relief when the door opened and he was standing fully dressed, smoking a Chesterfield cigarette.

"Let's have a cup of coffee in the lobby before we meet the bus. It's due at 6:00," said Dad. "When we get to Indianapolis, we'll have breakfast."

The departure time, deeply ingrained in our minds, was mentioned by Dad largely out of nervousness rather than from a need for verification. Coffee was indeed available free in the lobby in the early morning hours as a university courtesy, whereas breakfast was not served in the Union until 7:00 A.M. I was secretly pleased that he had thought out these minor details with kid-like enthusiasm.

The trip from West Lafayette to Indianapolis, a distance of 60 miles, took a little over an hour. As the Greyhound bus pulled up near the state capitol, we saw an imposing structure that copied, on a much larger scale, the government buildings so typical of 19th century architecture, the commanding presence of which is as predictable in the town squares of Indiana as the location of silos are adjacent to barns on the neighboring farms. Opposite the capitol was a semi-circular cluster of two-story stores and offices undergoing rehabilitation and, a few blocks away, conveniently close by the refurbished shopping areas, were several hotels. We went in a restaurant on the first floor of the nearest hotel and ordered buckwheat cakes smothered in Vermont maple syrup. By the time we'd finished, it was time to take another bus and head for the speedway, as crowds were beginning to congregate around the departure point. We finally boarded a bus for Speedway (where the track is located) after several diesel-spewing behemoths had arrived and left packed with fans about to endure a thirty-minute trip. The ride reminded me of a recent Lowell Thomas travelogue I'd seen at the Music Hall that showed people in Calcutta sitting on the roofs of rickety buses while the drivers attempted to negotiate sharp turns in the road without hitting anyone. When we arrived at last in the midst of a swirling crowd of humanity, hundreds of people were parking their cars on front lawns for fees that paid off a month of the fortunate owners' mortgages. Passing under the gateway sign Indianapolis Motor Speedway, with its logo of a winged tire, we saw in front of us the wooden grandstand and paddock painted an eggnog white and trimmed in green where, in the former structure, our seats were located opposite the straightaway.

Walking beneath Grandstand A past the restrooms where, traditionally, during those rare days when it rains, the spectators retreat to munch on fried chicken, we found our seats in a section unprotected by a roof and close to the track. Before us, a panorama of ant-like activity unfolded consisting of several hundred colorfully-dressed people, who were moving around sleek racing cars parked in the pits on the other side of the track. Pit crews were chatting excitedly with drivers, car owners, and one another as they gave last minute directions to the mechanics armed with wrenches and screwdrivers who were fussing over their charges. Looming over this action, like a wooden Godzilla, was a multi-storied tower known as the *pagoda* that housed timing and communication equipment. Our seats, curved wooden slats designed to fit the average adult-sized, Hoosier rear end, were located in the bright sun only a few feet from the edge of the track behind a wire fence that wouldn't have stopped a driver's helmet thrown at us from the pits directly in front of us. Names, stenciled in large black letters against the oyster white background of the concrete pit wall, identified the designated chauffeurs of the race cars undergoing last minute preparations. The legends of motor racing—Ted Horn, Rex Mays, Tony Bettenhausen, George Robson, and Mauri Rose, last year's winner—were directly in front of us either in the cockpits or about to be seated. A driver with an unforgettable name, Tommy Hinnershitz, was about to climb into his race car, a strikingly beautiful, bright-red vehicle with gleaming, chrome-plated exhaust pipes emerging from either side of the hood and covering the front engine, that contributed to its overall aura of unbridled power. Moving frantically among the track officials, bustling pit crews, and scattered members of the now assembling Purdue University band, was the general manager of the Speedway, a three-time 500 winner and unquestionably the most beloved and skilled race car driver in America, Wilbur Shaw. By a strange coincidence I was attending a school picnic on Memorial Day at a classmate's home in Needham as a ninth grader when I learned of his first victory at Indianapolis in 1937. I have a vivid recollection of the incident because I ran into a wire clothesline and gashed my eyebrow while playing tag on that day and, when I went into her house to stop the bleeding with ice, my teenage hostess was jumping up and down upon hearing the news that Wilbur Shaw had won. Her name, incidentally, was Muriel Shaw! I had no idea she was related to him. Mr. Shaw had won in a car he'd built himself that was powered by a 4-cylinder Offenhauser "Offy" engine. In 1939 and 1940, he won again, but this time his car was a more powerful 8-cylinder Maserati.

Wilbur Shaw, the first person to win the Indy 500 two years in a row was also well known for another remarkable accomplishment. In 1945, he talked an Indiana businessman named Tony Hulman into purchasing the Speedway from Captain Eddie Rickenbacker and his associates, who had obtained the track from the Fisher family in 1927. The WWI flying ace had raced there in 1914 (coming in 10th) and fallen in love with America's most famous race track. World War II had closed the facility for four years and left the brick track, concrete walls, steel fencing, and wooden grandstands at the mercy of Mother Nature. When the war finally ended, Captain Rickenbacker realized that its restoration required more money than he cared to invest. Serendipitously, Mr. Shaw was aware of this once-in-a-lifetime opportunity to transfer ownership from the WWI hero to a Terre Haute millionaire sportsman (Hulman was an outstanding pole vaulter in high school and one of Yale's greatest all-around athletes), who had the finances to restore the Speedway not only to its prewar condition but, in the years ahead, to create the finest racing facility in the world. Captain Rickenbacker, after the consummation of the real estate transaction, became a frequent visitor to the Speedway and occasionally drove ancient race cars around the track to the delight of the time-trial spectators, some of whom actually recalled seeing these antiques perform in the pre-WWI 500's. As chairman of the board of Eastern Airlines, his plane crashed in the Pacific and he and several survivors clung to life for a few weeks by drinking rainwater and consuming an occasional flying fish that jumped aboard the rubber raft. Finally, when they were on the brink of starvation, a seagull alighted on Rickenbacker's head. He grabbed the bird by the legs, wrung its neck, and prepared to eat it uncooked. Several days later his party was rescued. He depended on the wings of aircraft when young and was saved by the wings of a bird when old.

First, the mournful sound of a trumpet playing taps, then the Purdue University band's rendition of the National Anthem, followed by Morton Downey's singing of "Back Home Again in Indiana" and, finally, the tearing open of the canvas roof of an Army tent in the infield with the release of thousands of balloons.

"Gentleman, start your engines," commanded Wilbur Shaw over the microphone. A symphonic roar emanating from 33 cars, almost in unison, filled the air with a noxious mix of castor oil, motor oil, and the exhaust from methanol combustion. Several hundred thousand people stood up to get a better look at the Chevrolet pace car which was slipping away from the pit area to take its position in front of the crawling race cars that were gradually getting into the correct position for the pace lap. Imperceptible at first, the pace quickened until, upon completion of the pace lap, the cars disappeared around turns 1 and 2 as they proceeded rapidly down the back straightaway. Suddenly, they came in sight at turn 4 and, as the pace car turned back into the pit area, they thundered down the main straightaway in response to the waving of the green flag by Seth Klein, the official starter.

As they roared past and entered turn 1, I made out the bane of the Speedway, Duke Nalon's powerful 8-cylinder yellow Novi whose twin had crashed during time trials only days before, killing Ralph Hepburn, a racing legend who was Shaw's old friend and racing competitor, and Lou Moore's controversial race cars, the Blue Crown Specials driven by Mauri Rose, the old pro and the rookie, Bill

Holland who, the year before was leading the 500 when his car owner pulled out a sign that said "E-Z". Most of Moore's money was invested in the two cars and he wanted them to finish in one piece. When Holland slowed, Rose whizzed by him to complete the few remaining laps and win his second 500. Holland's wife, in the meantime, was waiting to welcome her husband in the winner's circle! This race was certainly filled with intrigue, drama, and revenge, all of the baser elements of human endeavor. And there were 200 laps to go!

At the conclusion of an exciting race that ended without serious injuries, Mauri Rose was again the winner with Bill Holland coming in second in a *deja vu* situation that would be Rose's third and final win at the Speedway. The glamorous, powerful, and unwieldy 8-cylinder Novi, driven by the man with the semi-royal name, Duke Nalon, crossed the finish line in third place, and the driver with the most unforgettable name, Tommy Hinnershitz, managed to chauffeur his bright-red steed with the chrome-plated exhaust pipes into ninth place. A sad barometer of the intrinsic danger residing in auto racing in the late 1940's and early 1950's was indicated by the number of participants in the 1948 Indy 500 who lost their lives racing a few months (Ted Horn), and a few years later (Rex Mays, George Robson, and Tony Bettenhausen). All of these people were considered among the greatest race car drivers in the United States, having earned championship awards without having won the Indy 500, with the exception of George Robson. Their deaths were caused by mechanical failures, i.e., steering malfunctions, chassis misalignments, and axle breakage, not because of a lack of driving skills.

At the conclusion of the race, Dad put his hand on my shoulder and said:

"Son, that's the greatest race I ever saw. It made the old dirt track races we used to see in Massachusetts and New Hampshire look like kindergarten events! I want you to get tickets for next year. Maybe I can talk mother into coming with me."

We took the *Calcutta special* back to Indianapolis and climbed aboard the Greyhound bus for the sixty-mile trip back to Lafayette, where he boarded the Monon for Chicago and the New England States for home. The following year, he was unsuccessful in getting Mom to accompany him to the Speedway, but in 1950, she finally consented and the three of us saw a race that had to be stopped at 345 miles because of a deluge that sent the townsfolk into the restrooms to munch fried chicken and Johnny Parsons into the winner's circle in a race car whose engine was about to expire. We left the race track holding the local newspapers over our heads!

On June 13th, 1948, I received a Master of Science degree at the Music Hall after passing an oral examination on my cyanide research and answering questions in mammalian physiology and endocrinology. My mother, accompanied by her younger sister, Lillian, had arrived by train for the occasion and, after the ceremony, the three of us visited Chicago for a brief shopping spree, during which she bought a knotty pine gun cabinet at Marshall Field's for Dad ($50), before heading back to New England where I spent the summer doing odd jobs.

CHAPTER THIRTEEN

Pleased to meet you (future wife)!

The summer months of 1948 were occupied with occasional jobs provided by Les Cook that involved driving beat-up cars from Worcester dealerships to the towns around Needham where they were sold to fill the unprecedented postwar demand for cars that could run, regardless of their outward appearance. Studebaker and the Kaiser-Fraser vehicles were among the first new models to appear after WWII, while General Motors, Chrysler, and Ford struggled with the conversion of their enormous facilities in the Midwest from the expertise, equipment, and materials required for the production of tanks, aircraft, and other forms of ordnance to the designing, manufacturing, advertising, and shipping steps involved before the successful introduction of new models, acceptable to a new- vehicle-deprived public, could be accomplished.

The new Studebaker, usually available in green, was a cute little car that looked like a modified, two-man Japanese tank, with an almost 360-degree, glass-enclosed cockpit prompting envious jokesters to call it a *mobile greenhouse.* Others complained that they didn't know whether it was coming or going, an attribute that could conceivably be attractive to potential bank-robbers. As for the Kaiser-Fraser automobiles produced by Henry Kaiser, whose shipyards launched a Liberty ship almost daily thanks to revolutionary prefabrication techniques he helped to develop, they were of two models: the Kaiser and the Fraser. I was with Dad who was making a house call when we saw our first Kaiser.

"What's the most striking feature of this car?" he had asked.

Without giving it a minute's thought, I answered: "The paint finish. I've never seen such a glistening surface."

"Right! But the shortage of chromium is evident in the small bumpers, and the car's body looks fragile."

In the market for a new car himself, Dad bought a brown, four-door Cadillac that turned out to be the wrong color—it was a lemon. He had an altercation with the dealer and vowed never to buy a Caddy again. In retrospect, the car was a victim of the conversion process at the GM factories, where the employees were overworked to bring new products to the marketplace. After that unpleasant incident, Buick became the car of choice for Dad and Mom through the next decade, although he did buy a Chrysler sedan for Mom in the late forties that was a superb car. Of course, the dealer of that car was our dear friend—Les Cook.

In late August, I headed back to West Lafayette, not by train this time, but in my high school graduation present, the black Plymouth convertible with the bright red leather upholstery that Dad had driven in the yard with Les in the spring of 1939, acting like a couple of giddy schoolboys. As I got in the car and waved goodbye to the folks, Dad's pronouncement: "Don't drive it to school!" came to mind. (I smile when I think of those *good old days* when few parents, let alone kids, had cars. If he were alive today, how he'd laugh at the antics of the kids at high schools in California who flaunt their Mercedes' convertibles, BMWs, and Ferraris, and park their cars next to the vehicles of the faculty, which resemble survivors of a demolition derby!). My how times have changed! Although Specky and I trudged through snowdrifts, icy rain, and near-hurricane winds to avoid offending anyone in the Needham community, those spartan journeys on foot saved wear and tear on the car, which was in showroom condition and ready for the long trek to Indiana.

The trip south to Pennsylvania in the absence of a New Jersey turnpike meant innumerable stoplights and heavy traffic once I'd left the confines of Southbridge, Massachusetts, where I refueled the car and devoured Howard Johnson's specialty, a fried clam roll and strawberry frappe. When I finally zoomed west on the Pennsylvania turnpike, after seeing the hills of Connecticut, the urban sprawl of New York City, and the oil refineries of New Jersey, I continued for several hours until I reached Washington, Pa., where I stayed overnight. The next morning, I entered Ohio with its unique brick-surfaced roads and stopped for gas and breakfast. I was extremely lucky to discover a place that dispensed gas and edible food. And there was also entertainment—behind the station was a functioning oil well. In those days, Ohio had as many restaurants on its major highways as ducks have molars, and the worst had the name Ma or Mom in their signs. By late afternoon, I drove over the Wabash and up the hill to West Lafayette and onto Sheetz street as the sun was setting. The Jacksons were just finishing supper when I parked the car in front of the house and started up the backstairs with a loaded suitcase.

"Welcome back, Mr. Mann," said Mrs. Jackson. "There's a registered letter I slipped under the door of your room. It looks mighty important."

"Thanks, and nice to see you," I replied, wondering if she had given it the steam treatment.

Upstairs, amid the dust from several months of ignored cleaning, I picked up the letter and sliced it open with a penknife. It was from the Secretary of the Board of Trustees, Frank C. Hockema, informing me that I was the recipient of a Purdue Alumni Research Foundation award supporting my doctoral research. It was also a welcome supplement to my G-I Bill, the income from which had previously provided funds for food, clothing, and housing.

Dr. Hiestand, having just returned from vacationing at his cabin at Lake Mendota in Wisconsin, was pleased to see me and asked that I sit in the chair next to his desk. He reached in his jacket pocket and handed me a photograph. It was a picture of him standing next to a beached rowboat with a fishing rod in one hand, which he was using as a cane, and a string of largemouth bass dangling from the other.

"Congratulations! You hit the jackpot! How do you explain your reversal of fortune?" I inquired.

"Serendipity. I made a couple of flies from Sheila's shed hair which did the trick. Incidentally, I took her to the vet before we left for Wisconsin, and he discovered she had diabetes. She didn't wet once on our old hooked rugs after her blood sugar level was normalized with regular insulin. And Jimmy Hatlo used another cartoon idea I sent him. It's been quite a summer! How was your summer?"

I told him about the highlights, the trip back by car, and the letter I had received informing me of the Purdue Alumni Research Foundation award.

"Well, congratulations are in order! I recommended you for the award based on your ability to perform original research, in addition to your enthusiastic approach to your course work. You're no longer a probationary graduate student."

Just then someone knocked on the door. "Come in," said Dr. Hiestand.

A slightly overweight, round-faced, swarthy, neatly-dressed man in his early forties entered the room with an apprehensive look on his face that disappeared when he saw me.

"I didn't mean to interrupt your meeting. I'll come back later."

"No, wait, I'd like you to meet my doctoral student, David Mann, who just arrived from Massachusetts." Turning in my direction, he said: "And this is Dr. M. X. Zarrow who is starting a department of endocrinology. He's also from Massachusetts with advanced degrees from Harvard."

We shook hands and exchanged hometown origins. He grew up in Millbury, and was familiar with the amusement park called White City, on the outskirts of Worcester, where I had taken dates on two-seater motorboat rides on a man-made pond created like Mirror Lake by damming a small stream late in the 19th century. Dr. Zarrow was likewise acquainted with Needham's Paramount theatre and its 35-cent double features.

When I asked him about the nature of his doctoral research, he answered: "I investigated the physiologic actions of relaxin, the hormone of pregnancy discovered by my mentor, Fred Hisaw. Have you heard of him?"

"I've heard of both of them. I had Dr. Hisaw for introductory zoology following Dr. Darrah's botanical course. He lectured about relaxin and its effects on the pubic symphysis of the 13-line ground squirrel. It's not only produced during pregnancy by the ovaries, but also by the placenta."

"I hope you'll sign up for my endocrinology course this semester. I'm now in the process of getting my lecture material together, and there'll be a laboratory session once a week. Did you take his endo course as well?" inquired Dr. Zarrow.

"Yes, and I didn't enjoy it. Too much emphasis on cutting the spinal cords of female macaques to study menstruation. The whole course covered hormones that acted below the waist." I responded.

"Well, this one I'm giving this fall involves all of the endocrines, their evolution, their hormones, and their effects. It is very practical and will provide a solid basis for your physiology major."

Dr. Hiestand interrupted our conversation, having listened intently to the brief discussion of our hometown and educational reminisces:

"Why did you knock on my door, Dr. Zarrow? Do you have a problem you can divulge in front of Mr. Mann? If not, I will arrange another meeting date with him."

"It's not that important. I was going to ask you if the university golf courses were challenging. My wife and I love to play. She's better than I, but that's beside the point."

"You're asking the wrong person, Dr. Zarrow," said Dr. Hiestand. "You obviously aren't aware of my medical problem. I have multiple sclerosis. In fact, I've had it for about fifteen years. That's how long it's been since I've played golf."

Dr. Zarrow's round visage gradually reddened as he attempted to speak with a throat that had suddenly become dry.

"I'm embarrassed," he eventually replied. "I had no idea. And I'm an endocrinologist," he added.

"I don't think it has anything to do with hormones," said Dr. Hiestand, a hint of indignity in his voice. Then, in a lighter tone, "Ray Cable, the parasitologist, whose office and laboratory are in the basement, should be able to answer your question. He was around when the university laid out the golf course."

Dr. Zarrow thanked him, shook my hand again with a sweaty palm, and retreated quickly, like a dancer finishing a gavotte, his complexion gradually returning to its former swarthy appearance.

As the sound of his footsteps in the corridor became imperceptible through the closed office door, Dr. Hiestand turned to me and said:

"Purdue is very fortunate to have attracted a man of his brilliance. By all means, sign up for his course right away. You should also enroll in Professor Edwards' pharmacology course at the Pharmacy school. You'll be with undergraduate pharmacy students, but your grading will be more stringent than theirs because of your graduate status. Have you given any thought to your research?"

"I'd like to do something more disease-oriented than toxicology."

"I agree. How about a study pertaining to some aspect of *diabetes mellitus?* Discovering that Sheila was suffering from it reminded me of Banting and Best's landmark research on dogs that led to the introduction of insulin for its treatment in 1922."

"That's the year I was born. Maybe that's a good omen. I'll do a literature survey to find out what's been done and what's yet to be done."

"Are you familiar with the biochemical method for determining blood glucose levels in mammals?" asked Dr. Hiestand.

"Dr. Corley's course reviewed the procedures of Folin Wu and Folin Malmros, but we didn't perform them because there was no laboratory."

"They aren't difficult, as biochemical procedures go. All you need is 0.1 ml. of blood by aspiration in a glass tube and a few, inexpensive chemicals. A spectrophotometer, which we have in our graduate lab, is used to make the final determinations. It's tedious until you've done it a few times. I envision a project that involves glucose determinations in small mammals such as the rat or rabbit. I'd like you to discuss any ideas you have with Dr. Zarrow as well as with me. In fact, he may offer some original approaches and even suggest a collaborative effort relative to his own research interests you could incorporate in your thesis."

I thanked Dr. Hiestand for his advice and walked to the Pharmacy school, located near the central architectural and dual-purpose gem of the university, the impressive Administrative/Music Hall edifice. A Quonset hut adjacent to the school housed laboratory facilities of an expanding pharmacology department and reminded recently discharged Navy veterans of their boot camp days. Professor Edwards' office on the first floor of the Pharmacy school building was locked and in darkness. Across the hall, however, an observant secretary, seeing my frustration through an opened office door, asked if she could be of assistance.

"Professor Edwards is on vacation. If you'd like a course outline and school curricula, see Tom Miya, his graduate assistant," she suggested.

"And where do I find this gentleman?" I inquired.

"His office is in the funny-looking building next to this one. If he's not there, you'll probably find him puttering in the pharmacology lab in the same building."

"Thanks," I replied, and walked to the Quonset Hut, the Navy's inexpensive solution to a barracks' housing shortage during WWII that had its innovative, taxpayer-saving counterpart in the Army's ubiquitous general purpose *(GP=jeep)* vehicle created for cheap transportation.

A young man of Nisei descent, Tom Miya, was on his knees unpacking glass jars in the undergraduate lab when his head suddenly appeared above a cardboard carton. He immediately noticed me and asked:

"Are you looking for me?"

"Yes, if you're Tom Miya. I'm David Mann from Dr. Hiestand's graduate lab. I'm was looking for Dr. Edwards to get more information about his pharmacology course but he's away. I intend to enroll as a graduate student."

"Great," he said, standing up and shaking my hand. "I'll give you a course outline and lecture and lab schedules. Dr. Edwards, incidentally, is a friend of Sheila's master. He'll be pleased to meet you next week."

That dog must be as famous as Lassie around here, I thought as I pocketed the papers, thanked Mr. Miya, and headed for the library to begin a literature survey.

The following week the semester began with physiology seminar (assignments), endocrinology (lecture and lab), pharmacology (lecture and lab), and the remaining hours for research. Pharmacology, as introduced by Dr. Edwards, was especially interesting. Before the course started, I met the professor, a stocky, middle-aged man, who seemed curious as to why I was taking the course.

"It was Dr. Hiestand's suggestion," I explained. "He felt with my undergraduate biology background and physiology major, it was a logical adjunct."

"He's right!" he said with a chuckle. "Pharmacology is the pragmatic side of physiology. I hope you'll like it well enough to make it a minor."

In the first lecture, I learned the difference between a drug and what Dr. Edwards called a *laboratory tool.* The former was a single substance or mixture that mitigated, prevented, or cured disease in man and animals. A *lab tool* was a chemical that was used to study a biological reaction. Nicotine to map pathways in the autonomic nervous system by blocking impulses at ganglia was an example. When it was given in a solution as an enema to get rid of intestinal worms (and frequently caused death), it became a drug. I was surprised to learn that most drugs mitigate or prevent disease, yet very few cure.

Dr. Zarrow's enthusiasm permeated his endocrinology lectures and mixed philosophical with evolutionary concepts. The handful of graduate students (less than a dozen) became hypochondriacs as the course progressed. Reading assignments of various endocrine deficiencies confirmed aches and pains that hitherto went unnoticed, but now were recognized to be due to hypofunction of the thyroid, adrenals, and gonads. It was a miracle that everyone was alive when the course ended.

As for my research, I embarked on a study of the effects of chemical agents on blood sugar levels in rats. When the research was well underway, Dr. Zarrow became interested in the normal blood sugar of sheep, having read that it is consistently low compared to that of man. Throughout the months of November and December, blood was taken from the jugular or ear vein of 4 normal sheep (Dorset and Shropshire), and blood sugar determinations were made by me. It was found that though there is a tendency towards a low sugar level, many have values comparable to lower limits of other mammalian species. Nursing lambs showed wide variations with overall high blood sugar levels due to the drawing of blood at different intervals after nursing. George Neher, a fellow graduate student, owned the sheep and the farm where this activity took place. His wife, a superb cook, always treated us to rich pastries, cake, and pie at the conclusion of each blood letting.

The daily lives of some predoctoral students were frequently as exciting as those of guards in an abstract art gallery. Occasionally, the monotony of routine existence was interrupted by strange happenings. During the height of the football season, when wild crowds poured into Ross-Ade stadium in anticipation of a Purdue win over Notre Dame (about once in every five meetings under coach Stu Holcomb), I was busily aspirating blood from rats' tails, or coating galvanized funnels with liquid paraffin. After the paraffin dried, the funnels were placed under circular cages, each containing a young rat, which was given a normal diet of Purina biscuits, vitamin supplements, and high concentrations of glucose or sucrose in the drinking water. The purpose of the experiment was to determine if ingestion of large quantities of sugar for prolonged periods (months) could cause *diabetes mellitus* in normal rats. The presence of sugar in the urine, one of the symptoms of human diabetes, a histologic study of the pancreas and other organs, and a significantly elevated blood sugar might confirm this hypothesis. The experiment was an ancillary study performed as insurance against possible negative results that might occur from my blood sugar experiments with various chemical agents. Because rat urine is extremely corrosive to galvanized metal objects, the funnels had to be coated with paraffin. After applying a thin layer of paraffin inside the funnel with a paintbrush, it was placed in a bucket of cold water to harden evenly. Unfortunately, the paraffin container, being warmed on a metal shelf with a Bunsen burner, suddenly tipped over from outside vibration and burst into flame. I quickly grabbed the flaming container (I was wearing thick gloves) and dropped it in the bucket of cold water. A column of flame the diameter of the bucket shot up to the ceiling accompanied by a loud explosion that jolted the laboratory door off its hinges.

"You should have used the carbon dioxide extinguisher," said Dr. Hiestand the following Monday when he asked why my eyebrows were missing. "It's a good thing there were no classes in the annex across the way. The psychology lecturer would have had a heart attack seeing volcanic activity."

At a party held for his graduate students and the Stullkens at the Hiestand home the following Saturday, (to celebrate a Purdue win over ND), I met Betty Hiestand for the first time. She was in her late thirties, her hair swept back like a gypsy, and exuding Midwestern charm that set everyone at ease. Each student had brought a gift of wine or candy. One by one she carefully opened each gift. When she opened mine, she gasped as her eyes fell upon chocolates covered with colorful fungi. The top layer looked like a novelty mineral garden that grows when vinegar is added! She thought it was a joke until I explained I'd just bought the box from a local drugstore and was completely innocent of being a prankster. It was my most embarrassing incident during three and a half years at Purdue.

My two friends in adjoining rooms at the Jackson residence, during the years 1948-50, were undergraduate students in liberal arts who were Hoosiers, yet from different social backgrounds. Bill Smith, the son of a famous scientist, was not following in his father's footsteps which was a good thing, because his father, a geology professor, was credited with being the first person in the world to explain satisfactorily how quicksand is formed. Bill, nicknamed *Rocky,* was a brilliant conversationalist who conducted bullsessions each weekend on every subject from girls to religion into the wee hours of Sunday morning. The last member of this triad was the occupant of the room in which the *Crossbones* Kid was packing to return to Indianapolis when I first arrived at Sheetz street in the fall of 1947. Its other resident, a student who had majored in wildlife, but not the kind presented academically at Purdue, had flunked out. A young man, whose physique resembled a pyramid standing on its apex, was Indiana's top student wrestler (in his class) from Terre Haute (Tony Hulman's hometown). I had never seen a person with such wide shoulders since seeing Johnny Pesky of the Boston Red Sox signing autographs at Sears-Roebuck near Fenway Park in Boston. I have forgotten his name but recall his frequent request to me *"to become involved in sports. Don't be a spectator when you can be a participant."* It was hard to explain that being a graduate student didn't leave much time for frivolity. If I had a lot of free time, I wouldn't be wasting it on a sport other than auto racing. For some unexplained reason, he never urged *Rocky* to indulge in a sport. When they both found out I enjoyed seeing auto racing and that one of the reasons I had selected Purdue was its proximity to the Speedway, they looked at me with disdain and disbelief.

Now with a car available, I had an opportunity to attend midget auto races once spring arrived and the rain-soaked oval dirt tracks dried out. The closest one was only a few miles away. Toward the end of one race or heat, a spectator who had been sitting at the edge of the track stood up to get a better view just as a midget car veered off course and struck him. I gasped as I saw his body twist completely around in a wheel and flop motionless on the ground. He was badly injured with many broken bones, but miraculously survived and, according to the monthly reports that appeared in the local papers, he recovered completely within a year. The incident brought back ugly memories of a Northeast Racing Association event in Massachusetts where a similar incident occurred. A driver climbed out from under his overturned race car and waved to the crowd that he was all right. Suddenly, as he was climbing over an embankment to safety, another race car, swerving out of control, struck and killed him. I had to attend these races alone, for the Hoosiers I knew preferred seeing the artful dodging of basketball to the mayhem of auto racing. When May finally arrived and with it time trials at the Speedway, I was accompanied by P.V. Hammond, a black graduate student in pharmacology, who wanted to visit his mother in Indianapolis and also get a haircut, an impossibility in town. Appreciative of my kindness, his mother asked me to dinner, an invitation I gratefully accepted despite my dislike for chicken, having regarded them not as food but as pets since my childhood in Needham and adolescence in Norfolk.

P.V. Hammond not only accompanied me on my weekend trips to Indianapolis to see the time trials (while he visited his mother), but also became a member of a study group I formed whose singular purpose was to get A's on major pharmacology tests. The group originally consisted of P.V. and me and was founded on the premise that our thought processes were not much different from those of Dr. Edwards and Tom Miya *(having been exposed to the same lecture and lab content although from different perspectives)*. Therefore, why couldn't we come up with the same questions they did in constructing the exam? The outcome proved to be so successful that P.V. and I considered ourselves either psychic, or pharmacologists in a former life. Our first meeting place was a dark corner of the university library. After our fellow grad students, who fared poorly on the test, heard about our review sessions, we relocated to a larger location, a basement room of the Bio Annex, to accommodate the new participants. With respect to my research, in addition to the sheep work with Dr. Zarrow at George Neher's farm, the rodent studies were going well. In a search for chemicals that would elevate, depress, and antagonize these effects on the blood sugar levels of rats, I ran across a paper in the library which described how the injection of a rat poison, ANTU, or alpha-naphthylthiourea, produced elevated blood sugar levels in rats that was prevented by the injection of potassium iodide. I modified the procedure by substituting thiourea for ANTU and administering potassium iodate for potassium iodide in drinking water for 2 days, whereupon the thiourea was injected. The elevation in blood sugar was prevented by the ingested potassium iodate, an interesting observation that was of sufficient interest to Dr. Hiestand that he asked me to prepare a paper for publication. We were both surprised when the manuscript was accepted without any changes by a prestigious biology journal (Proc. Soc. Exper. Biol. Med., 73: 657-658, 1950).

Arthur G. Zupko, a doctoral student under Dr. Edwards, aware that I was evaluating various agents on blood sugar levels, gave me an aqueous solution of *Veratrum viride* to inject in rats. This drug was once used to treat high blood pressure, but soon discarded for agents of lesser toxicity. The problem arose from the fact that V.v. is not a single entity, but a complex of plant chemicals (alkaloids) whose actions are hard to assess pharmacologically. Zupko was studying certain aspects of the problem of predictability and was curious to know what effect the complex had on mammalian blood sugar levels. I found that freshly prepared solutions of V.v. caused a drop in blood sugar of 30.6 mg% 20 minutes after being injected into rats fasted for 12 hours. When the solutions were refrigerated for 3 days prior to injection, sugar levels rose 41.8 mg%. The fall in blood sugar with freshly prepared aqueous solutions was similar to the action noted years before with synthelin (synthetic insulin). Further study, however, revealed that synthelin produced its insulin-like action by destroying the liver! The results of my V.v. investigation were published in the Proceedings of the Indiana Academy of Science, volume 59, 1950 as an abstract. My Ph.D. research was going well, but my sugar engorgement study in rats to induce diabetes was a failure, producing obesity and little else.

At the end of May it was time for the running of the Indy 500 and a visit from Dad. The reader will note that my father came a third of the way across the United States in 1948-49 to attend the famous race with me, but did not go to my high school graduation on June 13th, 1940 *(the day, incidentally, when France fell to the Nazis)*, or to the graduation ceremony at Purdue, when I received a Master of Science degree in physiology, also on the 13th of June (1948). In both situations, mother and her mother attended high school ceremonies, and mother with sister Lillian attended the festivities at Purdue. I realized many years later why he preferred not to attend. During his teenage years, while attending high school in Boston he picked up a cigarette smoking habit, which should more accurately be termed addiction. Throughout his years at medical school, his early years at TB sanatoriums in Rutland, Vermont, and Johnson City, Tennessee, he continued smoking and picked up tubercle bacilli along the way. During a hiatus at the farm in Norfolk, he gradually recovered from pulmonary TB with the help of a carefully controlled diet, plenty of fresh air, and rest. But he continued smoking. When he began his practice in Needham after an unsuccessful start in Medford, he ran a pulmonary clinic every Thursday evening in nearby Framingham, and continued to smoke cigarettes. By the time I began smoking, after being introduced to the habit (addiction) at the Harvard Freshman smoker in 1940, he was smoking several packs daily. At that time, he could not go beyond one hour without reaching for a cigarette. When I brought home my little corncob pipe with the red H painted on its side, he laughed, thinking it was a passing fancy. It wasn't. Puffing on a pipe required too much effort, so I purchased a pack of cigarettes for 25 cents. By experimenting, I found that Camels were the strongest and Old Golds the weakest.

Thanks to Dr. Edwards' brilliant course, I learned a great deal about smoking and nicotine. Having worked with potassium cyanide for my master's degree, I was shocked when I learned that nicotine has the same fatal dose in human beings as potassium and sodium cyanide and the gas, hydrocyanic acid, *I mg/kg. This means that a person weighing 150 pounds (a non-smoker) will die when subjected to 70 mg. of either poison (1 kilogram = 2.2 pounds)!* Of the two, nicotine kills more quickly. In the old days, when gas executions were routine at California's San Quentin prison, cigarette, pipe, and cigar smokers took longer to die than non-users because their habit produced methemoglobin which acted as a partial antidote to hydrocyanic acid (cyanide gas) inhalation. On the other hand, if a chronic tobacco user were given 1 mg/kg of nicotine, it would not kill because of the degree of tolerance or resistance that would be present. Accordingly, cyanide docs not produce tolerance because it acts on primitive respiratory processes within the cell to cause death, whereas the mammalian body develops tolerance to the effects of nicotine on more complex, and more recently formed respiratory centers of the brain. Nicotine is addicting; the cyanides are not. A consistent withdrawal symptom from nicotine usage is the onset of diarrhea within 4 days of stoppage. In contrast to cyanide, which acts only at cellular levels, nicotine acts at both the psychic (the mind) and cellular (physiologic) level. Thus, it is truly addicting.

Dad and I loved the Indy 500 of 1949 for two major reasons. One was the excitement engendered by Duke Nalon's powerful yellow Novi becoming unmanageable and ending up having a spectacular love affair with the wall. It burst into flames spewing burning fuel in a narrow path 60 feet long that forced drivers either to race through it like a carnival act or try their luck maneuvering on slippery grass. The Duke jumped over the wall to safety and probably prayed he was conscious and the car had not overturned. Today's steel fencing provides a protective barrier for spectators in grandstands around the track. If the fence and grandstands had been in place there in 1949, Mr. Nalon would have been trapped. The second reason was the brilliant driving by Bill Holland who, outfoxed the year before by the sly driving of the other driver of the Lou Moore-owned race cars, Mauri Rose, captured first place and got to hold the gleaming Borg-Warner trophy with its silver plaques bearing the likenesses of former winners, including three of his nemesis.

This excitement was nothing compared to the grand, once- in- a-lifetime happening about to unfold in my home during the month of July. The stage was originally set back in 1943 when cousin Willard (my favorite cousin as stated previously) joined the Air Force and was sent overseas to Scotland where he fell in love with a lassie named Sophie Jarvis, whom he had met at a dance. After the war ended, he invited Sophie to come to the United States where they were married and settled down at 555 Main Street in Stoneham, Massachusetts. One day in July, mother informed me that Willard and his new bride were coming to Needham to introduce Sophie and her younger sister, Mary, who had just arrived a few days earlier by ship.

On the appointed day, a new maroon-colored Oldsmobile convertible arrived in the driveway. I was in the kitchen with mother when the doorbell sounded. She rushed to the front door where Dad was already greeting the guests, and inviting them into the living-room. When I heard the commotion, I walked in and headed straight to the pretty, young lady in the dark blue suit. Ignoring the others, I shook her hand saying:

"Pleased to meet you!"

I greeted Willard, kissed his mother, Aunt Carol, and his wife, Sophie. Suddenly, mother called me to the kitchen:

"How do you like her? she asked, while slicing yellow cake with white frosting.

"Nice, but her accent is funny," I replied.

Before eating the cake and enjoying tea, mother asked Mary and me to:

"Skip down to Morgan's store for a quart of pineapple sherbet."
And that's how I met my wife.

CHAPTER FOURTEEN

Dr. Livingston, I presume!

It didn't take long for me to arrange a date with Mary after our guests had left to return to Stoneham. It was Dad who suggested the occasion, a midget auto race to be held in Norwood the following Saturday. A phone call confirmed her willingness to see something unique on this side of the Atlantic. Motor racing in Britain was not miniaturized. Indeed, the stars of the motor circuit at Silverstone and lesser raceways were giants in the Formula 1 league: men named Stirling Moss, Jackie Stewart, and Jimmy Clark, who, if not large in stature, were driving race cars of far greater size and participating in far longer races than those in this country who raced in midgets.

When I picked her up at 555 Main Street in my black Plymouth convertible and returned to Needham, Dad, Mom, and an additional race fan, Les Cook, were waiting to accompany us to the oval dirt track several miles away. I introduced Mary to Mr. Cook, mentioning that he was the father of two daughters and two sons, and that he and his wife Maisie had been friends of the family for many years. At the track, the cars vied for strategic positions in the qualifying heats, skidding sideways and frequently into each other while raising clouds of dust in an effort to determine who'd compete in the final and longest race for the big money. In that event, a car lost a tire that sailed into the grandstand and landed without striking anyone. I recalled the first Jimmy Hatlo cartoon idea I had sent him in which "poor old Tremblechin" had attended 2,428 baseball games without ever catching a fly ball, yet getting hit by a tire while watching his first midget auto race. Truth is stranger than fiction or is it *art imitates life!*

On the second date, I had the privacy of just the two of us and a chance to find out how she liked this country. In the brief time since her arrival, she'd seen a Red Sox game at Fenway Park, sung Loch Lomond on the radio (WBZ), and been on several dates with acquaintances of cousins Willard and his sister, Carolyn. We drove from Stoneham to route 128, and then to Onset on Buzzards Bay at the entrance to Cape Cod. We crossed the Cape Cod canal on the

Bourne bridge and headed for Falmouth, Quisset, and the oceanographic ambience of Woods Hole. At the Spindrift restaurant overlooking the harbor, I had a delicious plate of fried clams (with stomachs) while Mary savored a swordfish steak anointed with lemon juice. Before dining, we had spent a few minutes watching fishermen unload swordfish carcasses at the pier next to Sam Cahoon's fishmarket. The dark black bodies, devoid of tails and swords, revealed light colored meat in cross-section that was selling for 55 cents a pound. After lunch, we took a ride along a shore road, with Martha's Vineyard looming across the water 5 miles away, inhaling the mixed aroma of salty breezes and the fragrance of flowers blossoming near the sandy beaches of the harbor where Coast Guard vessels were loading and unloading newly painted and barnacle-encrusted steel buoys. The day ended too soon as we headed back to Stoneham and into a setting sun.

Before Mary returned to Scotland, my parents and I invited her to ride in Dad's blue-and-white Buick roadmaster through the neighboring countryside beyond Needham. We drove down Chestnut Street where I pointed out the site of my third grade class in the old Kimball school, now occupied by the Fire and Police Department, past the original brick building of the new Glover Memorial Hospital (original cost: $50,000), and crossed the bridge over a torpid Charles River into the Wallace Nutting setting of Dover. Within minutes we drove past the brilliant white barn and sculpture studio of Mrs. Peabody, a superb equestrienne, sculptor, and breeder of magnificent Arabians and on the opposite side of the street was Mr. Parker's mansion. Mary said it reminded her of a British aristocrat's estate with its manicured lawn, trimmed shrubbery, and carefully positioned trees standing like sentinels along the circular asphalt driveway.

"The gentleman who lives there is my patient," said Dad, noticing in his rearview mirror Mary's interest in the place.

"You'd think a person who owned property like that would have everything, wouldn't you? During WWII his only son dipped the wings of his airplane when he flew over the house on his way to Newfoundland and his father, standing outside on the lawn, waved, and that was the last time he was aware of his existence! I imagine that happened many times with RAF pilots," he added.

Just beyond Mr. Parker's estate, Dad followed the road to the left past a small pine-forested hill upon which Dover's tiny brick library stood.

"Just beyond that hill is the home of Mr. Westcott, another patient of mine and a true genius who helped to perfect technicolor film. He's a dropout from MIT. And nearby is the farm estate of Senator Saltonstall who was formerly the governor of Massachusetts. How do you like this country?"

"I'm very impressed. From what I've seen, it's very beautiful and the distances between cities are misleading. The United States is huge."

We passed through Dover square and headed for the horse country of Medfield where the grazing pastures of fox-hunting clubs and their stables were demarcated by walls constructed by stone masons long dead.

"This looks more like the countryside in England," noted Mary. "The farther north you travel in Britain, the fewer the trees and the more rugged the terrain. Now I know why this part of the country is called New England."

In the center of Medfield, where the road is shaped like a "T", Dad parked the car in front of Clement's and the four of us entered an old-fashioned drugstore with a marble-sided soda fountain at one end near a front door and a row of several wooden booths facing the side door through which we had just entered. A young girl took our order and soon we were enjoying a variety of ice cream dishes.

"We always stopped here on our way to *the Cedars* or the camp at Mirror Lake," said mother, slowly chipping away at a blob of coffee ice cream now the size of a golf ball.

"What a lovely little town," commented Mary. "What do the people do for a living? There must be something else besides fox-hunting and a few stores to provide work."

"That's an astute observation," said Dad. "Across the street is a hat factory that's going out of business. When we turn right, you'll see a large building next to the railroad tracks where farm equipment and animal feed are sold. My father used to come here by horse and buggy from Norfolk to sell milk, eggs, and other produce and buy animal feed. Norfolk was nearer, but he got a better deal here. Massachusetts was dotted with small farms when I was a kid. Each farm had a few cows, chickens, pigs, and a couple of horses to pull the plows and provide transportation. Even today you can see the remnants of farm equipment—plows, harrows, and mowers—rusting outside the unpainted, weathered barns, typically without silos. Today, the farmland has been consolidated into much larger acreage with more efficient agricultural techniques that yield bigger crops. The construction and real estate businesses have reshuffled the land by merging small farms into a few goliaths and building expensive homes on what's left."

We finished our ice cream and drove on through Medfield past a pond beside which the ruins of a stone foundation suggested the presence of a small building.

"That's where a grist mill once stood," said Dad. "Many of the ponds and lakes around here provided ice for the old wooden refrigerators. This one didn't because the water quality was poor. A railroad trestle spans the water behind that island in the middle of the pond. When a train crossed over it, coal and other debris fell into the water."

Presently, we went by a large farm with two beautiful barns on the left side of the road and a lovely farmhouse on the other.

"That's the Kelly estate. She's a Texas oil heiress who prefers living here. That's Noon Hill in the distance which King Philip, the famous Indian warrior, used as a lookout to spy on the colonists and plan sneak attacks."

Suddenly, Dad stopped the car and pointed at something at the edge of a hayfield: "That's where I shot a red fox years ago with a 0.22 rifle. I had it mounted and placed on top of the gun cabinet in the waiting room. Must have been back in the twenties."

He drove slowly to a small bridge, turned off the motor, and pointed at the dark, slowly-flowing water beneath.

"That's the Stop River at the Medfield-Norfolk line, Mary. I want you to see something interesting."

We all got out of the car and followed Dad to the right side of the tiny bridge. Peering down at a ledge of stone that partially supported the bridge, he pointed and said:

"Aha! It's there. Do you see it?"

"I see a coiled snake," replied Mary.

"You're right, it's a watersnake like the one that swam over me when I was swimming here as a kid. Do you have snakes in Britain?"

"They're very rare. There's a poisonous viper found in England. Scotland is like Ireland, no snakes," answered Mary, with a boastful tone in her voice.

"This river is really a creek. It's only 6 feet wide and probably 3 or 4 feet deep. The fishing is lousy but in the fall it provides a nice income for muskrat trappers," said Dad.

"It's also a good source of leeches for biology supply houses," I added.

A mile ahead, Dad slowed the car so Mary could see the old homestead where he was born. Then, we continued on to a small restaurant (now called Boomer's Real Estate Agency) where we had lunch before visiting Mirror Lake, the Bok dairy farm and the high concrete-walled prison in Walpole, before returning home through the beautiful flower and tree-strewn countryside surrounding Norwood and Dover. Mary stayed overnight and the next day I drove her back to Stoneham where she prepared to depart from Boston on the Furness-Withy ship "Newfoundland" for Scotland within the week, expecting never to set eyes on the Mann family again. Time and an almost daily barrage of letters arriving in Scotland from Indiana would soon prove her grossly mistaken!

Back at Purdue in the fall of 1949, my time was largely occupied with the blood sugar research, which included weekly jaunts to George Neher's farm for sheep work with Dr. Zarrow, in addition to rat studies in the Bio Annex lab with Dr. Hiestand. Searching for a fill-in course to supplement my rapidly dwindling choices in the biological sciences, I enrolled in Jimmy James' course in histology, sat through a couple of his lectures and several lab sessions, and then dropped out of the class when it dawned on me that the subject matter was too familiar. At Tufts College and its medical school, I had stared bleary-eyed for hours at glass strips containing slivers of rat and human livers, pancreases, and other major organs in cross- and longitudinal sections, and those purposely sliced at weird angles to confuse the observer. I liked Jimmy James as a gentleman and enthusiastic teacher. (He was never called Dr. and rarely *professor,* for he was an associate professor with a Master's who had not pursued predoctoral studies beyond that advanced graduate degree because he was a financial victim of the great depression). For that reason, he was one of the most popular teachers in biology, for the students could relate to one who was closer to them academically than the more auspicious academics who possessed what the students irreverently referred to as something that was piled high and deep *(the formidable Ph.D.)* Mr. James was highly approachable; no appointment secretary had to be notified days in advance to see him; if he were not lecturing, he was available to answer any question within his domain, which included personal problems. Mr. James was a teacher/chaplain with a talent, demonstrated years before, for using parental-like approaches to find solutions to academic and other problems, which undoubtedly saved many a disturbed student from suicide. As one would anticipate, many members of the faculty were jealous of his extraordinary advisory talent and the associative respect it engendered among the student body.

There was another professor, who shall remain unnamed, who deserved the appellation—*character.* When his lecture duties were over on Friday afternoons, he would go home, don old clothes, and head for downtown Lafayette where he would associate with the local boys for a beer or two between visits to antique stores where he searched for unusual clocks and wristwatches. This was his hobby, and rarely did a student shopping downtown on a Friday night realize that the hobo-like character walking along the street in the company of several similarly-clothed men and perhaps heading for the local pub, was the same neatly-dressed professor he or she had been listening to a few hours earlier. I never encountered him wandering the streets of Lafayette with his colleagues. His reputation as a *character* had been related to me by Dr. Hiestand who was trying to incorporate his strange behavior in a Jimmy Hatlo cartoon (They'll Do It Every Time) without being sued by the victim. He gave up in exasperation when he realized that to have a cartoon idea accepted by Mr. Hatlo, the name of the sender not only had to be known, it also had to appear, along with the contributor's address, in the finished cartoon.

The *evening hobo* professor was not nearly as well known for aberrant behavior as one of the engineering professors who had lost a leg and wore a leather prosthesis that was ill-fitting. He had been known to sit for an hour or so in a movie theatre while ushers struggled to extricate the artificial limb wedged between the seats. Once while walking through the campus, he kicked at a dog that got in his way and the leg went flying into a tree! Of course, the incident made the *Purdue Exponent* and was soon picked up by the Associated Press to be read throughout the country. The fact that the poor soul was a distinguished professor in the engineering school made it even more humiliating.

Every school and college has its unusual characters who deviate from the so-called *norm.* In retrospect, president Conant of Harvard wished to rid the college curriculum of geography because, in his own words, "it more rightfully belonged at the high school level and not as a college subject." When I took the course during the second half of the senior year, the professor, an astute observer of human cultures and behavior, said, during the first lecture, that "I love Chinese women. Do you know they don't wear anything under those long dresses with the slits on the sides?"

I think this observation got back to the president and he was embarrassed by the ethnicity of the remark. What amazes me about this is my crystal-clear memory of it more than fifty years later! If scholarly knowledge presented in other courses at Harvard were as vividly recalled, and its reason for retaining it for so lengthy a time were understood, neurobiologists could discard their present theories about improving the brain's retentive processes.

The geography course was entertaining, especially when the professor occasionally referred to one of the world's authorities in geography who happened to be Dad's patient, Mr. Leonard Packard. He resided next to my boyhood best friend, Jimmy Davidson, on Warren Street. Once, when I was walking down Great Plain Avenue, having been to Charles Stevens' ice cream emporium (without Specky), I met Mr. Packard and graciously asked him: "How is your wife?"

"She died thirty years ago," he replied.

"I'm sorry to hear that," I answered.

That was the last time I saw the author of one of the most widely read (at the high school level) geography books in the English language.

A major requisite toward fulfilling the requirements of the degree of Doctor of Philosophy, in addition to residency, coursework, and original research, was the satisfactory completion of reading assignments in two foreign languages. Those acceptable for testing by the linguistic department of the university, overseen by professor Gunn *(known affectionately as Pop Gunn)* were French, German, and Russian. Presumably, these languages were representative of the nations most actively engaged in scientific research, whereas Italian and Spanish were the languages of countries widely known as being representative of the Fine Arts, whose master practitioners were Michelangelo, Da Vinci, Raphael, and Goya, Velazquez, El Greco, Picasso, and Miro, respectively.

Professor Gunn had established a routine procedure for the evaluation of each language. For example, a student's prowess in reading scientific French was evaluated as passable if 100 lines of a monograph were translated with 100% accuracy. A dictionary was provided and the time required for completion was one hour. German, on the other hand, involved translating only 50 lines during the same time period, also with a dictionary. I didn't know anyone who chose the Russian examination in the days before Sputnik focused attention on the pioneering Soviet effort in space science. The candidate's department chairman was responsible for choosing a foreign article that reflected the student's interest. Not only the student's major areas of research had to be known (mine was physiology/diabetes/mammalian blood sugars), but also the minors (1st-endocrinology) and (2-pharmacology). Professor Gunn's obsession with linguistic perfection became legendary after foreign students, who spoke and wrote the language of their homeland, routinely failed examinations because they made small grammatical errors or misspellings during the translations into English! I was extremely lucky, for I passed the French test the first time. It took three attempts before I successfully completed the German requirement after spending many months studying examples of the language and putting a sizeable dent in my monthly budget. The introductory course I sat through at Sever Hall in 1940-41 was comparable to that taken by Hitler in the first grade. Today, computer skills replace one of the language requirements as part of the *dumming down* of American education.

Upon completion of the sheep study in the late fall of 1949, I began preparing an abstract by condensing several months' work with

Dr. Zarrow at George Neher's farm for publication in *Federation Proceedings* which, along with hundreds of others, would be distributed to those in attendance at the annual meeting of the Federated Societies of Biology to be held in the huge Convention Hall at Atlantic City in March, 1950. At that time, I planned to be interviewed for a teaching position in physiology, having fulfilled almost all of my predoctoral requiremerrts. When the day finally arrived, I flowed into the cavernous airdrome-like hall with hundreds of other budding biologists heading for the interview booths. There I signed a card and was handed the number 142.

"When this number appears on the bulletin board, check the time and you will be interviewed by a potential employer," said the young girl with a smile.

I spent the next two hours wandering about the huge convention hall gawking at exhibits of the latest advances in biotechnology and the newest revised editions of corpulent medical books, that now served a dual purpose of inculcating knowledge in the reader and serving as defensive weapons when hurled at an attacker, Every ten minutes I made my way back through the hordes of impeccably-dressed young people and the established scientists, resembling dust-bowl farmers heading West, to the job-wanted bulletin board searching for the magic number 142 among the hundreds of other cards, pushpinned haphazardly on the cork surface like darts tossed at a target in an English pub. At last, after the 12th visit I spotted my number and rushed to the young girl seated behind a collapsible table laden with brochures, After a lengthy wait, I finally received information about the time and place of the interview. It was scheduled for early afternoon in a tiny booth furnished with two chairs, a small table, and a wooden coathanger, The interviewer's name was Dr. Livingston, I smiled when I read the name because many years before, Dad's sister, Aunt Agnes, had told me that the paternal side of the Mann family was related to the Moffetts of England who, in turn, were related to the Scottish African explorer, Dr. David Livingstone. Maybe I was about to meet a distant relative!

When one o'clock arrived, I entered the small booth and shook hands with a gray-haired, distinguished-looking gentleman.

"I'm David Mann," I announced, surprised at the strength in his grip. "I'm Dr. Livingston without the 'e'," he said. "Please sit down."

In the next 15 minutes, I learned he had an educational background that was remarkably similar to mine: B.S. and M.S. degrees from Ohio State and a doctorate from Cornell in physiology. His Ivy League experience had come at the end of his degree chain, whereas mine had begun at the beginning, and both of us had acquired some of our education from Midwestern universities. After seven years of government employment as a research pharmacologist in the Department of Agriculture and the embryonic F.D.A., he decided on a teaching career and accepted a position as assistant professor of pharmacology at the medical school of the University of Pennsylvania under department chairman, Dr. Carl Schmidt, co-discoverer of ephedrine. In 1929, following an eight year tenure at Penn, he accepted an invitation from Dr. Parkinson, dean of Temple University's medical school, to reorganize the former department of materia medica into the department of pharmacology. Almost two decades later, he headed a similar department at Temple's school of pharmacy when a new administration at the medical school initiated chairperson rotation. Sixty-one and in the twilight of an outstanding research and teaching career, he opted to head the department of pharmacology at the pharmacy school for four years. Upon reaching the mandated retirement age of 65, he was given the choice of leaving academia or continuing at another institution with a later retirement date. Because the departments of physiology and pharmacology at the pharmacy school had a vacancy at the assistant professorial level, Dr. Livingston graciously volunteered to drive 60 miles from Philadelphia to Atlantic City to interview potential candidates and also to renew acquaintances with his peers. Devoted scientist that he was, he had also submitted his abstract, *Effect of epinephrine on intestinal volume,* that was published in Federation Proceedings. To my amazement during our brief discussion, he had inquired about my blood sugar research in sheep, obviously having read my abstract before driving to the New Jersey shore! Interestingly, his epinephrine research was conducted during his tenure as a faculty member of the medical school, suggesting that he had not yet assumed his academic duties at the pharmacy school. If I accepted his invitation to come to Temple, both of us would be newcomers to the facilities and faculty of the school of pharmacy.

Our half hour interview ended with a polite handshake (still firm) and an invitation to visit Temple if I were interested in the teaching position. When I answered affirmatively, he handed me a card embossed with his name, school address, and telephone number.

"We can discuss details over the telephone when you return to Indiana. Send me a telegram to confirm your arrival date in Philadelphia, At that time, I'll arrange a meeting with the provost of the university and the dean of the pharmacy school."

On June 13th, 1950 the following *Western Union* message was sent:

PA386 DEA54S
DE. LAA195 PD=UN LAFAYETTE IND 13 115P

=DR A E LIVINGSTON= DEPT OF PHARMACOLOGY
SCH OF PHARMACY TEMPLE UNIV PHILA=

ARRIVING THURSDAY AS PLANNED SEE YOU THEN=
=DAVID MANN=

CHAPTER FIFTEEN

The Temple Owls and Joe Sprowls!

The flight from Indianapolis to Philadelphia was uneventful save for the roar of four huge engines mounted on the silvery wings and the cabin vibration emanating from propellers slicing rarefied air. At the airport, I went by taxi across the Penrose Avenue bridge, past the disarray of an oil refinery with its bulbous storage tanks and steel candles spouting flames, to Broad Street, one of the longest city streets in the United States. We circled the concrete island where Philadelphia's City Hall, an architectural hybrid of Egypt's Pharos of Alexandria and the Louvre, thrust its awesome mass skyward while serving as a pedestal for William Penn, and continued several miles north on Broad Street.

"That's Temple on the right," said the driver, pointing to several stone buildings beyond Montgomery Avenue. "On the other side of the street behind that stone wall is a Civil War cemetery. And we just passed the faculty club that used to be a funeral home. I expect they serve a lot of box lunches there," he joked. "Your pharmacy school is a mile or so from here. In the forties it was the old Packard dealership before Temple picked it up as war surplus for 110 grand."

Minutes later, he parked in front of a 4-story building with a bronze plaque placed at eye-level on the wall at the entrance on each side of two huge glass doors. The one on the left identified SCHOOL OF PHARMACY and the other, SCHOOL OF DENTISTRY. I paid the taxi driver, thanked him, took a deep breath and entered the building. Inside, the lobby glistened with a freshly polished linoleum floor that continued as brightly colored tiles up a stairway to the second floor. Opposite the stairs was a small elevator, whose operator suddenly appeared as the cage-like door slid into the wall.

"Can I help you?" asked the young black man whose name was Roy Wright.

"I'm looking for Dr. Livingston or Dean Sprowls. I have an appointment with them."

"The dean's office is over to the left. Dr. Livingston has an office on the 3rd floor."

I thanked Roy for his help and headed to the former location. Inside the office was a bank-like wooden counter behind which three young women were busily typing. One of them looked up and asked if she could help me.

"Are you a student?" she asked.

"No, I have a meeting with dean Sprowls and Dr. Livingston about a teaching position."

"Dr. Sprowls is at Broad and Montgomery at another meeting right now. Are you Dr. Mann?"

"That's Mr. Mann. I expect my doctorate next January. Is Dr. Livingston in?"

She rang his office and soon the professor arrived wearing a white lab coat.

"So nice to see you, Mr. Mann. I want you to come to my office and meet Dr. Munch. He's the professor of pharmacology in the dental school, and Dr. Larson is the professor of physiology in both schools whose office is adjacent to the laboratory."

We entered the elevator and Dr. Livingston introduced me to Roy Wright, the operator, whose face broke into a big smile.

"I ran into this gentleman a few minutes ago. He's the first person I met at Temple," I said, shaking his hand.

On the third floor, we walked down a green-walled corridor past microbiology and histology/pathology laboratories to an office door marked in black letters Pharmacology adjacent to another door designated Pharmacology/Physiology Laboratory. Opening the office door, I saw a cluttered rectangular room with two oak desks on the left side, a narrow corridor, and a long, continuous table on the right above which were several shelves loaded with jars and bottles filled with chemicals. At the end of this room was a door that opened into Dr. Livingston's bailiwick.

"The second desk belongs to Dr. Munch," said Dr. Livingston, removing his lab coat and replacing it with a suitcoat. "He's in and out like a jackrabbit. We'll have lunch at the faculty club and then meet with the provost, Dr. Gladfelter, and dean Sprowls later in the afternoon."

As we climbed into his green Studebaker in the parking lot and headed down Broad Street, I wondered if the box lunches at the faculty club contained chicken legs in a crossed position.

There was little doubt as to the previous function of the faculty club: at one end of the dining-room a stage, adorned with red-velvet curtains and barely large enough to accommodate a couple of Radio City Rockettes, suggested a viewing area and, at the left of the front of the stage, a light-colored patch revealed the former location of an electric organ. The taxi driver, probably a part-time teacher, was also correct. Box lunches were being served in white cardboard containers containing condiments, war surplus metal eating utensils neatly wrapped in paper napkins and, of course, food. Seated across from Dr. Livingston, I noticed for the first time his enormous goitre which required a partial unbuttoning of his collar.

"You've discovered my goitre," he suddenly remarked, his face reddening. "One of the unfortunate aspects of growing up in Ohio at the start of the twentieth century. I should have worn a turtle-neck, even though it makes me look like an ancient movie director."

"My landlady at Purdue has the same condition. It's amazing that the lack of a simple element like iodine in the diet can be so detrimental to health."

We talked about endocrinology for a few minutes before leaving for Conwell Hall, where I was scheduled for a half hour interview with the provost, Dr. Gladfelter, before being whisked back to the pharmacy school in Dr. Livingston's green mini-tank -like Studebaker to meet dean Joseph B. Sprowls. While he went to his car to wait, I sat in a waiting room reading a university brochure before a tall man opened a door and invited me into his office.

"I'm Dr. Gladfelter," he said in a distinguished voice with an accent I'd never heard before, "the provost. Please have a seat."

During the next 30 minutes, I learned that *Gladdy,* as he was called by the faculty at informal occasions, was a farm boy from Pennsylvania Dutch country who became the principal of a high school affiliated with Temple prior to joining the Temple University family. The present ruling hierarchy consisted of president Robert L. Johnson, a successful businessman and one of the founders of Time magazine, vice president Tomlinson, treasurer Sterling Atkinson, secretary John Roads, and provost Dr. Gladfelter. Gladdy was especially interested in my educational background and research interests and was intrigued with my early days spent on farms in New England and Canada. We even exchanged viewpoints about which cows were the best milkers! I departed from his office sensing that Temple University had an extraordinary individual in Dr. Millard Gladfelter who possessed Lincolnesque attributes with a face that called to mind the features of Harry S. Truman.

"How did you make out with Dr. Gladfelter?" asked Dr. Livingston on the return trip to the school of pharmacy.

"He's a very intelligent man who seemed to be interested in my background and qualifications," I replied.

"He always supported me at the medical school when certain factions were on the opposite side of the fence. When a stranger is brought in from outside the university community, who occupies a position that prevents nepotism, trouble will ensue. There's always someone on the faculty with a relative who thinks he or she can do your job better than you. Academia is no different in this respect from what goes on in the military, business world, or among theologians. All have a common denominator—they're people."

He turned right at Allegheny Avenue, then left, and left again as we entered the fenced-in parking lot immediately south of the school. As we got out of the car, Dr, Livingston continued:

"I didn't mean to lecture to you about human relations. Now let's see if Dr. Sprowls is in his office. I only met him a week ago. He told me he was in a class with Hubert Humphrey, who was a pharmacist before entering politics. And his cousin is married to Errol Flynn! He visited her when she was in a play at New Hope. I don't recall her name."

"I think it was Lili Damita."

We entered the lobby and the little waiting room opposite the three secretaries who were still typing behind the counter. A loud voice from the inner office boomed:

"Is that Dr. Livingston with Dr. Mann?" When his personal secretary, Jessie Smith, responded to the affirmative, he said: "Please come in, both of you."

The man sitting behind a beautiful wood desk stood up and shook hands with both of us. He was in his late thirties, probably six feet tall, with a boyish smile and the look of an outdoorsman.

"Do you have teaching experience?" he inquired, getting immediately to the heart of the matter.

"None at all," I answered. "At least not professionally. I took over an introductory German class at college when the professor failed to show up."

"What are your research interests?"

"I like toxicology, but I'm also interested in endocrinology, especially diabetes."

"We have a position open with the title: *assistant professor of pharmacology and physiology,* paying 4500 dollars for 9 months. The initial contract is for one year which, if satisfactorily executed, is renewed for three years. After four years of satisfactory academic achievement, tenure is granted and is usually accompanied by a promotion. The person selected will teach in the undergraduate laboratory of Drs. Livingston and Larson in the pharmacy school, but will only be responsible for a few lectures in the department of pharmacology. Are you flying back to Indianapolis tonight?"

"Yes. I'm taking a taxi to the airport within the hour. It's been a pleasure making your acquaintance, Dean Sprowls."

He rose and extended his hand, saying: "We'll let you know as soon as we finalize our decision." As he walked to the door with Dr. Livingston and me, I observed a limp which he acknowledged after seeing the look on my face.

"I had TB of the pelvic bone from drinking unpasteurized milk during my boyhood days in Colorado. Have you ever been to Colorado?"

"Yes, in 1938, when I went past Pike's Peak on the Santa Fe railroad. Unfortunately, it was at night. All I recall is a huge black blob on the horizon."

I said goodbye to the dean and Dr. Livingston and, two hours later, was in the air with a head swirling with thoughts about Temple University, the city of Philadelphia, and several people who were unlike any I had known before. There was Dr. Livingston, the distinguished-looking, gentleman-scientist; Dr. Millard Gladfelter of the commanding Lincolnesque presence; and Dean Joseph B. Sprowls, a cowboy look-alike, who spoke with Midwestern honesty. I looked forward to seeing them again under more professional circumstances.

The academic events leading up to my flight to Philadelphia involved taking oral and written examinations covering lecture and/or laboratory material from courses in physiology (my major), biochemistry, pharmacology, and endocrinology (my minors), and finally completing my German translation (third time). The written tests were first, taken with pen in hand, in Dr. Hiestand's office over a two-hour period. The night before I wrote down the items I anticipated, i.e., blood clotting factors, drugs affecting body temperature, and the physiology of the kidney and respiration. Sitting at a table in front of Dr. Hiestand when he gave me the examination sheets, I almost fell off the chair when I saw the questions! I had foreseen all of them correctly! Consequently, when he handed me the results of the first test, he congratulated me: "You have a perfect paper."

The pharmacology, biochemistry, and endocrinology tests had a slightly different outcome. Dr. Edwards had asked about cyanide poisoning, which I knew from my Master's research, but when he asked for an example of *"slow adrenaline,"* all I could think of was the sluggish effect ephedrine had on the blood pressure in contrast to *epinephrine (adrenaline),* which acted rapidly and was far more potent on a comparative dosage basis. The answer he was looking for was *"epinephrine in oil"*, which was a preparation given intramuscularly for a slow and sustained action. *Adrenaline* incidentally is the so-called trade or proprietary name for the generic drug *epinephrine* in the United States whereas, in the United Kingdom, *adrenaline* is the generic equivalent of *epinephrine.* In other words, when an athlete boasts after an event that he or she has experienced an *adrenaline rush,* an *epinephrine rush* is more accurately being experienced. Another way of putting this is that *adrenaline* is a commercial preparation in this country while *epinephrine* is the natural hormone being produced by the adrenal glands of athletes. In Britain, the same athlete would be producing *adrenaline* as a natural hormone, such are the vagaries of pharmacologic and endocrinologic nomenclature.

Trying to sleep before the oral examination was an impossibility. I spent the night worrying about what would be asked and kept reviewing the highlights of each course. And then I could envision the professorial line-up: Drs. Hiestand, Edwards, Andrews, and Corley, each asking a humdinger of a question which I failed to answer correctly over and over again. Then, I pictured them rising as one and leaving the room with looks of disgust.

In reality, the oral examination was a pleasure. I knew the answers and even enjoyed some of the levity thrown in to relieve tension:

"Did you know that *diabetes mellitus* can occur in plants?"

"No," I answered.

"Ever heard of sweet pea?"

Earlier, during the Memorial Day weekend, I had entertained my parents by driving them to Indiana's beautiful park, Turkey Run, halfway between Lafayette and Indianapolis, in the Plymouth convertible (with the top down). The two-lane highway, surveyed arrow straight through some of the richest farmland in the United States, held Dad's attention as he contrasted these chocolate-brown tilled fields with the boulder-strewn land of his father's Norfolk farm. At Turkey Run, we walked for an hour or so along sylvan trails and over log-formed bridges spanning shallow streams until, exhausted from the heat and exercise, we headed back to the parking lot, had a soft drink at a log cabin restaurant, and then departed for Lafayette. On the trip back, I decided to pass a slow-moving car in front of me. The coast appeared clear ahead so I got into the left lane when, to my surprise, a car came directly toward us, having been obscured by a deep dip in the road. I quickly ran off the road to the left which, fortunately, was an open space. Never again would I pass a car unless there was at least a mile of open, car-free road surface at least a mile ahead. Apparently, the flat land of Indiana takes on a gradually undulating topography as one drives south of Lafayette. Like Dad, I hadn't noticed the roller-coaster dips while admiring the countryside. I was expecting a reprimand, but instead got a: "Nice going, son!"

At the 1950 Indy 500, mother was enthralled with the tremendous spectacle surrounding the start of the race and experienced the awesome moment when 33 cars appeared around the fourth turn and began their fast run past the green flag and into turn one. At 350 miles, the dark clouds overhead released tons of rain and abruptly stopped the race with an ecstatic Johnny Parsons waving vigorously as he passed by the checkered flag. We left the grandstand with newspapers over our heads and mud on our shoes looking for an empty bus for the short ride back to Indianapolis. A black limousine passed us and inside we gaped at Clark Gable and Barbara Stanwyck sitting in the back. They had been filming MGM's "To Please a Lady," in which Gable played the daring race driver, Joie Chitwood, and Miss Stanwyck, a newspaper reporter. Despite the climatic conditions, that day is etched firmly in my mind as though it had happened last week. Mom, an inveterate film buff, always seemed to be in the right place to see a movie star. Once, we visited a friend in New Hampshire who turned out to be the widow of the late Samuel Hines, one of Hollywood's great character actors, who usually played an admiral or general, but occasionally appeared as a parson in Westerns. He had received a degree from Harvard Business School, decided to go into acting, and lost a fortune during the depression. He quickly regained his wealth as his fame grew as a character actor in the same league enjoyed by J. Carroll Naish. Many years later, mother asked if I could drive a friend, Mrs. Lehr, to a store in Jenkintown. When she stepped in the car, I asked her if she were related to Lew Lehr, the famous comedian who always appeared at the end of Fox Movietone News in a filmed short subject about chimpanzees that ended with **"Monkeys are the craziest people!"**

"You remember my husband?" she exclaimed. *Mom was attracted to film stars like silver nitrate was attracted to gelatin.*

In late June, I received the good news that I had been appointed to the pharmacy school faculty as an assistant professor of pharmacology and physiology with teaching duties to start shortly after Labor Day. Having also received word that I had successfully passed the oral and written preliminary examinations, there was a final course commitment to fulfill before heading to Philadelphia— enrollment in *invertebrate zoology* at Woods Hole, Massachusetts, spanning a six-week period beginning in mid-July and ending in late August. The credits obtained from sociobiologic exposure to this scientific mecca on Cape Cod, the home of the world-famous Woods Hole Oceanographic Institution (WHOI) and the Marine Biological Laboratory (MBL), would complete my Ph.D. course requirements and also provide a splendid opportunity to meet and listen to the international superstars and Nobel laureates discuss their notable and award-winning contributions to humanity. Scientists from all over the world flocked to the red-brick buildings adjacent to the ocean to further their research and absorb the latest knowledge in embryology *(sea urchin eggs)*, study drug activity on giant axons and retinal cells of invertebrate eyes (the *squid* has the largest eyes in the animal kingdom), investigate fish and whale migrations and, lastly, to delve into the toxicology of such weird creatures as corals, coelenterates (jelly-fish), Portuguese men-of-war, and sponges. Where the knowledge acquired at the MBL and stored in its library was restricted to biology, the neighboring Oceanographic Institution utilized physics and chemistry to probe the mysteries of the oceanic depths. Submarine detection, for example, focused a great deal of attention on sonar, in association with M.I.T. and the Department of Defense, before, during, and after WWII. The dividends accruing from this research not only resulted in the eventual curtailment of German and Japanese U-boat activity in WWII, but also the discovery, in peacetime, of the long-lost graves of the *Titanic* and *Lusitania,* an additional testament to the ancillary benefits to be derived from this endeavor.

The time had come to invite my future wife, Mary, with whom I had been corresponding almost daily by mail and several times a month by telephone, to come to America and become my bride. She had originally planned to travel later by ship, but the ominous feeling that a Korean War was about to occur, prompted me to make a frantic call suggesting that she arrange to fly from Prestwick, Scotland, to the Logan airport in Boston. Thus, in mid-July, I drove mother's blue Dodge coupe to the airport in the wee hours of the morning and waited for the plane to arrive. After a few minutes, it dawned on me that I should be waiting at the overseas terminal instead of at the regular airport facility, so I hurried over to the former, peeked inside the building, where the Customs officers were checking suitcases, and there she was…dressed neat as a pin in a lavender-colored suit. Thank God I wasn't late! She had arrived in one of the greatest propeller aircraft ever built, the Howard Hughes- designed Lockheed Constellation of TWA.

CHAPTER SIXTEEN

squid pro quo

A year prior to Mary's arrival at Logan airport in July, I was invited by Clarence and John, two friends who shared my love for the outdoors, to spend a day trout fishing. I pulled on a pair of oversized rubber boots, anticipating wading into the stream, but neglected to wear socks. The only interesting thing that occurred during the day was the appearance of a baby weasel that sniffed at my boots while I stood on a grassy knoll wondering why my casting wasn't producing results. Several days later, another friend, Bob, and I decided to try our luck at saltwater fishing on Cape Cod. We expected to spend a couple of days on the Cape, rising early to catch the striped bass when they traversed the Cape Cod canal, slapping their silvery tails in piscatorial glee as they headed north toward the icy waters off the New England coast. We intended to find a place in Falmouth catering to itinerant travelers who, unlike the traditional summer tourists, stayed only a few days, not for a month or more. Bob and I stopped the Plymouth convertible in the center of town at Falmouth, enjoyed a plate of fried clams at the Town House restaurant, and then, as we walked toward the car, noticed on a bulletin board stuck on the side of a building, that several rooms were available at 117 Pin Oak Way. We asked a passerby in which direction Pin Oak Way was located.

"It's on the way to Quisset and Woods Hole. About a half mile from here. Turn left and drive down the entire length of the tree-lined street. It's the last house on the right next to the pond."

The house was a beautiful New England Colonial with gray shingles weathered by the air whose salty essence Edward Hopper had captured on canvases in the late forties. A middle-aged, distinguished-looking woman with cigarette in hand, appeared out of a side door as we drove in the yard.

"Are you looking for a room?"

"Yes, for several days. We want to try our luck fishing."

"My husband will love you," she smiled. "Hunting and fishing are his sole hobbies. And he's terrific at both of them. Please come in."

She introduced herself as Phoebe Crocker and invited us into the living-room, where I stared at a magnificent oil painting of a Spanish galleon engaged in combat with a British frigate, before sitting down next to Bob on a leather sofa perfectly positioned beneath its hand-carved golden frame.

"That painting takes my breath away," I gasped. "It's strikingly realistic!"

"My husband Stan loves it, too. Actually, he's my second husband. My first died several years ago from a heart attack. Now I want to show you the room. It's at the rear of the house with twin beds."

She snuffed out her half-smoked cigarette butt in a large quahog shell ashtray, took a deep breath, and proceeded to climb the hall stairs where we entered a large bedroom furnished with a maple dresser, hooked rug, several rattan chairs, and two comfortable looking beds.

"The bathroom is down the hall with a tub and shower. I charge ten dollars a night."

I looked at Bob, who nodded approval, "We'll take it for two nights."

As we each handed her a ten-dollar bill, I saw a strange expression on her face.

"Is something wrong?" I asked.

"Your face seems flushed. Are you all right?"

I looked in the mirror and also saw a slight redness.

"Maybe I'm allergic to fried clams. I had some for lunch an hour ago and they were delicious. I don't eat them very often."

Bob and I walked to the car to pick up our suitcases and were about to get a closer look at the pond when a large golden retriever ran in front of us as if to impede our passage.

"Wow Spar," said a young voice. "Don't block these people. You'll frighten them."

A teenage girl suddenly appeared and grabbed the dog by its thick leather collar.

"Hi! I'm Brenda and this is Spar. She hates to see people near water. If she saw you swimming in the pond, she'd jump in and try to rescue you!"

"That's nice to know, but I prefer swimming in the ocean. My name is Dave and this is my friend, Bob. We came down here to fish for blues and stripers."

We learned that Brenda was Phoebe's younger daughter and that her step-father, Stanley Crocker, was an avid fisherman and deer hunter. We met Stan that evening when Phoebe saw us in the hallway, returning from dinner at a Falmouth Howard Johnson's, and asked us to meet him in the kitchen.

"My wife tells me you're here for some saltwater fishing," remarked a heavy-set man in overalls, with a teddy-bearish appearance, while sipping coffee. "Have you fished on the Cape before?"

"Yes," I replied. "At the Canal last summer…without much luck."

"Well, you came to the right place. I've fished here for years using everything short of dynamite. And hunting, you should come here in the fall for pheasant hunting on Penzance Point. Do you hunt?"

When I related my early days hunting ruffed grouse (partridge), quail, and pheasant on the farm in Norfolk, his interest was roused. When I told him about Dad's moose hunting experience in the bogs of Quebec, he listened intently. The *coup de grace* occurred, however, when I told him about my article in *Hunting & Fishing* magazine and the exciting events that led up to the shooting of the monarch of Pocologan. He put down the coffee cup and gasped: "I read that article! You wrote it!…How to Hunt Whitetails?"

From that moment on, the lives of my family became intertwined with that of the Crockers. Unfortunately, Bob and I didn't get a chance to go fishing. The next day, while we were walking to Sam Cahoon's for bait, I became nauseous and vomited at the railroad station at Woods Hole. We drove back to Pin Oak Way and I went to bed with a couple of aspirin while Phoebe called a physician. His diagnosis was appendicitis after I complained when he dug his fingernails into my abdomen. "Does this hurt?" "Yes it does," I groaned.

Bob drove me home the next day and Dad discovered a black calloused area on my right heel. He sliced it with a scalpel and outpoured pus. The back of my right leg was inflamed from the heel to the thigh, the penalty of infection imposed by not wearing socks with rubber boots while trout fishing weeks before. A shot of penicillin arrested the infection and subsequent antibiotic therapy cured the condition, but also precipitated a mild case of psoriasis.

In the months that followed the thwarted Cape Cod fishing trip, my parents became acquainted with the Crockers, initially to thank them personally for recognizing I had a serious health condition that required an examination by a physician and, later, to spend a few days at their home, where Mom rested from her bookkeeping duties by playing bridge with Phoebe, Brenda, and her older sister, Pat, while Dad and I went saltwater fishing with Stan. On one of these occasions, I climbed into the back of Stan's pick-up truck, amid sawdust, left-over 2 X 4's, eelskin bait and rods, reminiscent of my bouncy ride with Dick Foley in Canada on the way to my meeting with the 12-point buck, and enjoyed the Cape landscape as Stan and Dad, seated comfortably in the cab, chatted like long lost relatives. Stan parked the truck at a cove where a number of small sailboats and rowboats were tied to a wooden pier, whose pilings were no longer recognizable because of barnacle infestations. We stepped one by one into the stern of an 18-foot Nova Scotia-built dory, causing the bow to rise abruptly into the air despite the presence of a heavy Danforth anchor attached to several yards of cordage. Like a see-saw ride between a thin and fat kid, the huge Johnson Seahorse outboard motor bolted to the stern transom and three passengers (two of whom weighed over 200 pounds) unequally counterbalanced the lighter contents of the bow.

Soon we were speeding past buoys attached to large yachts in the cove to the open ocean where waves, caressed by a light wind, became more noticeable as Stan carefully maneuvered the craft to avoid broadsiding.

"Throw the lines over now!" he shouted, above the noise and vibration, pointing to a flock of seagulls directly ahead, which were squawking and skimming a few feet above the ocean's surface.

When the eelskin-baited hooks, dragging a few feet beneath the surface, reached the place over which the seagulls had hovered, the rods suddenly bent sharply. Within ten minutes, Dad and I caught several bluefish, weighing between 3 and 5 pounds. At that time the world's record for bluefish was claimed by a 25-pounder, caught by a New York lawyer who happened to be fishing off Long Island.

I had never caught a bluefish before and was impressed by the piranha-like ferocity displayed by the creature. Carelessness in removing a bluefish from a hook could easily have resulted in the loss of fingers.

It was after this brief fishing trip that Dad, learning that Stan hunted deer each fall on the Cape (with a shotgun), invited the Crockers to join the Cooks and the Manns on their annual November hunting trip to Pocologan, New Brunswick. The amazing result of this invitation, a few years later after Phoebe's death, was Stan's relocation, after remarriage, to Pocologan where he retired from construction work to concentrate on hunting and fishing.

The events of mid-July 1949 on Cape Cod were serendipitous. A year later, when my bride-to-be arrived at Logan airport, she and my parents could visit me on weekends during my six-week course in marine biology at the MBL of Woods Hole. The wedding date, August 17th, had been set to coincide with that of my parents, although under much different circumstances, for Mary and I would have a modest church wedding at the oldest Episcopal church on Cape Cod, just across the railroad bridge at Woods Hole, with a few Needham friends, and several Purdue students, who were also taking the marine biology course, in attendance. By contrast, a mystery surrounded the marriage of my parents, who never revealed the circumstances prompting the services of a justice of the peace somewhere in Vermont or New Hampshire, and their short honeymoon in Wolfeboro, N.H. (revealed by two photographs of each standing on the eastern shore of Lake Winnipesaukee). The elopement mercifully prevented the hard feelings invariably arising in those who would have been overlooked if a traditional wedding had taken place. It's rather strange that the concept of weddings originated when money wasn't yet invented. With the invention of money came opportunities to wring wampum from almost every aspect of the ceremony, which became more complex with gradually added accoutrements, thus prompting those with meagre funds to choose elopement.

Our wedding plan originally called for my esteemed uncle, the reverend Mr. Harold Deacon, the distinguished pastor of Grace Episcopal church in Lawrence, Massachusetts, who was married to mother's younger sister, Lillian, to perform the ceremony. At the last minute, however, unforeseen circumstances arose that canceled his participation. While these preparations were going on in Needham and Lawrence, I had unpacked my suitcase in the spartan quarters of a wooden bunkhouse overlooking Eel Pond and been assigned a shaker-like chair in an annex of the Biological Laboratory, a single-story, barn-like structure, furnished with aquaria bubbling with sea water in which a variety of small creatures, captured from the neighboring ocean, lived. Dozens of male and female students, all in their mid-twenties, occupied similar positions to mine next to their aquaria throughout the large room, whose bare walls were constructed of wood. This was a temporary structure, built across the street from the main brick facility, and ideally suited for summer studies but, subjected to harsh New England winters, was shut down with all pipes drained, and year-round researchers moved to warmer quarters in the MBL building.

At the initial meeting of the course, whose schedule called for orientation and survey of content, Dr. L. H. Kleinholz, professor of zoology at Reed College in Oregon, was introduced as the person in charge of the course. A tall, muscular-looking man, already bronzed by the summer sun, named the lecturers who would discuss their areas of expertise and also described the nature of the local field trips (in the ocean) we would experience. The purpose of these oceanic field trips was to observe the local fauna and their relationship to one another and to their environment, and also to procure exemplary specimens for closer examination in the laboratory. Regardless of artistic ability, a pencil sketch was required of crustaceans, mollusks, coelenterates, and other small creatures collected from each mini-expedition. During the first week of the course, students were required to select a research project that could be conducted within the remaining time frame of 5 weeks. After watching the erratic activity of a small crab, that extended its stalk-supported eyes forward whenever I touched the dorsal part of its shell with a metal probe, I decided to record those areas of the carapace which, after tactile stimulation, caused the eyestalks, folded within the shell, to extend forward. My first thought was to explore in greater detail something I had noticed at Tufts summer school while dissecting a sandshark whose body had been embalmed with formaldehyde. After examining the heart of a male, I noticed that the female shark of a fellow student had a heart of a different color. One was darker than the other, but I don't recall which sex had which. At the time, I wondered if the formaldehyde had caused a color change in the shark's heart muscle based on the presence of a hormone that was found in one sex but not the other. Obviously, a few weeks was not enough time to investigate this intriguing mystery. The crab study, on the other hand, could easily be accomplished almost overnight. The results, incidentally, revealed that there were indeed sensitivity variations to tactile stimulation of the carapace: the posterior portion of the shell did not elicit a response, whereas the probe, when gently tapped beyond the mid-point of the shell (anteriorly), caused the eyestalks to respond. By using different colored squares to illustrate sensitivity variations, I mapped the results obtained from a consolidation of several experiments on different crabs.

Our first field trip was by rowboat (5 per boat) from Woods Hole to Hadley harbor, a short row away, where a glance beneath the surface of the shallow water revealed an abundance of quahogs or hardshell clams. We picked up a bushel or so for laboratory study and a clam chowder supper planned for that evening. It was a unique lesson that the out-of-state students learned that day: real New England clam chowder, prepared with fresh quahogs, potato, and milk, causes two days of diarrhea if the diner is not a resident Cape Codder.

At the end of the first week, the world-famous authority on invertebrates, Libby Hyman, lectured on her first love, microscopic animals with cilia that lived in the ocean. The possessor of one of the largest proboscises I had ever seen in a human being, Dr. Hyman spoke in *basso profundo* tones that were considerably embellished by a cold caught while arriving during an infamous Cape Cod sea turn. Despite the acoustical aberration of the lecture, her erudition shown through like the sun on a cloudy day, and she concluded her talk before standing applause.

One afternoon, while examining biological specimens obtained from a recent field trip, I observed an air of excitement brewing in the laboratory concerning a rumor of a visitation by Albert Einstein to a Woods Hole restaurant. A woman, undoubtedly a clam chowder-susceptible tourist, had observed a gentleman with bushy, gray hair emerge from the posh Spindrift and walk toward the Oceanographic

Institution. Almost overcome by emotion at seeing in person the world's greatest scientist/amateur violinist, she had quickly regained her composure and pointed to the hastily retreating figure, whose steps quickened to avoid confrontation when he heard someone blurt a name behind him. The incident was finally resolved when Dr. Kleinholz, hitting the vertical pin of a chrome-plated hotel bell used to summon bellboys, announced:

"May I have your attention, please. I'd like to introduce a visitor who is especially well-known to those who are majoring in or have studied endocrinology—Dr. Frohlich, after whom the syndrome, *adiposogenital dystrophy,* is named. The condition, you may recall, is characterized by hypogonadism and obesity due to a pituitary tumor."

A gasp was heard throughout the laboratory as an Einstein look-alike, though of shorter height, stepped forward and acknowledged the applause that followed. The thought was on everyone's mind: *no wonder he almost broke into a trot when he heard the woman's voice!*

During the first week of August, I boarded a WHOI-owned motorboat, with a third of the class, and sailed past the beautiful Elizabeth Islands to Cuttyhunk, an island lying several miles from the multi-colored clay cliffs of Gay Head, Martha's Vineyard. The vessel safely deposited its human cargo on a sand spit, beside which a riptide flowed, and then departed for Menemsha, Martha's Vineyard, where it had an errand to perform before returning a few hours later to pick up the students. Those left standing on the narrow sandy strip of land with a couple of instructors now climbed aboard a large dory and prepared to row the boat several hundred yards across the tide rip to the main part of the island where an abandoned Coast Guard station stood in a state of dereliction, its function of assisting the Navy in WWII submarine detection and coast guarding distant history. After a third of the students were comfortably seated in the boat, an instructor pushed the dory into the onrushing water, whereupon the hull tilted sideways dumping them into the foaming water. Fortunately, they were able to grab the gunwales and crawl to safety. If the boat had overturned in deeper water, the entire group would have been swept far out to sea beyond the Vineyard and No Mans Land, where the local sportsmen caught hammerheads and tiger sharks that swam out of the Gulf Stream and into the colder, food-laden waters off New England. Decades later, three actors would depart from Menemsha's harbor on Martha's Vineyard in search of the Great White shark during the filming of JAWS! Today, as one sails out of Menemsha's tiny harbor, the land that first appears on the distant horizon is that of Cuttyhunk.

By far the most interesting creature exhibited in the 100-gallon saltwater aquarium at the front of the laboratory, where each day a different member of the phylum under study was placed for closer examination, was the Portuguese man-of-war (Physalia). Actually, this coelenterate is not a single entity like its jellyfish relatives, but a colonial hydrozoan. Like piglets suckling at a sow's underbelly, hundreds of individuals called zooids are attached to the underside of a large, spindle-shaped float, with a sail-like structure, that bestows upon this strange colonial organism its European naval sobriquet. Instead of using hydrogen or helium to inflate its mini-blimp, the zooids direct nitrogen from the atmosphere into the balloon-like structure, thereby maintaining its integrity, which has a dual function: providing a means of locomotion at the mercy of wind and current, and preventing the horrible fate of sinking into the ocean depths where a paucity of food exists.

Normally, the Physalia obtains food from the presence of its tentacles, which trail in long strands behind and below the balloon, usually with a tiny decoy fish swimming among them, like eel larvae in the Sargasso Sea, to attract larger prey. Each tentacle contains poisonous cells with coiled needles called nematocysts, which stun those creatures attracted to the fish decoy *(strangely immune to the poison and a superb toxicological project for someone to find out why),* which, in turn, shares the remains of the victims with the ravenous zooids. Swimmers in tropical seas have been paralyzed and some have even succumbed from the onslaught of nematocyst attacks when they were unfortunate enough to have ventured into Portuguese man-of-war territory. Because their arrival, like that of many other coelenterates, is dependent upon such factors as tides,(moon phases), seasonal changes, water temperature and salinity, swimmers in tropical waters should familiarize themselves with the incidence of injuries attributed to these amazing organisms of delicate beauty and ephemeral appearance. As an example of seeing seasonal fluctuations in the incidence of jellyfish infestations from my own experience, I remember one February, when our ship anchored in the harbor of St.

Pierre, Martinique, the water was infested with thousands of small, parachute-like organisms (jellyfish). During a return visit a year later in April, I was hard-pressed to find even one.

The Portuguese man-of-war I was observing in its temporary 100-gallon saltwater home was scooped up from a much larger aquarium at the nearby Woods Hole Aquarium at the entrance to Penzance Point. Its tentacles, which attain lengths of 30 feet or more, were compressed in a transparent mass in which a tiny decoy fish, destined by nature to be in this unique symbiotic relationship, still swam. As an aside, when hurricane Diane struck the New England coast in the summer of 1954, the Woods Hole Aquarium was washed out to sea, returning its occupants to their original home.

Thursday morning, August 17th, our wedding day, opened with a warm, windless sunrise where I, and other student survivors of the overturned dory, who were snatched from the jaws of death at Cuttyhunk by quick-thinking instructors, had begun the study of phylum, *Mollusca* at the MBL. The phylum, *Echinodermata,* had been previously described as a fascinating potpourri of oceanic creatures, commonly found along the shores and in the waters off Cape Cod. Starfish, sea urchins, and sea cucumbers represented the majority of its members, whose characteristic features were a spiny skin, five-sided symmetry, and a peculiar water-pressurized system, which provides locomotion by the ingenious use of hollow, muscular, tubular structures known as tube feet. If one looks at the underside or ventral portion of a starfish, a tiny structure called a *madreporite,* looking like a miniaturized porous manhole cover, serves as the entry site for sea water which, in turn, maintains a complicated tubular system that acts as a microscopic tidal swell, the ebb and flow of which moves the tube feet and the starfish to its primary food source, an oyster bed.

Each student received a sea cucumber, widely known to scientists as a *holothuroidean,* to observe what happens when a drop of ammonia (pollution?) is added to a pan containing the animal floating in sea water. Everyone knew what to expect from an earlier lecture and, as anticipated, the animal regurgitated at once, turning itself inside out as protection against further exposure to a toxin. We never learned the name of the scientist responsible for this momentous discovery, but his or her contribution should not rank at the same level within the annals of questionable human achievement with that of the first person to eat raw oysters.

Now, listening to Dr. Kleinholz introduce the diversity of phylum *mollusca*, I became reacquainted with an old friend, *Venus mercenaria* (now called *Mercenaria mercenaria)*, the molluscan component of Dr. Welsh's experiment, using the isolated heart of the hard-shelled clam or New England quahog, a technique devised by physiologists to detect the presence of minute amounts of the neurotransmitter, acetylcholine, in biological fluids. Native Americans sought the purple-colored nacre within the shell to use as hard currency (wampum), and jewelry. Its close relative, the soft-shelled Ipswich clam, provided a young man named Howard Johnson with a delectable specialty, the fried clam roll, for his first restaurant at Wollaston Beach, Mass. Anointed with tartar sauce and held in a toasted hot dog roll, the fried clam was served with a choice of 28 flavors of delicious ice cream. Those who preferred steamed soft-shell clams removed the membrane from the *'neck'* (actually the suction apparatus) like a nylon stocking and used this appendage as a convenient handle to dip the clam in melted butter or clam juice prior to ingestion. The soft-shell's nacre, unlike its hard-shell relative, lacks a purple coloration, for the inside appearance is a drab-looking gray. Thus, to Native Americans, soft-shell clams were delicious to eat, but useless as a source of wampum. On the other hand, the empty shells were often used as tweezers to remove unwanted hair.

The morning was spent dissecting the morphologically simple molluscans that had been collected during earlier shore trips and preserved in alcohol. Gastropods, which seemed as plentiful as sand grains, were encased in shells, if they lived in salt water, whereas, if land-bound as slugs, lacked comparable calcareous vestments. The infamous oyster drill, *Urosalpinx cinerea,* a gastropod that practices molluscan cannibalism by eating its bivalve relative, the oyster, once wrecked havoc on the Chesapeake Bay economy before the cause of this devastating effect was supplanted by overdredging and the influx of toxic pollutants.

The antithesis of the lowly, sluggish, relatively uncomplicated animals-snails, slugs, chitons, clams, mussels, oysters, shipworms, and oyster drills—the awesome cephalopods, on the other hand, occupy the loftiest position among mollusks and also the highest niche among the animal rogues of literary folklore. The octopus and squid often appear in the world's literature as the embodiment of terrifying evil for several reasons: the presence of 8- (octopus) and 10- (squid) snake-like tentacles with peculiar (and frightening) suction cups, a chitinous beak (like that of an eagle or vulture), huge eyes (the largest in the animal kingdom, 12-14 inches in diameter), and an ink-sac that releases clouds of dark fluid, rendering the animal invisible in its marine environment. Furthermore, obtaining fragments of partially-digested squid, regurgitated by sperm whales (along with ambergris produced by squid beaks irritating the whale's intestinal lining), and extrapolating the dimensions of the original specimen from comparable parts of smaller members of the species, have revealed squid of enormous size. Incidentally, the architecturally perfect, multi-chambered nautilus, whose body occupies the terminal and largest "room" of the shell like an overseeing landlord, lacks the amazing ink-sac gland of the octopus and squid because its pearly fortress provides adequate protection. The jet locomotion of these advanced cephalopods evolved millions of years prior to the arrival of the ancestors of *Homo sapiens,* a fact that adds further to their mystique.

The giant axon of the squid and its stellate ganglion are two structures that have provided scientists with knowledge about how nerves generate impulses and what factors are involved in modifying nerve conduction. These studies, some of which were performed at Woods Hole, have produced commercially useful local anesthetics, and other agents that temporarily abolish pain during dental procedures and spinal surgery.

I carefully dissected a squid specimen, which each student received as an alcohol-soaked blob of grayish flesh, lying in a rubber-coated pan, from 1:00 to 3:00 p.m., whereupon, shaking nervously and numb with excitement, I departed from the laboratory and walked quickly to my room at the dorm. There, I shaved *(with a razor not a clamshell),* showered, and dressed in the clothes my folks, who were staying at the Crockers, had dropped off the previous evening, when Mary and I visited the minister for the wedding rehearsal.

Our wedding at the Church of the Messiah in Woods Hole, Cape Cod's oldest Episcopal church, started promptly at 6:00 p.m. and was attended by my parents, Les and Ruth Cook, the Crockers, and my

Purdue friends, the Vernbergs (who later became renowned and award-winning biologists at Duke University). The Rector, George H. Elliott, performed the sacrament of Holy Matrimony, substituting for Uncle Harold, about whom he expressed concern for his health. It was a simple, happy ceremony that lasted an hour. Mary, with a garland of flowers in her brown hair and wearing a stunning blue gown, looked lovely as she walked down the aisle with the best man, Stan, followed by bridesmaids, Brenda and Pat. The Manns, Cooks, Vernbergs, and Phoebe Crocker sat silently in the pews, the latter pondering the day when her daughters would also tie the knot.

Following the ceremony, everyone except the minister went by car to Falmouth's Coonamessett Inn for a dinner celebration. Situated beside a small pond, where guests could casually paddle kayaks and sail catamarans while awaiting confirmation of dinner reservations, the colonial-style, white-painted buildings of the Inn complex were operated by Edna Harris, a fastidious owner/manager whose expertise placed the restaurant and motel facilities in an upper echelon of quality denoted by an entrance hall lined with annually awarded plaques extolling excellence. Large banquet rooms and smaller rooms in alcoves for private parties, such as the one we were about to occupy, led to a centrally located lounge decorated with authentic furniture and oil paintings where elderly guests, too old to paddle kayaks or sail catamarans, sat on sofas and rocking chairs and smoked cigarettes while chatting with companions. Bringing up the rear of our post-wedding procession in the 1939 Plymouth convertible, Mary and I parked and walked around the grounds for a few minutes, enjoying the late afternoon sun and admiring the beautiful flowers and trees, before being seated at a long table amid handshakes, kisses, and congratulations. Most of the wedding party ordered the house specialty, baked lobster stuffed with bread crumbs and chunks of lobster claws, and French Fries, while Mary ordered Long Island duckling. Not to break the spell of togetherness, I went along with the duckling, too.

"When you said '*I do*'", Les Cook commented, between mouthfuls of lobster, *"you stood on your toes."*

"I was trying to look taller than the minister," I replied.

Halfway through the meal, a commotion outside attracted our attention. A stable across the street had caught fire and the owner was desperately trying to rescue his horses. Before the sun set on the most memorable day of our lives, Mary and I climbed into the Plymouth, said goodbye to our guests, and headed for Chatham where we spent our three-day honeymoon. The following week I completed my Woods Hole course and returned to 863 Great Plain Avenue in Needham, where I had left my bride of three days. The time had come to plan our future in a city unknown to both of us at a university the people of New England regarded as being a *southern* school because Temple sent its outstanding football team *north* to play Boston College in the nineteen thirties.

David E. Mann, Jr.

CHAPTER SEVENTEEN

Onward to Temple!

The morning of Friday, August 25th, I took public transportation from Needham to the station at Back Bay, where I stood on a crowded platform with a small suitcase and gawked at the Boston Red Sox team, who were making small talk and occasionally looking down the track for the train coming from South Station. I recognized the legendary catcher, Birdie Tebbets, from Nashua, New Hampshire, standing a few feet away, and moved closer to him to catch the sound of his high-pitched voice, the origin of his avian nickname. He was chatting about the forthcoming weekend series with the Philadelphia Athletics to a young player I didn't recognize. During the conversation with the young man (probably Maury McDermott their newly acquired pitcher), he mentioned the A's outstanding hitters, Gus Zernial and Ferris Fain, and, in the same breath, that he was looking forward to a meal of delicious snapper soup served at the Walnut Street Bookbinder's, a delicacy unobtainable at even the finest restaurants in the Boston area.

I looked around but didn't see the *'splendid splinter,'* Ted Williams, who obviously preferred flying on Northeast Airlines for a few minutes rather than spending several hours on a dusty train (*according to Pat Marston, my high school classmate, a flight attendant with the airline and familiar with his travel idiosyncrasies).*

Hours later, when the train finally arrived at Philadelphia's impressive Thirtieth Street Station, a competitor in architectural beauty with Union Station in Washington, D.C., I went by taxi down Market Street and across the Schuylkill River to Rittenhouse Square where I signed for two nights in a lovely room overlooking the arboreal and statuary splendor of the Square bordering Walnut Street. In retrospect, my accommodations were at least comparable, and perhaps even better, than those occupied by my fellow travelers who donned baseball uniforms that evening. It wasn't really a room it was a *suite* of rooms complete with a large stone fireplace and flowers.

In the *Philadelphia Bulletin,* I read about the availability of a three-room apartment at the corner of Oxford and Levick in Lawndale for 75 dollars per month. The next morning I went by taxi up North Broad Street and Rising Sun Avenue to the 2-story stucco-sided building, on Levick Street at Oxford Circle, which was divided into four similar apartments. It was owned by the De Marios, who occupied the house next door, and was located in a nice, clean, middle-class, German/Italian neighborhood. I examined the facilities (three rooms and bath) and, after calling Mary for her approval sight unseen, signed the lease in the presence of Mrs. De Mario, whose husband, Eugie, was working as a movie projectionist a few blocks away. In lieu of returning by taxi to Rittenhouse Square, I decided to follow the route I'd soon be taking to Temple. A short walk to Rising Sun Avenue, boarding a No. 25 trolley to Broad and Olney, down the subway steps for the quick ride to Allegheny Avenue and, finally, up subway steps to the pharmacy school.

Instead of entering the pharmacy school to look for Dr. Larson or Dr. Munch, I crossed the street and took the subway to Center City, where I walked down Walnut Street to Rittenhouse Square. Packing my suitcase, I hailed a taxi for 30th Street station and bought a ticket to Route 128, a few miles outside of Needham, where mother and Mary would be waiting. The next 6 hours were spent observing the detritus that invariably accompanies human endeavor in the cities of the Northeast Corridor. Outside the cities, however, breath-taking landscapes, with an occasional river or lake, appeared as if Mother Nature were offering a topographical contrast to balance the environmental degradation wrought by humankind. The return trip to Needham reminded me of the visiting aborigine who, when asked how far it was to Boston from Philadelphia, answered: "Two dead dogs!"

Mary was delighted when I described the apartment and the surrounding neighborhood and looked forward to meeting motherly Mrs. De Mario, her husband, Eugie, and their daughter, Mimi, whom I hadn't met, but was soon to become a bride herself. Mother, practical as a beaver in a creek that was drying up, began making a list of things to ship with Mary and, within a day or two, a truck pulled up in the driveway and began loading the more practical wedding gifts (toaster, orange juice maker, and waffle iron), furniture, cooking utensils, silverware, blankets, and winter clothing. Notified of the date of arrival of our belongings in Philadelphia, we waved goodbye to my folks the following day and drove west, following the route I had taken previously on trips to Purdue. Soon we were motoring along a goldenrod-laced highway of the Connecticut countryside, approaching the hills outside Hartford and then, thirty miles later, the ocean bay area of New Haven, an hour or so away from the outskirts of New York. Crossing the George Washington bridge, a structure that Paul Bunyon could have built if he'd had an erector set as a kid, we drove slowly down a hill on the New Jersey side where junk yards appeared in profusion on either side of the road. Freight trains, consisting mainly of tanker cars, crawled slowly over tracks that bisected the junk yards in every direction. In the distance we could see fires burning excess gas atop steel towers, below which gray-colored storage tanks of gasoline clustered like huge cakes behind barbed-wire fences. The odor of petroleum derivatives produced such a stench that, even with the windows tightly closed, the air inside the car reeked like that in a filthy gas station. Several hours later, thirsty, hungry, and bone weary, we pulled into the parking space of a restaurant off Roosevelt Boulevard called *Beck's*. Because one of Dad's patients was named Floyd Beck, a successful business man of German origin, I anticipated, as Mary and I entered the building, that this was going to be a *sausage and strudel* operation. Not our favorite cuisine. However, we were both happily surprised to discover that *Beck's* was one of Philadelphia's premier seafood restaurants, ranking up there with the two Bookbinder's in quality of seafood, if not in location and as a tourist magnet.

Revitalized by delicious meals, we departed for Levick Street, noting that traffic flow on Roosevelt Boulevard was growing noticeably heavier.

Minutes later, we were driving along Rising Sun Avenue, pondering why it bore a Japanese appellation, and wondering which Roosevelt had lent his name to the boulevard *(it was Teddie, not FDR)*. Maybe not so unusual for a city with two outstanding restaurants named Bookbinder's. Somewhere within the city limits there must be a bookbinding establishment where seafood is served while books are also being bound! Turning right, opposite Martin's drugstore, we drove slowly down Levick Street past Mr. Strobel's tiny grocery store and the German Protestant Home, where two Jersey cows were munching grass in a field surrounded by a high chain-link fence.

"There it is, Mary! That stone and stucco building on the right."

I parallel-parked the car under a sycamore tree, locked the doors, and walked with Mary to the De Mario's next door to introduce my bride and get the keys. The family was eating a late dinner, the aroma of which swept over us as Mrs. De Mario slowly opened the door. Recognizing who we were, she threw her arms around Mary, giving her a big hug, then, turning to me, she smiled and said:

"You've got a nice lady here. When you get settled, you must meet my daughter Mimi and Eugie my husband. I'll get your keys. You must be tired from your long trip."

The entrance to the first floor apartment was on the opposite side of the building, which required walking up several concrete steps. After fumbling with the key, the door, stuck from days of disuse, opened with a creak. I lifted Mary (8 stone or 112 lbs.) and carried her over the threshold, depositing her in an empty living-/dining-room, separated by a plaster ceiling arch. In the latter division, a kitchen, consisting of a sink, stove with oven, refrigerator, and several closets, was only a few feet from the last two rooms: a bedroom and bathroom.

"When the furniture arrives, hopefully tomorrow," said Mary, "it will get rid of the bleakness. We're close to stores and there's a Methodist church up the street where we'll probably meet people our own age. I like it!"

The moving van arrived the following day and unloaded its meager cargo in minutes. Now, with a couch, desk, bookcase, bed, dining-room table, lamps, living-room rug, and a red-leather chair from my Harvard days, the *bleakness* of the apartment was suddenly transformed into a comfortable domicile, large enough to accommodate friends who might visit from New England. During the two years we stayed there, my Canadian friend, Dick Foley, Firechief Dick Salamone from Needham, the Hiestands from Purdue and, of course, my parents, either visited for a day or stayed overnight. Happily, in 1960, after we had purchased our first home in the suburb of Cheltenham Township, Mary's parents, John and Sophia Jarvis, stayed with us for a week. They had sailed from Glasgow to Boston on the Furness-Withy ship, Newfoundland, *(6,800 gross tons;* 186 *passengers).*

I reported for work the day after Labor Day, taking the trolley/subway route to Broad Street and Allegheny Avenue, and walked into a crowd of pharmacy, dental, and dental hygiene students, who were entering the building to register (pharmacy students) or attend class (the rest). Instead of taking Roy Wright's elevator, I walked up the stairs to the third floor and, with my key, opened the door to *Pharmacology.* Dr. Livingston, clad in a white lab coat, was talking to a distinguished-looking, bespectacled gentleman with a tiny dark mustache, who was listening intently, though shifting restlessly on one foot and then the other.

"Oh, Dr. Mann," he remarked, as I closed the door, "I'd like you to meet Dr. Munch, professor of pharmacology of the Dental School."

Dr. James Munch, in his mid-sixties and unquestionably the most famous person at the dental-pharmacy school, was known to the pharmaceutical community throughout the United States as a creator of innovative drug combinations. For example, to protect a person from overdosing on barbiturates, he had come up with the fantastic idea of incorporating a small amount of the central nervous system stimulant, pentylenetetrazol (Metrazol), in each barbiturate capsule, an amount incapable of blocking the depressant action of the barbiturate but, when consumed beyond the recommended therapeutic dosage, the stimulant activity increased significantly to counteract the barbiturate-induced depression. Another original Munch idea resulted in the use of chlorophyll in pharmaceutical preparations to remove unwanted odors from body cavities. He was also an acknowledged genius in the development and improvement of bioassays, such as the biological assay that utilized a pigeon to determine the potency (and safety) of digitalis glycosides. In this procedure, the pigeon is anesthetized with ether and a needle is inserted in its alar (wing) vein through which the drug mixture (X) is infused until the heart stops beating. Using a digitalis mixture whose drug potency is known (K) or (standardized), the potency of the unknown can be calculated. The unpredictable amounts and mixtures of digitalis glycosides, obtained directly from the leaves of *Digitalis purpurea* and dissolved in alcoholic solutions for administration to heart patients, require a bioassay because a chemical assay cannot be used.

"I'm pleased to meet you, Dr. Munch," I said, shaking his hand. "Dr. Zarrow, the endocrinologist at Purdue, asked me to say 'hello' if I ran into you. Do you remember him?"

"Yes, I do. He has a better looking mustache than mine," he replied with a smile.

The morning was spent discussing lecture and laboratory schedules with Dr. Livingston and exploring the 25' X 12' stockroom located several steps from my desk. A Dutch door opened from the stockroom into a large laboratory where pharmacology and physiology classes were held in both the pharmacy and dental schools. The walls of the stockroom were lined with green-painted steel shelving crammed with Bunsen burners, rubber and plastic tubing, clamps, tripods, wood and metal cages, Ohaus spring balances, and Harvard and Livingston kymographs. In one corner, brown glass jars of various sizes, containing light-sensitive drugs and chemicals, (one held several hundred grams of acetylsalicylic acid *(aspirin)* with its original cost-35 cents-scrawled in ink on the label), were placed a few feet from steel cylinders of nitrous oxide, oxygen, and carbon dioxide that stood like silent sentinels, tightly secured by leather straps on two-wheeled dollies, awaiting the call to duty. The worn linoleum floor was cluttered with metal drums, the labels of which incorrectly revealed their true contents.

"I have a code book that translates *sulfuric acid* into *ethyl alcohol,* etc.," explained Dr. Livingston, proud of his anti-pilfering system. "There is an *Achilles' heel* in the system, however. If Dr. James from chemistry comes into the stockroom after hours to borrow sulfuric acid, he could blow up the place! I hate to reveal the code to him because, once someone knows about it, its integrity is jeopardized."

"What's this?" I asked, pointing to a large, camouflaged wooden box on the floor.

"That's a field telephone from WWII that was left behind when the Bendix Corporation sold the building to Temple. The Packard Motor Car Company was the previous owner, which explains why there's so much colored tile in the lobby, the former showroom."

"It looks exactly like the building on Commonwealth Avenue in Boston which ex-governor Alvin T. Fuller of Massachusetts used for his Packard dealership. When I was a kid, my parents and I used to shop on Thursday evenings in Central Square. Before crossing the

Charles to Cambridge, Dad would park the car in front of the William Moreton showroom, next to the Packard building, where we'd admire the gleaming white yachts, while mother sat in the car reading a magazine. Those wooden yachts cost $10,000 for a 38-footer and $15,500 for the 50-foot Queen of the Fleet. They were built by the Matthews Company of Port Clinton, Ohio."

"Ohio," repeated Dr. Livingston. "That's my neck of the woods. I was born there."

Just then, a heavy-set, balding man wearing a white lab coat, stuck his head through the upper part of the Dutch door, and asked:

"Is this Dr. Mann?"

"Yes, Dr. Larson," replied Dr. Livingston. "You mean you haven't met Dr. Mann? An introduction is in order."

"My duties in the dental school begin in late August. I've been busy with lectures and I've already had a week of laboratories. Pleased to meet you. I understand you'll be assisting me in the physiology laboratory in the pharmacy school. When you have a few minutes, I'd like to discuss the schedule and nature of the experiments."

"I have time right now. Dr. Livingston has been familiarizing me with the inventory of the stockroom. I ran across a paper by you on blood sugars in rodents which coincidentally was my area of research interest at Purdue"

"The author of that paper is *Ed* Larson, who is now a professor of physiology at the University of Miami medical school. He used to be at Temple med as Dr. Livingston well knows. My first name is *Evert.*"

Embarrassed by the error of authorship recognition, I excused myself from Dr. Livingston's presence and silently followed Dr. Larson to his office.

"I understand you're from Massachusetts and went to Harvard. I went to Yale as an undergraduate and studied at Clark University in Worcester for my advanced degrees."

Oh God A Yale man I thought! Then replied: "Clark University is where Robert Goddard launched the first American rocket into space when I was in kindergarten in Needham. It looked like a coat hangar. Have you ever been to White City?"

"Many times. I loved to ride in those little motor boats."

A rapport based on mutual interests gradually emerged as the months passed. I was impressed with the academic skills of the people he called his close friends: *Mo* Leach, a brilliant illustrator and biology teacher who taught in the pharmacy school and Ralph Wichterman, the world's foremost authority on *paramecia* (the slipper animal) and brilliant biologist and histologist at the main university, whose daughter married Albert Saint-Gyorgyi, winner of a Nobel prize for discovering vitamin C. These three people were a sort of *rat pack,* the academic equivalent of Frank Sinatra's talented entertainment buddies on the West coast.

Having received my teaching assignments from Drs. Livingston and Larson, I began preparing for the laboratories that began the following week. Tuesday afternoons and Thursday mornings were devoted to simple pharmacologic procedures illustrated by the experiment demonstrating how individual variation in response to the fixed dose of a drug occurs. Mice, of comparable age, sex, weight, and strain, were restrained, one per container, actually in an empty mayonnaise jar that was sealed with a metal lid. A fixed volume of liquid general anesthetic (ether or chloroform) was squirted from an insulin syringe through a tiny aperture that was then sealed with tape, whereupon the jar was gently rotated to prevent stratification of vapors. At 5-minute intervals, bodily reactions were noted for 30 minutes. When the drug-induced responses were compiled at the end of the study, they were categorized in percentages as follows: no response, ataxia or loss of balance, unconsciousness, cessation of breathing, etc. Among Swiss laboratory mice, which are far more homogeneous in genetic make-up than a comparable numerical group randomly selected from the human population, the responses proved to be unpredictable. The pharmacy student became strikingly aware of a fundamental problem with clinical therapeutics—that *no one can predict the reaction of a patient to the administration of a drug.* Thus, a therapeutic dose of aspirin can relieve headache or muscle pain in one person and kill another person by anaphylactic shock within ten minutes. Furthermore, this latter example of an unpredictable acute or short-term side effect has a lethal counterpart in the chronic or long-term side effects of certain drugs. The literature records a tragic incident involving patients who had received a drug to visualize the gall-bladder for subsequent x-irradiation. *Thirty years later, each patient developed terminal cancer!*

Wednesday mornings and afternoons belonged to physiology and involved experiments on the circulatory system. At one corner of the lab, Dr. Larson had installed a tiny room, enclosed in wire screening, where students sat to record their EKGs. One student, who had experienced electroshock therapy, fainted when the electrodes were strapped on his arms. Oddly, he fainted again in a pharmacology class when each student was given an opportunity to sniff a tiny bottle of ether. This general anesthetic requires several minutes of inhalation to reach the end of the first stage of anesthesia, loss of consciousness. In his case, however, he attained it in seconds!

In addition to EKG recordings, Dr. Larson required students to take blood pressure readings on each other, along with bleeding and coagulation times, and red and white cell counts. To supplement and expand their knowledge of the heart and circulation, he presented interesting lectures, and requested reading assignments in Best and Taylor's classic text, *The Physiological Basis of Medical Practice.* As a recent graduate student of the subject, the material presented in the course seemed to be second nature. To the students, however, the subject ranked in complexity up there with advanced calculus.

My duties in both laboratories required helping students, unfamiliar with scientific terminology, to interpret the cookbook-like directions, use the equipment, answer pertinent questions, and summarize the results. Also, important aspects of each experiment were reviewed to prepare students for quizzes and hour exams looming in their immediate future. In those days, before the introduction of electronic devices that corrected a hundred multiple choice questions in a fraction of a second, papers were laboriously corrected by hand. First, a template was prepared by incising tiny holes or squares representing the correct answers with a scalpel or razor. Secondly, using a red-colored pencil, the correct answers were marked on the student's sheet, and lastly, close visual examination of the location of the red mark for each answer was accomplished. If a dark pencil or ink mark appeared in the row on either side of the red color, the answer was incorrect (marked by an X in the appropriate row). Although forewarned not to use a red- colored pencil, occasionally a student would slip up, requiring that the paper be redone by the instructor using a different colored crayon.

The worst-case scenario occurred when essay-type questions were asked on an examination, requiring countless hours to correct. I remember two answers I shall never forget. One dealt with the answer to the anatomical request: *Describe the route of blood flow in the lower human extremities:*

Answer: It flows down one leg and up the other!
To the question, What is a side effect of testosterone?
Answer: Pregnancy.

In pharmacology, I was scheduled to give only two lectures: *cyanide poisoning* and *malaria and antimalarial drugs.* Dr. Livingston aptly handled the rest of the lectures, which he invited me to attend to embellish my knowledge and also to study his podium technique. Concerning the latter, he told me the story of the minister who always pounded the lectern when he came to a weak point in his sermon, "It works in academia, too!" he said proudly. When my turn arrived in late fall, he also attended my two lectures, noticing the students' reactions to my unfamiliar New England (Boston) accent as well as taking notes on the lecture content. At the conclusion of my cyanide lecture, he commented: "When you say *sulfur,* emphasize the r. It sounded like *sulfa.*"

To prepare for criticism of my accent by students, I was ready with the following repartee: "Do you realize that two New Englanders came to Philadelphia from Boston and Springfield, Massachusetts, years ago to create two great universities? One, Benjamin Franklin, founded the University of Pennsylvania; the other, Russell Conwell, founded Temple University. Without these two great educators, God only knows what your accents would be like today!"

Thanks to the intelligence, decency, and understanding of the students, I never had to use this rebuttal.

One Tuesday morning, while preparing for the afternoon lab in pharmacology, I discovered a small wooden box hidden deep in the recesses of a stockroom shelf, where metal clamps were strewn in disarray. Without opening it, I showed it to Dr. Livingston, hoping it belonged to him and was not a left-over from the Bendix Corporation's departure.

"So that's where it ended up" he commented, admiring the casket-shaped mahogany box, like an antique dealer encountering a rare find. "While I was packing my books and mementoes at the medical school a month ago, for delivery to the pharmacy school, I received a call from Dean Sprowls to attend an emergency faculty meeting. A med student graciously offered to help pack the rest of my belongings and unpack them at the pharmacy school. Somehow this ended up in the stockroom instead of in my office. If I'd had the time, I'd have inventoried what went into the cardboard boxes. Thanks for discovering this. Would you like to see the contents?"

When I answered affirmatively, he ran his thumbnail along the groove below the cover and gradually opened the box. Inside, the interior was lined with red velvet material and contained a grooved wooden block in which tiny glass rods were embedded in hair-thin slots, all pointing in the same direction like bullets in a machine-gun belt.

"These look like glass rods," I commented. "Do they have any significance?"

"They belong in a museum," answered Dr. Livingston, somewhat smugly. "Thank God you found them."

"They were specially prepared by Mr. Graham, the glass-blower at the University of Pennsylvania medical school, for a colleague of mine, Dr. A. N. Richards. He inserted them into the kidneys of frogs to obtain glomerular filtration fluid, sealed them, and sent them to a lab where the contents were revealed. Thus, he was able to determine which substances were excreted by the kidney. You've heard of high and low threshold substances—those reabsorbed by the kidney and those that aren't?"

"Of course!"

"When he presented his results at a Federation Meeting, a young scientist asked: 'Why did you use a frog?'"

"Better to have used the frog than nothing!"

Attempts to repeat his experiment by others consistently failed because the glass capillary tubes broke during insertion. Eventually, he divulged the secret of his success. Mr. Graham, the glass-blower, had prepared tubes made of quartz not glass!

"What he accomplished sounds like the kind of basic knowledge that warrants a Nobel Prize," I commented, as Dr. Livingston gave the contents a final glance before closing the lid.

"I agree with you. But if your research, basic as it may be, deals with excretion, it is taboo to the Nobel committee."

(*I'm pleased to report, however, that the University of Pennsylvania rectified this deficiency by honoring Professor Richards with a building bearing his name. Unfortunately, Dr. Livingston passed away before his colleague and friend received this recognition*)

Between lectures and laboratories, I spent a number of hours per week filling in the gaps of my doctoral research, which required injecting rats with agents that prevented or modified the onset of diabetes mellitus after the administration of alloxan. This chemical destroys the *Isles of Langerhans* in the pancreas, which are the source of insulin. If this hormone is in short supply or absent, diabetes occurs. The discovery of alloxan advanced the study of the origin of diabetes mellitus by providing researchers with a simple method of inducing the condition. In large mammals, diabetes is produced experimentally by removing the pancreas, which also causes trauma during surgery, a lengthy post-operative recovery period, and entails the expense of anesthetics, sterilization, and the expertise of veterinarians and ancillary personnel. Complete removal of a rat's pancreas is almost impossible because it is diffuse and spread like a membrane over intestines and other structures. On the other hand, when alloxan is injected into mammals in appropriate dosage, the chemical destroys insulin-producing cells wherever they may be, producing a diabetic state within a few hours (depending upon the species). Furthermore, some scientists believe that an alloxan-like substance may be produced in diabetic-prone people, which makes the search for substances that block or prevent its Islet-destroying action a sensible approach to treating or preventing the condition.

By late fall, I had laboriously typed the thesis in multiple copies and forwarded them to Dr. Hiestand. On January 28th, 1951, after a wintry ride through western Pennsylvania, Ohio, and Indiana, Mary and I were seated in the Hall of Music at Purdue, she, a few rows behind me in the audience, and I, in the front row with dozens of other candidates, awaiting President Hovde's command to rise and receive my hood and degree. Dr. Fred Andrews placed the hood on my shoulders, handed me the black folder containing the doctorate degree, and, with a firm handshake, sent me forth into the world of scholars.

CHAPTER EIGHTEEN

Weirs can be Weird

The acquisition of a doctorate degree had a remarkable effect on my psyche and attitude toward teaching. During the months I taught at Temple before the Purdue commencement, I had felt like an impostor when someone addressed me as "Doctor." Now, I had academic verification of my entitlement to "the rights and privileges appertaining thereunto" the holder of a doctorate degree. I have yet to determine what these rights and privileges are, but they certainly sounded legitimate when delivered by the President of the University!

The trip to Philadelphia, following the gracious hospitality of our hosts, the Vernberg's, who had attended our wedding ceremony at Woods Hole, was nightmarish. In Ohio, where we had searched in vain for a decent restaurant on the trip out and settled for *Ma's Dinner Bell (ugh!),* we slid off the road into a snow-encrusted meadow and was quickly rescued by a truck driver who gallantly attached a rope to a bumper and hauled us back on the highway. My attitude toward truck drivers changed dramatically on that day from highway nuisances to heroic individuals. Without that tow, Mary and I would have frozen to death after trying to keep warm by burning my doctorate degree!

At my first Temple commencement in June (1951), held in the cavernous Convention Hall in West Philadelphia, I watched intently while Eddie Cantor, the famous comedian, movie actor, and former Ziegfeld Follies star, received an honorary degree and was asked by President Johnson (of Temple) to say a few words. He spoke for ten minutes about his early childhood days growing up in New York City, where he was raised by his grandmother. George Jessel, known as the "Toastmaster General" of the United States and a comedian in his own right, gave him a bath every Saturday night and those days when they swam together in the East River during the summer months. At the conclusion of these brief anecdotal remarks, Mr. Cantor sat down amid thunderous, standing applause! The principal speaker, Wellington Koo, the U.S. Ambassador to China, was the next degree recipient and spoke for half an hour. At the conclusion of his talk, everyone felt sorry that such a distinguished person had to follow Eddie Cantor! No one present at that commencement in 1951 could possibly recall what Dr. Koo said, yet those in attendance, when asked about Eddie Cantor's contribution to the program, conjure a mental picture of a young Georgie Jessel giving *Mr. Banjo Eyes* a bath.

In the commencements during the next 36 years, when I served as a member of the faculty procession or as a marshal (1958-1987), the following outstanding personages received honorary degrees: Margaret Bourke-White (photographic journalist), Richard M. Nixon (vice-president), Clare Boothe Luce (playwright/reporter), Cecil B. DeMille (movie director), Eric Severeid (TV journalist), Roy Wilkins (head of the NAACP), Allan Dulles (CIA head), and Hubert Humphrey (vice-president). Mr. DeMille, a giant in the movie industry, was the shortest (5 ft. 4 inches) and Mr. Humphrey was the tallest (6 ft. 2 inches) of the male recipients of honorary degrees.

Mr. DeMille's imposing presence at Temple's commencement in the seventies reminded me of the first time I'd seen the famous Hollywood director. It was in September, 1950, when Dr. Larson, and his charming wife Edna, had invited Mary and me to attend the Ringling Brothers Barnum & Bailey circus, whose spanking-new, flame-proof tents were set up in Philadelphia's Lighthouse field. When we were seated in the grandstand with the opening procession about to begin, having observed the complicated ropes and wires strung overhead to accommodate the high-wire artists, our attention was suddenly focused on a vehicle that supported a crane on which two people were seated, one behind what I immediately recognized as a technicolor movie camera.(Thanks to Mr. Westcott's discussion back in Dover about developing technicolor film, he had pointed out the external differences between the small, conventional type of movie camera, using a single roll of black and white film, and the monstrous technicolor camera, which required a protective hood and three rolls of film for recording red, yellow, and blue, the primary colors. The extremely complicated chemical processing technique, although very expensive, produced amazingly authentic images in realistic colors that could be stored for years when confined to a dry, cool environment.) Suddenly, the distinguished-looking ringmaster announced over the loudspeaker that "the gentleman sitting next to the cameraman is Mr. Cecil B. DeMille, who is going to film the opening procession for a segment of his new movie 'The Greatest Show on Earth.'" The next voice, in the stentorian tones everyone knew from his years as host of the *Lux Radio Theatre,* directed that "when the clown riding a tiny car goes by, give him a big hand. He's a hometown boy named Jimmy Stewart (actually from Indiana, Pa.). And on the parade floats you may be able to distinguish in costume Betty Hutton and Cornel Wilde, who play high-wire artists. Finally, although there will be no close-ups of the audience, I'd appreciate it if you would disregard our presence by looking at the performers and not at us. Thank you!"

Sure enough as the procession unfolded, we were able to see Miss Hutton and Mr. Wilde waving at the crowd from gold and silver-painted floats bedecked with brightly colored feathers. And that lanky guy with a face hidden behind zinc oxide and a bright red nose, who was maneuvering a weird-looking car, was our favorite movie actor, Jimmy Stewart a.k.a. Buttons the clown, *What a publicity photograph that would have made—Jimmy at 6 ft. 4 inches standing next to Cecil B. at 5 ft. 4 inches!* How did I know Cecil's height was five feet 4 inches? Well, when he walked down the corridor past me at the Temple commencement, I realized he was the same height as Mary.

Two years later, this film, one of my favorites, won an *Oscar* for Mr. De Mille as the best movie of 1952. It was a thrill to have witnessed a small part of its production, whose viewing today on videotape brings back vivid memories of a happy evening spent with Evert and Edna Larson.

Traditionally, the Thursday commencement at Temple in mid-June marks the end of the teaching year for an assistant professor who, if non-tenured and completing a first-year commitment, hurries home to see whether the letter on the kitchen table is thick or thin. *Thick,* of course, means that it contains a three-year contract awaiting a signature, while *thin* is the worst-case scenario of apologetic prose that ultimately means goodbye and good luck! Regardless of the outcome, I had accepted a summer teaching assignment at Purdue, which Dr. Hiestand had offered me following my January commencement, that was scheduled to begin the last week in June and end 8 weeks later on August 21st. I breathed a sigh of relief when I opened the fat envelope and read the following words from President Robert L. Johnson:

"Please sign the new contract as soon as possible, if you are amenable to the conditions stated therein."

Before the week ended, we had packed our summer necessities into a new four-door, green Plymouth (a gift from the folks who had sold my venerable 1939 Plymouth coupe with 22,000 miles on the odometer), and headed for West Lafayette via the identical route taken in January on icy roads when the surrounding landscape was covered in deep snow. Now, crossing the flat fields of Ohio on a hot summer afternoon, the asphalt road frequently revealed red bricks underneath where the tarry surface layers had worn away. The thought entered my mind that these bricks were probably laid in the early 1900's after the two-and-half mile Indianapolis Motor Speedway, similarly paved, proved to be such a sturdy and endurable road surface. Chalk up another significant advancement of this great sports' spectacle that has benefited the American motorist!

Stopping at a gas station in Ohio, we noticed an oil well slowly pumping in the backyard.

"You folks from Pennsylvania?" asked the grizzled attendant. "Always wanted to visit Titusville and ask the oil men there how I can increase production from a daily dribble. Ever been there?"

When I answered negatively, he changed the subject. "How'bout them Phillies?"

"We're Red Sox fans. The Phillies could use a Ted Williams."

"No argument there," he replied, scratching his chin with a greasy finger as he handed me the change. "Have a nice day."

When Mary and I arrived at the Biology Annex the next morning, Dr. Hiestand's secretary informed us that he and his wife had departed for their lakeside cabin in Wisconsin, where they expected to remain until September.

"He left you these," she said, handing me a thick packet, "and wished you good luck in your first teaching venture at Purdue." (Illegitimus non Carborundum was scrawled in India ink across the front of the packet. These Latin words were also burnt into a wooden panel screwed on the wall above his office door frame. I recognized the varnished white pine wood and lettering style as the kind sold at sportsman and crafts shows in the East where the wording usually read: Home Sweet Home.)

"And before you go," she continued, "here are the keys to your efficiency apartment. He even drew a little map showing where the place is located—behind the Ross-Aide stadium. In the summertime, things are pretty quiet, unlike the fall and winter months when it's like living next to an elementary school!"

"I know the noise, having listened to the crowds walking to the stadium on Saturday afternoons, when I was measuring blood sugar levels in rats. After the game got underway, a roar would occasionally be heard as Purdue scored a touchdown against Notre Dame or Michigan."

Mary took the map and apartment keys, while I tucked the thick packet of instructions inside a jacket pocket, thanked the secretary for her assistance, and started the car. Within minutes, we were walking up the steps of our home-away-from-home, one of many look-alike brick dwellings, that seemed to be fused into a single, gigantic hunk of masonry, Inside, there was a sparsely-furnished living-room, tiny kitchenette, bathroom, and closet. The central focus of the living-room was a large, upholstered couch, whose gaudy colors clashed with those in the reproduction of Georges Rouault's ugly painting of The *King,* which hung over the audacious piece of furniture. I had read in an art magazine that Rouault was regarded by his peers as the *least* talented member of the impressionist movement. Others were more direct, critiquing his work as *pure schlock.* Regardless of these appraisals, if this painting were the original, its value would probably have been equal to the original cost of half a dozen of the type of apartments we were now occupying! The other memorable thing about this apartment was the closet which contained a half-empty bottle of Four Roses on a shelf. Looking closely at the label, one could make out a tiny pencil mark that denoted the whiskey level at the time of the owner's departure! I could picture the former occupants, upon their return in the fall, using an hydrometer to test the specific gravity of their precious liquor! Ah, the ingenuity of the modern-day chemist!

After dinner that evening, I made myself comfortable on the couch beneath THE KING and carefully unfolded the papers from the thick packet that Dr. Hiestand had scrupulously prepared and vested with an appropriate Latin warning, when a loud knock vibrated the door. Mary, who was busily putting clothes on hangers for availability in the "Four Roses" closet, glanced at the papers spread over my lap and decided she'd open the door. A girl in her mid-twenties was standing on the worn rubber welcome mat, awaiting the invitation to enter.

"Hi, I'm your upstairs neighbor, Barbara. My husband and I watched you arrive today and he suggested I drop in to welcome you and ask if you needed anything. He's a graduate student in the engineering school. Right now he's in the library picking up some assigned books."

"I'm pleased to meet you. Come in, Barbara, but excuse my husband for not standing. He's preparing to teach a summer course in physiology and has notes spread on his lap."

Barbara entered the living-room behind Mary, who introduced her to me as an upstairs neighbor. Unable to rise without sending papers in all directions, I shook her hand and asked that she take a seat.

"You're a professor?" she asked, when comfortably seated. "You look the same age as Harry, my husband. Are you a Purdue professor?"

"No, I teach at Temple University in Philadelphia. My major professor, Dr. Hiestand, invited me to teach a summer course in physiology while he's vacationing in Wisconsin. The assignment for summer lectures by biology professors in the school of science is on a rotation basis, so he's not stuck with it every year. It's fortunate that I'm on a 9-month teaching schedule at Temple and had the opportunity to fill in for him."

"Is Temple a Jewish university?" inquired Barbara, after accepting an ice cold glass of Royal Crown cola offered by Mary. "I'm not familiar with it."

"No, the name Temple comes from the Baptist Temple in Philadelphia where its founder, Dr. Russell Conwell, preached. You may have heard about his famous "Acres of Diamonds" sermon. He repeated it all over the world thousands of times."

"Mary," I noticed a foreign accent. "Are you of Irish descent?"

"No, I was born in Scotland. Where are you and Harry from?"

"I'm from Terre Haute and Harry's from Chicago. Dave, are you Scottish, too? You don't have a Hoosier accent."

"No, I grew up in a little town outside of Boston. You probably never heard of Needham. It's next to Wellesley, the home of Wellesley College."

Barbara nodded recognition of the latter, paused a moment from her questioning and stared at Mary, who, after adding an ice cube to her guest's cola, sat down beside me on the couch. "When are you expecting?"

"The end of October when we'll be back in Philadelphia with cooler weather. I see you're also expecting. You can tolerate the heat much better than I. In Scotland, a hot day is one when the temperature is in the low seventies!"

"Our mutual condition is one of the reasons I had to see you, Mary. I have a wonderful young obstetrician named Dr. Payton. I've written down his office address and telephone number in case there's an emergency. Dave, you probably know where the hospital is in the north end of town."

"I sure do. I had to get a shot of penicillin there when I got stuck with a dirty needle in the lab a couple of years ago. It's on a little hill overlooking the town."

"Mary, there's one other thing I must tell you before I go to meet Harry. A terrible tragedy occurred here several weeks ago. One of my close friends put Toni hair-conditioner in a milk bottle and mistakenly fed it to her baby. Of course, the baby died, and she had to be hospitalized!…It's been nice meeting both of you. I look forward to introducing you to Harry and learning how Boston and Scotland got together."

"She seemed very nice," said Mary. "Did you get a chance to look at Dr. Hiestand's instructions before she arrived?"

"Yes, the lecture notes were printed in ink and couldn't be more legible if a monasterial scribe had done them. He even indicated when the mid-term and final tests are to be given on a small calendar. Imagine, a whole year's course in physiology in 8 weeks with 2 lecture hours per day. By the time I return to Temple, I'll be a veteran lecturer. And it starts tomorrow! Will Barbara be coming back with Harry tonight?"

"I don't think so. She conveyed her message and realized that we're both busy."

David E. Mann, Jr.

Dr. Hiestand's truncated *introduction to physiology* unfolded the next day before a class of 55 students, the majority of whom were female, who were satisfying the requirements of their pharmacy curricula, but with fewer credits because of the absence of laboratory and review sessions. The one-hour lectures, one in mid-morning and the other in mid-afternoon, unfolded without a hitch, although a barely perceptible groan was heard when the test dates were announced. The world's most predictable academic question was also asked: "Will the final examination include information from *all* the lectures or only the material given since the mid-term?" The universal answer (perhaps originally proposed by a cagey law professor): "The emphasis will be on the material *since* the mid-term, but you will also be responsible for the content of the earlier lectures." Which is a subtle way of saying: "You'll be responsible for EVERYTHING."

The interval between the daily lectures (three hours) permitted me to have lunch at the apartment with Mary and return in time to review the content of the afternoon lecture in Dr. Hiestand's office. Dr. Zarrow, who had become a close friend during our blood sugar studies on sheep at George Neher's farm, and was pursuing his endocrine research during the summer months, frequently dropped in to see how I was doing. One day he invited Mary and me to join him on the university golf course the following Saturday.

"I know Mary is well along gestation-wise, but the fresh air and exercise will be good for her. My wife will keep her company while she plays a few tees."

"I brought tennis racquets, but forgot my golf clubs," I replied, fumbling for an excuse to avoid undue exercise in the summer heat.

"You can rent them at the clubhouse," suggested Dr. Zarrow. "Half price for faculty."

The following hot, humid Saturday morning, three golfers, gifted with varying degrees of athletic expertise and accompanied by my agreeable pregnant wife, started around the course. Everything was going well when all of a sudden Mary let out a little cry. Within twenty minutes, she was at the Lafayette hospital undergoing an examination by Dr. Payton.

"I'm glad you got here when you did. You were about to have a miscarriage!"

That was the last time I played golf in Indiana. Dr. Zarrow, whose only form of exercise, other than walking to class, was chasing after the little white ball, succumbed to a heart attack during the next decade.

The weeks passed by uneventfully with frequent visits to Dr. Payton's office for hormonal therapy that prevented further signs of miscarriage. One incident, however, almost upset the endocrine apple cart. We both loved seafood, especially lobster fresh from the Atlantic, preferably from the frigid waters off the coast of Maine. During one excursion to the local market, we decided to purchase a couple of frozen lobster tails, which come from crustaceans native to the tropical waters off the coast of South Africa and the Caribbean and lack the meaty claws of their Maine relative. Mary prepared them that Friday evening and the next morning we awakened with a joint case of violent morning sickness. In retrospect, because of the prevalence of thunderstorms in the area during the past week, we reasoned that the power to the freezer had probably been cut off for hours causing the food to thaw and deteriorate. Ever since that experience, we have shunned seafood that emits an odor, and avoided cheeses that don't.

I had prepared a surprise for my students who were studying for the mid-term examination. To the question as to whether the test was to be essay, multiple choice, or other, I replied: "other."

When the day arrived, they were handed an examination in the form of a crossword puzzle that had taken me many hours to prepare.

"Now, you have one hour to complete this examination. All you need to know is half the information and the other half can be easily figured out if you know how to spell."

Amazingly, the students enjoyed the innovation and did well. I looked forward to comparing the grades from the mid-term test with those from the traditional multiple choice, final examination only a month away. While I was contemplating this intriguing evaluation of student adaptability to testing techniques, I received a letter from the Hiestands inviting us to spend the last week in August with them at the lake. It would have been great fun, but unfortunately the folks had requested that we spend our same date wedding anniversary (August 17th), belatedly with them, on a short trip to the Foley's residence in Canada.

Thus, when I had completed my summer teaching commitment and we had bid farewell to the Zarrows, our upstairs neighbors, Harry and Barbara, whom we spent many pleasant hours with and, of course, Dr. Payton, we took two days driving back to Philadelphia where, after a day's rest, we headed for Needham and another two-day drive to Pocologan, New Brunswick.

Leaving Needham in the early hours of the morning, driving along the 12-mile coastal route of New Hampshire while inhaling the cool, salty breezes sweeping off the ocean from the distant Isles of Shoals, crossing into Maine at Kittery, past the rock-strewn beach at York (like that of the French Riviera), required the remainder of the day to reach the U.S.-Canadian Customs at Calais.

"Because your wife is not a naturalized citizen of the United States, you will need a photograph of her to enter Canada. Otherwise, she will not be able to return to the United States," said the Customs officer. "There's a professional photographer named Tracy who lives in Calais. I'm sure his studio is closed now. But you can call him at home. I have his telephone number on his business card."

Dad called Mr. Tracy, who was enjoying dinner, and explained the situation. As a fellow Rotarian, Mr. Tracy graciously consented to meet in half an hour at his studio, which was across the street from the hotel where we had stayed a few years earlier in our deer hunting days.

Dad parked and told Mom to remain in the car when Mr. Tracy appeared. After the photograph was taken and developed in the darkroom, Dad paid Mr. Tracy and thanked him for his special courtesy. When we stepped outside, there was Mom with a distressed look, pacing up and down.

"What happened?" asked Dad.

"I locked myself out."

During the short ride to St. Stephen, New Brunswick, where we stayed overnight, the argument continued on and on: "Why did you get out of the car? I told you to stay in the car, etc., etc. You've done this before, blah, blah, blah."

The air finally cleared the next morning when Dad's Buick crunched the gravel in the oval driveway and, at the sound, the Foley's emerged from the front door, like little carved figures on a toy Bavarian weatherstation, with welcoming smiles.

That evening, after the usual gourmet dinner of clam chowder, boiled lobster, and blueberry pie, we sat on the porch overlooking the Bay of Fundy.

"Can you see the herring weirs strung out in the Bay?" asked Dick. When I answered affirmatively, he continued: "Early this spring, the fishermen used pile drivers to drive posts into the mud so they could fasten nets to them. Tomorrow, we're going to go down to the beach and row out to a weir to see how many fish are caught. It's quite a sight."

The next morning, leaving mother with Marge to recover from her recent verbal chastisement, Dad, Dick, Mary and I went by car to a nearby rocky beach where access to a weir, several hundred yards offshore, was afforded by a beached dory. As we looked out at the calm surface of the weir, glazed by the rising sun, I noticed a strange sight.

"Look, there's something sticking out of the water inside the weir!" I said excitedly, attracting everyone's attention to a triangular form breaking the silvery surface.

"That's probably a skate," explained Dick. "They occasionally follow the herring into the weir and have a banquet. Let's row out and have a look. Maybe we can scoop it up in this handnet before it eats up all the profits. There's one in the dory. I know the owner. He lives over there on the point. Makes enough money in the summer from herring to take care of his food and liquid needs, if you know what I mean, the rest of the year. He also owns the weir and will appreciate our taking care of this uninvited guest."

Mary chose to remain ashore while the three of us stepped carefully into the dory, wiping the damp seats with a burlap bag flecked with pearly fish scales, and bailing out the excess seawater with a rusty cup, left to prevent shrinkage of the wood bottom from dryness. (After use, last one out scooped up a cupfull of ocean and splashed it over the bottom.)

"Let me row," I requested. "I've never rowed a dory. I'll feel like a Gloucester fisherman on the Grand Banks."

It took a minute to synchronize the heavy oak oars so that the boat moved in the right direction in the presence of a strong tidal influence. When we reached the periphery of the weir, we gasped collectively. A huge shark glided past our 12-foot dory just beneath the surface and several inches away, separated from us by the delicate filigree of the net.

"My God!" exclaimed Dick. "I've never seen anything like this! It's at least two feet longer than us!"

"It looks like a torpedo with teeth! It's either a Greenland shark or a Great White," I postulated, finally putting into practical use my recently acquired marine biological knowledge. "But I'd bet it's the latter, because Great Whites are much larger. If we could borrow a motorboat, we could enter the weir, lasso its tail, and drag it ashore."

"No way," said Dick. "I still have a few years to live."

That evening, Dick received a telephone call from John, the owner of the dory and weir, that was overheard by 12 other people on the same line. (In the nineteen fifties, when residents of rural Canada made a local call from home, clicks could be heard as the party line sprung into action. Obviously, highly confidential information was reserved for distribution either by mail or face-to face discussion.) At the conclusion of the brief conversation, Dick announced that tomorrow the "seiners" would remove the herring from the weirs for sale to the canneries as sardines. We were invited to witness the activity from his motorboat.

"Did he mention the fate of the shark?" inquired Dad.

"Apparently, it found the exit. Which is truly amazing given the size of its brain," answered Dick. "Thank God the ocean is too cold and the tidal swell too formidable for swimming, otherwise there'd be some casualties every summer."

The next morning, Dick, Dad, and I went to the rocky beach and boarded John's 20-foot motorboat, a cedar-planked, oak-ribbed, open-deck craft with a modified seawater-cooled Buick engine that burned amber-colored gasoline.

"In Canada, gasoline used by fishermen is cheaper than the fuel sold at service stations because there's no road tax," said Dick. "The dye is added to marine gas to keep everyone honest. The motorist who tries to bilk the government by using marine gas pays a heavy fine, if caught."

John, a tall, lanky, fortyish gentleman suddenly appeared wearing rubber boots, faded blue overalls, and an old L.L. Bean shirt. He was lugging a WWII petrol container, designed by the Nazis for transporting fuel to their tanks, and a school lunchbox illustrated on its cover with a lithograph of Buster Crabbe as "Flash Gordon." Dick introduced Dad and me as we helped him over the gunwale, getting a whiff of alcohol and almost a hernia from the tank's weight. I spotted a half-filled package of cigarettes tucked in a pocket of his shirt as he crawled over the side and wondered if we might meet the shark again if he lit up at sea. After priming the fuel pump, he flipped a switch and the engine started with a peculiar throbbing noise.

"That's Bessie sucking up seawater. It takes a minute or so before she gets cool enough to put into high gear."

When the sound subsided, he put the propeller into reverse, and we backed a few yards into the bay, whereupon he spun the wheel and moved the gear forward. The boat responded by kicking up frothy waves at its stern and pointing its bow toward a number of rowboats and wooden barges loaded with fishermen, who were preparing to drop nets inside the weir and harvest the finny crop for deposition in the barges waiting outside.

"There's Max," shouted Dick, waving at a pot-bellied man who was standing amid wooden barrels crammed in the open cockpit of a Maine lobsterboat. A rusty steel pole, formerly used to yank lobster pots from the ocean floor, held a block and tackle from which a huge brass-rimmed scale hung.

"Max isn't a fisherman," explained Dick. "He buys fish scales that flake off when the netted fish are unloaded on the barges. The plastics industry uses them in everything from cosmetics to those pearly-looking fountain pens. Max, how much are you paying this year?"

"Thirty Canadian dollars per gallon," he replied. "Up two dollars from last year."

"Reminds me of a pig farmer, all of whose product is used except the squeal!" I commented, recalling a *bon mot* from an old Farmer's Almanac.

"John's *take* from the fish he sells to the cannery will be at least ten thousand dollars, which is more than twice the annual salary of the average worker in the United States. Not bad for a couple of months work. But it has its downside," said Dick, whose expression suddenly changed from serene to scorn. "Occasionally, whales swim into the bay and slam into the weirs, destroying costly nets and snapping the poles on which they're strung. No one knows whether they destroy the weirs intentionally or are unable to locate their presence on radar.

I suspect it's the latter because the whales, playful as they appear, don't have any innate desire to free trapped herring or consume them as food. They're plankton eaters."

With only two days left before returning to the States, the Foley's suggested a trip to St. John twenty-five miles away, where our family could shop for fall clothing imported from Scotland and see the famous reversing falls at the mouth of the St. John river. This remarkable phenomenon of water flowing simultaneously in two directions is caused by the exceptionally high tides of the Bay of Fundy. The city reminded Mary of Aberdeen, the *granite city* on the eastern shore of Scotland where, five years later, boarding a double-decker bus in that city with our two young children, we gasped at the photograph on the front page of a passenger's newspaper. It was the *Andrea Doria's* stern about to disappear in the ocean off Nantucket.

"At first glance, I thought it was our ship, the *Homeric,*" said Mary, "but then I realized that she has a white hull. This one is black."

CHAPTER NINETEEN

publish or perish!

My second year at Temple, with a three-year contract for 9 months of annual teaching, began in September, 1951, and provided me with a greater sense of financial stability, scholarly accomplishment, and academic preparedness than I had experienced the year before, the latter feelings gained from my January commencement and the summer school teaching experience at Purdue. The prior year, my research had been restricted to completing my laboratory studies and recording the results in my thesis. Now, it was time to get down to business and decide in which direction I wanted my future research efforts to go. Would I continue investigating agents that affect blood sugar levels of the white rat? (Dr. Livingston had graciously provided me with the necessary equipment to accomplish this when he learned that I needed additional biological information to complete my thesis.) I was also fascinated with drug reactions on the isolated clam heart, a project I had resurrected, on the spur of the moment, from a dim recollection of Dr. Welsh's experiment in invertebrate biology at Harvard, in response to a university request for a *Sigma Xi* science day exhibit.

Each spring the university chapter of the *Society of the Sigma Xi,* the national honorary scientific research society, co-sponsored with Temple a science day for hundreds of high school students of the Delaware Valley, who arrived by bus and spent several hours admiring a variety of scientific exhibits provided by science professors from the professional schools of medicine, dentistry, and pharmacy, and by those who taught chemistry, physics, biology, and anthropology at the main university. An outstanding scientist (also a *Sigma Xi* member) was invited to attend and give a one-hour lecture whereupon, the president of the university, amid pageantry, presented the speaker with an honorary degree. As secretary of the local chapter one year, my duties included finding an appropriate lecturer. I admired Admiral Hyman Rickover, father of the nuclear submarine, and believed him to be an excellent choice. I made a major mistake, however, when I wrote in my invitational letter that "because you are in the twilight of a great career, you are now able to visit Philadelphia from Washington in March, and dazzle the local youths with your brilliant accomplishments!" The return letter tersely informed me that "I am busier than ever in developing nuclear propulsion systems for a variety of ships." To emphasize his point, the admiral had listed the names and types of ships at the end of his letter! My second choice, Dr. Paul Siple, the Eagle scout who had accompanied Admiral Byrd on several expeditions to the South Pole, graciously accepted my invitation and gave an inspiring talk before six hundred fascinated students. He arrived from the airport by taxi in mid-March and was greeted by several professors who were covered in snow from a vernal equinox storm while awaiting his arrival on the steps of the pharmacy school.

As for the new direction of my research, I settled on *both* additional blood sugar studies in rats *and* observing drug actions on the isolated clam heart.

During my first year at Temple, regardless of the weather, Dr. Livingston and I walked a few hundred yards up North Broad Street, five days a week, to Fisher's seafood restaurant, where the two of us chatted over a lunch of roast beef, deviled clams, or shrimp bisque (served only on Fridays). Seated on solid oak chairs in the ambience of a Hollywood-style German castle, with an impressive interior crafted in painted plaster, we discussed various topics, ranging from youthful memories to contemporary events. Once in a while, Dr. Livingston would tell a joke, usually mildly humorous in nature, acquired from a newspaper or magazine, that was rarely off color. When the subject turned to research, he showed me a photograph taken during his graduate school days at Cornell. It was a black-and-white photograph of several students standing beside long wooden tables that were packed with early twentieth-century laboratory equipment. I glanced at the young faces and took a closer look.

"I recognize you, Dr. Livingston, standing beside a man with a familiar face. That's Ed Boring, my psychology professor at Harvard! And what a head of hair he sported in those days!"

"That's my class in physiology…my major. And I remember Ed Boring who was the best known member of the class. His doctoral thesis dealt with nerve regeneration. He'd sever superficial nerves in his arms with a razor and then observe how long it took before the sensation of a needle prick returned."

"That seems to be a prerequisite for working at Harvard," I commented. "Become totally immersed physically in your area of teaching expertise. My parasitology professor, Dr. Cleveland, was a walking laboratory of parasites, especially malarial parasites."

"No wonder you enjoy giving lectures on malaria! You admired his unstinting dedication to the educational process, which is sadly missing in today's society. I wonder what unique sacrifices Harvard's geology professors had to make when they were graduate students? Take their lectures for granite?" joked Dr. Livingston.

At one of these luncheon conversations, I asked him what he thought was his most significant contribution to pharmacology. After an illustrious academic career instructing young physicians-to-be in the skills of prescribing drugs and understanding their actions and limitations, and the publication of numerous research articles in prestigious journals, I was astonished at his reply:

"During my graduate school days at Cornell, while strolling along the edge of a deep ravine one fall afternoon, I heard a cry for help. Peering into the ravine, I saw a young man standing on a rock ledge, about a hundred feet below, waving frantically and yelling "help!" I quickly summoned the university fire department who rescued him within half an hour. The young man was named Eddy who, a few years later, became the world's foremost authority on the pharmacology of the opium alkaloids, which include morphine and codeine. That was my most significant contribution, saving his life!"

On November 1st, our first child, a 7 lb. 4 oz. healthy boy, was born on the 13th floor of Hahnemann hospital in Philadelphia after a long, difficult labor watched over by Dr. Hunter, a firm believer in the natural, breathe and bear-it, technique of childbirth. Drenched to the skin from a rainstorm that had begun the previous day and continued throughout the next day, I walked into the room to congratulate Mary and see our firstborn when a nurse entered, gave me the evil eye, and exclaimed: "What's he doing here?"

In today's *la-de-da* society, a new husband is treated as a decent human being and often assumes a passive role in the birthing ritual. In the foreseeable future, it is likely that he'll also be allowed to take up position at an appropriate location in the delivery room to catch his offspring in an oversized baseball glove as it emerges from the birth canal! The memory of that nurse's obnoxious behavior, occurring at what should have been one of the happiest events in my life, infuriated me at the time. But after some introspection, I resolved that her behavior was probably based on some primeval reflex harkening back to the days when her ancestors were arboreal. If there were another reason behind it, I didn't want to know about it!

A happier day arrived when we brought David home and were greeted by the DeMarios, our Methodist minister and his wife, Frank and Rena Davis, and our neighbors, the Hackers, a childless couple who reacted like close relatives and happily pitched in to help Mary while I was at the university.

Dad, examining a patient in his office when Mom received Mary's telephone call, emitted a restrained, but exuberant "hooray." Within a week, my parents arrived at our apartment to meet their first grandchild. Mary's parents in Scotland received the good news by telegram, a less expensive means of notification than an overseas telephone call. (Five years later, they would meet David, our second child, Caryn, and me when we sailed from Quebec to Southampton, journeyed from London to Edinburgh on Britain's crack train, the *Flying Scotsman,* and crossed the Firth of Forth bridge to Ballingry.)

In August of 1952, I participated in the first appearance of color television at a pharmaceutical meeting, the American Pharmaceutical Association's Centennial Convention, held in Philadelphia. Sponsored by Smith, Kline and French Laboratories and presented by the A.Ph.A, in cooperation with the Temple University School of Pharmacy and America's oldest pharmacy school, the Philadelphia College of Pharmacy and Science, the program was entitled *Pharmacy in color television,* and consisted of experiments and demonstrations in pharmacology and pharmaceutics presented live on television in 35- minute segments over a three-day period.

Monday morning, August 18th, Dr. Livingston and I demonstrated the modern pharmacology research team in action by testing agents that block the actions of epinephrine-like drugs before an audience of several hundred pharmacists, pharmaceutical scientists, and educators, who stared at colored images on television screens set up in the lobbies of participating hotels throughout the city. Tuesday morning, in a segment entitled *Evaluation of Potency and Safety,* I joined Dr.

James C. Munch, professor of pharmacology, on leave from Temple as medical director of Strong-Cobb and Co., of Cleveland, to demonstrate acute and chronic toxicity studies. Finally, on Thursday afternoon, Dr. Fritz O. Laquer (M.D.), associate professor of biochemistry at Temple's school of pharmacy and I demonstrated vitamin deficiency states with the help of colorful (and colorless) charts. Dr. Laquer had sneaked out of Nazi Germany in the thirties and arrived by freighter in South America where he was reunited with his wife and children who had separated to avoid suspicion and expedite movement in tumultuous times. Settling in Philadelphia and joining the faculty, Dr. Laquer brought Old World dignity and charm to the school and a high level of teaching and research skills to the disciplines of organic chemistry, pharmacognosy, and biochemistry. His greatest accomplishment was the isolation of digoxin, a cardiac glycoside, used to treat congestive heart failure. A distant relative, E. Laqueur, had been the first to isolate the male hormone, testosterone, in milligram amounts from several tons of bull testicles!

Three other participants in the historic program were Dr. Herbert M. Cobe, professor of bacteriology (Temple pharmacy), Dr. Alfred N. Martin, assistant professor of pharmacy (Temple), and G. Victor Rossi, M.Sc., instructor in pharmacology (PCPS). Dr. Cobe, a graduate of Stanford University and heir to the Lewandos dry cleaning fortune in Massachusetts, was a superb lecturer and researcher. When he appeared on the stage of the school's auditorium each May with the rest of the faculty during pre-Memorial Day ceremonies, he was invariably introduced as an ambulance driver in WWI. Then, the dean, in the same breath, would introduce Dr. Laquer and add:

"And Dr. Laquer was in the same war, but on the opposite side!"

Dr. Martin was a contemporary Purdue classmate, whom I never met there, although in biochemistry class he sat directly behind me. "You were next to a blond in the front row of Dr. Corley's class and never turned around," he remarked when we met at a faculty meeting during the first week of school. Al and his wife, Mary, became close friends as a successful career in pharmaceutical research and education blossomed, eventually earning him the sobriquet *father of physical pharmacy* after the publication of his landmark text that transformed the *art of pharmacy* into a science. He quickly advanced to a full professorship and became dean of Temple's pharmacy school before joining the faculty of the prestigious University of Texas at Austin.

G. Victor Rossi received his doctorate degree in pharmacology at Purdue and became a vice president of the Philadelphia College of Pharmacy and Science (America's first pharmacy school), after directing its pharmacology department for several decades. Among the many innovative ideas he introduced to improve undergraduate education was an annual lecture series given by world-renowned pharmaceutical scientists, who discussed circumstances leading to therapeutic breakthroughs. Many founders of today's successful pharmaceutical houses, i.e., Eli Lilly, Rorer, and McNeil, had attended the school as adolescents.

Co-sponsors of the television program were the deans of the two local schools, Temple, represented by Joseph Sprowls and PCP, by Linwood Tice. Despite an obvious academic rivalry between the two institutions, these educators were close friends who respected each other and worked together to advance their profession. Dean Sprowls was the co-author of a widely used pharmacy text and recognized by his peers as an outstanding teacher and researcher. Long plagued with illness, he succumbed to cancer in his sixties. Dean Tice, on the other hand, lived to be 87, authored over 150 papers, served on editorial boards of the prestigious *United States Dispensatory* and *Remington's Pharmaceutical Sciences* and, as a culmination of this, received the highest award in American Pharmacy, the Remington Medal. He was also my friend.

After lunch at the downtown Bookbinder's a few years ago, Mary and I ran into Linwood at the cashier's counter.

"Hello, Dave," said the man who looked like an older version of Dick Clark, the ageless host of American Bandstand. "I didn't know you could afford to eat here!" He also had a great sense of humor.

In early 1953, I received an envelope from Dr. Zarrow containing a dozen reprints of our joint publication on blood sugar levels entitled: *Effect of Insulin and Epinephrine on the Eosinophil and Blood Glucose Levels, in Sheep; Lack of Diurnal Rhythm from* The American Journal of Physiology, 171:636-640 (1952) by M. X. Zarrow, M.E. Denison, B. Rosenberg, D.E. Mann, Jr., and G. M. Neher. My portfolio of publications had now reached two with one abstract. And I had just completed another blood sugar study in rats that would be presented at a pharmacology meeting at Yale University in New Haven. I was on my way to not perishing. A grant application, which I had requested from the Tobacco Industry Research Committee, had arrived by mail and a young graduate student in pharmacology, Joseph Shanfeld, and I were in the throes of planning a study of alleged cancer-producing effects of cigarette smoking.

During my second year at Temple, I often accompanied Dr. Livingston when he visited the medical school across the street to borrow equipment unavailable in our stockroom. In a basement room that reeked of machine oil, I was introduced to Charlie Thor, the master technician and inventive genius, who had assembled two of the shiny, stainless steel and aluminum Livingston electric kymographs that had replaced the commonly used, smaller, wind-up Harvard types. Under the supervision of Drs. Henry Wycis and the Spiegel-Adolphs (husband and wife), neurosurgeon and neurologists respectively, Charlie had constructed the first stereotaxic apparatus. This device, a metal frame supporting a slender scalpel, was used in delicate brain operations in a manner comparable to an offshore platform that sends, with remarkable accuracy, a pipe into a sea bed where oil deposits have been discovered by seismographic soundings. Instead of seismographic recordings, however, x-rays of the patient's head, taken from different angles, provided the precise locus. The device was placed on the patient's shoulders and the probe was positioned over the site of brain pathology by turning tiny, serrated wheels. Once correctly positioned, another wheel was turned and the scalpel/probe entered the brain and destroyed the offending area. Mr. Thor informed me that the original wooden model was now on view at the Smithsonian Institution in Washington, never having been used on politicians there to the best of his knowledge. I asked him if he were related to the washing-machine company by the same name, recalling the gray monster in a room off our kitchen that vibrated the whole house.

"No, but I was taken to the cleaners during a visit to Las Vegas!"

On one visit we stopped at the physiology laboratory at the medical school when Dr. Livingston spotted the department head, Dr.

Morton J. Oppenheimer, pouring mercury into glass containers in preparation for gas analysis experiments.

"Dr. Oppenheimer, I'd like you to meet Dr. Mann, a physiologist from Purdue who went astray and became a pharmacologist! Actually, he's with both Dr. Larson and me."

A slightly heavy-set man, who resembled the Hit Parade singer, Snooky Lanson, wiped his right hand with a wool cloth and grasped mine.

"It's a pleasure to meet a fellow physiologist, even though he's a hybrid. You and Dr. Livingston might be interested in attending an international meeting of physiologists to be held at Convention Hall this December. Coincidentally, I'll be chairing the affair."

At Dr. Oppenheimer's suggestion, I joined the Physiological Society of Philadelphia for local medical and pharmaceutical scientists, which met every 4-6 weeks at one of the city's five medical schools. The agenda began with a 30- to 60-minute slide presentation of a member's research, followed by a question-answer period, usually initiated by an elderly, renowned medical scientist, whose contributions to physiology were worthy of a Nobel prize, who critiqued the validity of the work and frequently uncovered the Achilles' heal in its execution, thus rendering it valueless, to the embarrassment of the red-faced speaker. Wine and hors d'oeuvres were then served prior to a gourmet meal to allay ruffled feathers. Those new to the *modus operandi* of the Society were well advised to rehearse giving their paper with at least one dry run before their peers and mentors to minimize the chances of being labeled *scientifically incompetent.*

Despite this important negative aspect, the social advantages of Society membership could not be overlooked. In addition to conferring with renowned scientists from the area, a fine opportunity existed to meet younger members of the various faculties to discuss the pros and cons of their research, or perhaps, more realistically, to foster relationships with the opposite sex for those who were either unmarried or about to leave their mates.

At the medical school of the University of Pennsylvania (historically, the oldest and academically one of the finest in the United States), Dr. Livingston introduced me to Mr. Graham, the glass-blowing artisan, who demonstrated how to transform ordinary glass tubing into cannulas for insertion into blood vessels (the main reason why we were in his laboratory because students routinely broke the tips). This is the same genius who created the quartz capillary tubes for Dr. A. N. Richards' pioneer kidney experiments. In a sense, Mr. Graham was the counterpart of Temple's skilled Charlie Thor, although the former worked his magic in the more restricted media of glass and quartz rather than in metal. Furthermore, to my knowledge, Mr. Graham's magnificent creations of laboratory glassware were never displayed at the Smithsonian in contrast to Mr. Thor's wooden stereotaxic apparatus, which occupied a place of honor in the medical section alongside such grotesque antiques as the blood-letting equipment used to treat the ague of George Washington and Thomas Jefferson (the first president to die from chronic diarrhea), and the crude wooden cabinet that housed a prototype of the modern device that records electrocardiograms.

Joining the Physiology Society of Philadelphia had the additional benefit of permitting Dr. Livingston and me to attend Dr. Oppenheimer's chaired international meeting at reduced rates where, in the lobby of Convention Hall, we observed two Nobel laureates (in medicine), Drs. Otto Loewi and Sir Henry Dale, from both sides of the Atlantic, chatting with the great Scottish pharmacologist, Professor Gaddum. About what? Perhaps the relative merits of Philadelphia scrapple and Scottish haggis!

In October, 1952, Mary, son David, and I moved from our Philadelphia apartment to a four-bedroom home, the Lanfair, in nearby Cheltenham township, the smallest of three models that included the intermediate-size Plymouth, for families of four, and the large Standish, ideally suited for families of six or more. Across the street from the homesite was a large vacant lot where, according to the builder, colonial-style stores were to be built to provide services and products in addition to those offered by a delicatessen, pharmacy, barbershop, bank, and franchised grocery store on an adjacent street.

In the spring of 1952, we had scoured the surrounding countryside for land where a home within our financial capability could be built. In Cheltenham, several miles from our apartment, we stopped one day to admire single brick homes in various stages of construction rising from the dust of a freshly bulldozed hill. At the bottom of the hill, a sign in front of a completed Lanfair home invited interested persons to inspect its innovative features. After a brief tour, we pocketed a brochure and informed the salesman we were definitely interested. After living in a tiny apartment for two years, the Lanfair seemed like a palace—compact, well-built, and glistening with new plumbing and lighting fixtures. It had only one drawback: a single garage located beneath one of the bedrooms posed a deadly hazard of carbon monoxide poisoning if an automobile engine were inadvertently left running. Also, the garage door lacked insulation, which meant that cold air would refrigerate the bedroom floor above during the winter months, causing gas heating bills to rise dramatically. To eliminate these disadvantages, most Lanfair owners converted their garage spaces into recreation rooms or libraries within a year or so after purchase.

We spent the summer months watching our house rise from the dusty soil literally brick by brick. After settlement was made in October, my parents arrived the following weekend to inspect the property, congratulate us for purchasing a home that would cost a few thousand dollars more in Massachusetts, and help with scraping paint off the windows. In the following months, friends who had never visited the Philadelphia area came for an overnight stay at 49 Dewey Road: Dick Foley, Helen Linane (head nurse at Needham's Glover Hospital), Firechief Dick and Mary Salamone, to name a few.

True to his word, the builder constructed a series of single-story stores whose plain stucco rearwalls were hidden by a high wooden fence. Tall lighting towers were installed to prevent parking lot accidents and discourage crime. A short distance from the stores was the small Melrose Carmel Presbyterian church, whose membership was mushrooming with the influx of young Protestants attracted to the fine, reasonably priced homes and a superb public school system.

A faculty/wives' get-acquainted party, held at the Drake hotel in the fall, was the social affair that brought John and Elizabeth Lynch into our lives. John, an assistant professor of pharmacy, had graduated from Temple with honors and married Elizabeth Helm, also an honors graduate from the Philadelphia College of Pharmacy and Science (PCP), whose father owned a pharmacy in North Philadelphia. When Mr. Helm passed away, the Lynch's sold the business and moved to a brick home in Cheltenham township, approximately nine miles from the school of pharmacy, and a two-minute walk from the sylvan setting of the Presbyterian church where John served as an elder.

I first saw John Lynch while descending the hotel stairway from the second floor looking for Mary, who had accompanied Mary Martin to the ladies' room, when I noticed a man who resembled my favorite character actor, Donald Meek, carrying a shapeless pile of mink stoles to the coatroom clerk. Later, seated next to John and Elizabeth at the dinner, a multi-decade friendship began while the after-dinner speaker, an alumnus of the school of pharmacy, interspersed his serious talk with some of the corniest jokes I had ever heard:

"I fell into my coal bin the other day, but I didn't get hurt. It was soft coal!"

"When I started out as a pharmacist I had only half a shirt on my back. The other half was in front!"

(Why do I still recall this trivia?)

Shortly after this incident, Mary and I invited dean Joseph Sprowls and his wife, Rosalie, to dine with John and Elizabeth and to show off our modest dwelling. After the meal, Elizabeth, sitting beside Rosalie on a couch in the living-room, innocently commented: "Wasn't it nice of that pharmaceutical manufacturer to give a bottle of perfume to each faculty member?"

Rosalie gave Elizabeth a glassy stare for a couple of seconds and then, turning to her husband, asked: "What bottle of perfume? Joe, did you get a bottle of perfume?"

With a sheepish look, the dean meekly responded: "I gave it to Jessie Smith (his secretary)."

This *faux pas* quickly terminated the conviviality of the evening. One can only conjecture what the homeward-bound conversation in each car was about!

"Whew," gasped Mary, as our guests departed. "Thanks for giving me the bottle of perfume! It was very thoughtful of you!"

CHAPTER TWENTY

gimme that oldtime religion!

Mary and I became members of the Melrose Carmel Presbyterian church after being impressed with the quality of the sermons given by its young, dynamic minister, George Munzing. We joined the church a few weeks after moving into our new home and were promptly informed that, because a new addition was needed to accommodate the rapidly growing attendance in Sunday School, an assessment of at least ten dollars per month was required in addition to our original pledge. Older astute church-goers, who contemplate joining another church always check two things that are financially oriented: a) the pastor's salary (has he or she had a raise recently?) and b) is the mortgage paid in full with no further need for expansion? Our two-year association with the Methodist church of Reverend Frank Davis had fulfilled these two requirements and, in addition, provided meaningful religious services. Mary and I attended many of the Sunday morning sermons, during one of which our infant son, David, was christened without an unanticipated incident.

Reflecting on my own religious upbringing, I remember that mother and I faithfully attended the service at the Evangelical Congregational Church each Sunday, while Dad visited his patients at the Glover Memorial hospital, or made housecalls. During my elementary school days, I had to attend Sunday school presided over by an elderly lady named Mrs. Lord, who passed around a glass globe reinforced with copper rods that kids, sitting in tiny chairs at wooden desks, happily dropped pennies into, under the false assumption that Mrs. Lord had the same relationship to her husband that Mrs. Santa had to hers. The penny fund provided biblical material and crayons that helped the instructor, a diabetic businessman with a ketone breath, who had been shanghaied from the choir for a teaching stint, describe events such as Noah and the flood for 45 minutes, whereupon the remaining 15 minutes were spent coloring the passengers as they boarded the ark. Having been on a cruise before attending my first Sunday school class, I wondered why the ark had neither life preservers nor lifeboats! Youngsters are quite inquisitive and perceptive as Mary learned one Sunday during her class at Melrose Carmel when a youngster innocently asked: "Where did God stand when he created the Universe?"

Dr. Harry Woods Kimball was the distinguished minister of the Congregational Church in Needham and the founder of a national youth organization called *The Comrades of the Way.* I joined the *Comrades* in my teens and quickly discovered that its primary purpose was to promote Christian fellowship among youths in the surrounding Boston area. Once a month our group boarded a bus and spent several hours at another Congregational church. Someone would read from the Bible, followed by a brief discussion of the passage, whereupon our *Comrades* would engage the host members in a jeopardy-like game, and then refreshments would be served.

I remember one game when I broke the tie: To the question: "Who is the emperor of India?" After the opposition shook their heads, I answered: "King George VI."

During my high school days, I occasionally sang in the choir with young people and several elderly women who could have belted out "Ol' Man River" if called upon to replace Paul Robeson in Showboat! I asked the organist, a middle-aged lady, if she had begun her music career on a hand organ. "Yes, and the monkey is now the minister!" she responded with a hint of annoyance in her tone.

Eventually, I became an usher, biding time in the church basement with my close friend, aviation instructor, and head usher, Bill Patton, as we exchanged jokes with the other two ushers with one ear attuned to a ventilating shaft, waiting for the organ's *basso profoundo* tones that would signal two minutes remained for us to ascend the rear stairs and stand at attention prior to our military-like entry. At every service there is always someone who makes change during the collection. These annoying transactions wreck havoc on the synchronous gait of the ushers, who work in symmetry to collect from the forward half rows while the other pair passes plates to the remaining rows. The organist, under these circumstances, has to improvise like a pianist at a nickelodeon. When this situation arose, faithful movie-going parishioners often recognized melodic excerpts from Biblical epics, produced during the twenties and thirties, as the organist patiently waited for the ushers to reassemble and walk down the central aisle singing: "Praise God from Whom all blessings flow!"

309

David E. Mann, Jr.

After the mahogany plates, their felt linings covered with change and greenbacks, had been carefully stacked by the minister, placed on a table, and the prayer said before a standing, bowed congregation, the ushers did an about-face and walked smartly down the aisle, glancing at the person who dared to make change, before disappearing again into the basement. Sometimes, when the church was half full, members observed that only the hard of hearing occupied the front pews while the younger members positioned themselves near the air-conditioning duct, which carried the conversations of those in the basement!

During the summer of 1938, I spent several weeks at a religious conference in Northfield, Massachusetts, that was held annually in memory of its founder, Dwight L. Moody. A dozen youths from my church mingled with several hundred from all over New England to study the Bible, attend seminars, and enjoy the beauty of Nature. The Northfield School for Girls, in view of our campsite, comprised a number of Georgian style buildings in one of which, a magnificent Olympic-size swimming pool, the finest in the Commonwealth, had been inspected and approved by Johnny Weissmuller. The Olympic swimmer had recently signed a new movie contract to continue his starring role as *Tarzan.* Although the school was officially closed in the summer months, its trustees had graciously allowed the conference students to use its physical education facilities, which included the pool. The alternative to the pool was swimming in the nearby Connecticut River, which was the coldest and deepest in New England.

Reverend Mr. Edward Condit replaced Dr. Harry Woods Kimball, who retired after a distinguished literary and theological career in 1938, when Specky and I were enjoying our all-weather, 5,280-foot jaunts to the high school atop the hill of Memorial Park. Mr. Condit was a tall, slender, fortyish man with a commanding voice, whose preaching and vocal techniques were indistinguishable from those of Frederic March, the star of the movie *One Foot in Heaven,* which was featured at the Needham Paramount theatre soon after he arrived in town. Parishioners noticed the remarkable vocal transformation that occurred after he had seen and loved the movie. If one shut his or her eyes during a Condit sermon, the rich, deep voice of Mr. March would miraculously emerge from the speakers. One sermon was

310

especially memorable—when Mr. Condit ascended the pulpit one Sunday with a roll of toilet paper, held the loose end tightly between his fingers, flung the remainder of the roll over the podium and, watching it as it rolled down the aisle, commented: "those seated on one side of the paper represent the followers of Jesus, while those on the opposite side are non-believers." There is more to this parable that time has completely erased from my brain. After all, it was almost 60 years ago!

The church experienced new life when Mr. Condit hired Mabel Friswell, (Frizzie), an ebullient music teacher, to inject adrenaline into the torpid youth organization and revitalize the choir. Thursday evenings were reserved for youth choir rehearsals and several Sundays each month were set aside for the performances, which gave the older throats in the adult choir a much-needed rest. She frequently arranged for her young choir members to participate in unusual events, such as ushering at professional tennis matches. I remember watching Bill Tilden, Donald Budge, and Alice Marble, the greatest tennis players of the decade, while directing late-comers to their seats at Chestnut Hill. She sponsored hay rides in the fall, dances in the spring at Norumbega Park to the exciting tunes of Blue Baron's band with Buddy Moreno as the male vocalist, and treated us to evening snacks of fried clam rolls and strawberry ice cream sodas at Mary Hartigan's or Howard Johnson's near Norwood when the rehearsals went exceptionally well prior to important religious holidays.

Frizzie continued her work with young people at the church while teaching music appreciation and voice at a local college. She never married and died in the late 1940's, leaving many young adults with memories of a happy pre-World War II America. There was only one disadvantage I experienced in being a Congregationalist. When I was transferred to the Bureau of Supplies and Accounts (Busanda) in Washington, D.C., after 12 weeks of boot camp at Bainbridge, the FBI interviewed my friend, Dr. Chester Mills, as to whether I was a security risk. My membership in Dr. Kimball's innocent Christian organization, *the Comrades of the Way,* written on my naval application, had suggested a Soviet connection!

CHAPTER TWENTY-ONE

On the road to tenure

There is no better way to attain mastery of an academic discipline than to stand before a group of young, inquiring minds and emphasize the important aspects of your subject with the ultimate goal of providing students with information that will remain with them throughout their working lives as a sound basis of core knowledge that is indispensable to the successful practice of their profession. In pharmacology, the principles of drug action are invaluable in making judgments about new chemicals about to be introduced into therapeutics as drugs. Every drug belongs to a class of agents whose scope of action is well known even though the chemical structure may be different. Cocaine, for example, is a botanically-derived agent that is a member of the same class of drugs to which procaine, the first of a series of synthetic drugs known as local anesthetics, belongs. All of these agents interfere with the membrane flow of ions in structures that exhibit electrical activity, such as nerves and heart muscle. The former attribute makes them extremely useful in dentistry to eliminate pain. On the other hand, the inadvertent injection of a small dose into a blood vessel can arrest a beating heart. The unique structure of cocaine, a benzoic acid ester of the base ecgonine, which is additionally esterified with methyl alcohol (it's a double ester chemically), conveys upon its pharmacology a central stimulating action in the brain (like the amphetamines), responsible for its habituating and rarely addicting actions, and an epinephrine-like action that is observed as dilated pupils and blanching of the skin from vasoconstriction. The student is expected to assimilate all of the pharmacologic actions pertinent to each class of drugs and also to know the exceptions. Furthermore, if one is aware that a drug has local anesthetic properties, the assumption should also be made that this agent, which possesses the most common pharmacologic activity of most therapeutic agents, also has antihistaminic activity, i.e., the ability to block histamine receptors and thereby reduce allergic reactions.

Dr. Livingston, eventually satisfied that I was getting the essence of my malaria and cyanide lectures across to the students the first year, expanded my repertoire to include cathartics, emollients, demulcents, and digestants, subjects he found either distasteful, uninteresting, or uninspiring. His colleague, Dr. James Munch, who discussed cathartics in his dental lectures, proudly mentioned his basic research with yellow and white phenolphthalein, the former being the active ingredient of Ex-Lax. Yellow phenolphthalein was unique among therapeutic agents insofar as a single dose exerted its cathartic effect on the intestinal tract, was then absorbed within the body to act again at the same site. Recently, it was implicated as a potential human carcinogen based upon its cancer-producing effects in rats in massive doses and quickly removed from the Ex-Lax formulation. I wish Dr. Munch were alive today to comment on this public health measure.

According to a slender brochure that arrived by mail in September, 1952, Dr. Livingston's old friend and colleague, Dr. Carl Schmidt, chairman of the pharmacology department at Penn's prestigious medical school, was scheduled to give six Thursday evening lectures entitled: *New Developments in Pharmacology* in the auditorium of the Philadelphia College of Pharmacy and Science, commencing in October and terminating just before the Thanksgiving holiday.

Seated in the front row, surrounded by several hundred pharmaceutical scientists from the Delaware Valley, Dr. Livingston and I listened intently to one of the giants of pharmacology describe, without lecture notes, recent advances in therapeutics in an authoritative manner with crisp words that rolled smoothly off a silver tongue:

"Drugs designed to treat obesity have encountered a variety of major problems initially unforeseen. Gordon Alles' amphetamines seemed to be a panacea for weight loss by acting upon areas of the brain that brought about a loss of appetite. Unfortunately, this action was stimulatory and not confined to the hypothalamic area, but also involved other areas in the cerebral cortex and the vasomotor and respiratory centers of the more primitive medulla oblongata. As a result, sleeplessness, habituation, and other detrimental physiologic

effects ensued. Ironically, the amphetamines closely mimicked the actions of ephedrine, which Dr. Chen and I isolated from Ephedra leaves in China, but lacked significant therapeutic effects in patients with bronchial asthma (causing only slight bronchodilation), and induced insignificant circulatory effects in contrast to ephedrine which caused fatalities in patients with cardiac asthma by arresting the heartbeat. The most remarkable antiobese preparation that has come to my attention was conceived by a physician in Quincy, Massachusetts. He prescribed the oral administration of an innocuous-looking white capsule which caused the recipient to lose weight in several weeks. When inquisitive medical colleagues examined the capsule to ascertain the nature of the wonder ingredient, they were astonished to see the scolex of a tapeworm. After ingestion, the capsule dissolved in the stomach juices and passed into the duodenum where the exposed scolex (head) attached itself to the intestinal walls, whereupon its body segments (proglottids) arose and soon deprived the host of essential nutrients.

The biggest problem with present weight reduction schemes is that the weight loss cannot be targeted in restricted areas, such as the buttocks, hips, stomach, etc., without affecting other areas such as the brain. To affect the brain is to invite irreparable damage manifested as mental aberrations, retardation, and even death."

I learned a lot from the 6 lectures, returning to the classroom revitalized and enthusiastic about my dynamic discipline and also anxious to test a new lecture technique.

In addition to teaching and research activities, the duties of a college teacher also include attendance at regular meetings when student's academic problems, departmental budget allocations, replacement of aging equipment, and modifications of physical plant are discussed at faculty meetings, and faculty and student grievances, reorganization proposals, departmental mergers, union and budgetary problems, and outlook for salary increases routinely comprise the agenda of faculty senate meetings. The latter also foster social relationships of both a personal and professional nature and encourage joint research projects that require the expertise of more than one discipline. Also, several weeks prior to the end of the second semester in May or early June, a special meeting is called by the dean to review the names of recipients of the multitudinous commencement awards, some of which had been previously selected, oftentimes after lengthy and tedious deliberations by fickle department members, who let a student's personality interfere with the evaluation of scholarly accomplishment. Our department teamed up with Dr. Eby's pharmacognosy department to award the *Pharmacology-Pharmacognosy Award* to the student achieving the highest average in these courses over a two-year period. The award was named for a former *materia medica* professor, who taught a meld of both disciplines in the days when the majority of drugs of the late 19th century were of botanical origin and not produced synthetically in a chemical laboratory. (Ironically, although chemically-derived drugs today have far greater purity and dosage stability than those of plant origin, they also possess a greater potential for causing allergic reactions, while the botanically-derived, ancient drug digitalis, derived from the leaves of *Digitalis purpurea,* treats congestive heart failure and other cardiac irregularities with an allergic potential that is practically non-existent.) The monetary value of the Pharm/Pharm award was substantial in the early years of its inception but, following decades of inflation, had degenerated into barely enough money to purchase a lobster dinner for one at Bookbinders. Regardless, it served as an excellent predictor of academic success in professional school as illustrated by the black minority student recipient who entered Jefferson Medical school and, four years later, graduated number one in his class!

At my first faculty senate meeting held in the basement of Mitten Hall and presided over by Dr. Millard Gladfelter, the provost of the university, I sat next to a heavy-set, distinguished-looking woman who wore a name card that denoted her appellation was *Miss Peabody.*

"Miss Peabeddy, pleased to meet you. I'm David Mann from the pharmacy school."

"And you're from New England," she added. "Only New Englanders know how to pronounce my name correctly. The rest pronounce it PeaBODDY! I'm from Maine."

That was the last time I saw her, but her legacy to the university must have been considerable, for a dormitory bears her name,

Peabody Hall, undoubtedly the most mispronounced building at Temple.

In the years 1953-1954, several significant events occurred: on the warm, cloudless morning of April 23rd, Mary and I welcomed a daughter into our family at Hahnemann hospital, where a new obstetric nurse had replaced the nasty one I had encountered on the monsoon-like morning of November 1, 1951; the Temple University News reported that the Tobacco Industry Research Committee had approved the grant Mr. Shanfeld and I submitted on behalf of the school of pharmacy's department of pharmacology for *"the study of alleged cancer-producing effects of smoking,"* and, lastly, Drs. Livingston, Munch, and I attended a fall Federation meeting in New Haven at Yale University where I presented a paper entitled: *Effect of potassium iodide and potassium iodate on blood sugar response to thiourea.*

Our baby daughter, weighing 7 lb. 6 oz. (2 oz. heavier than her older brother) was named Caryn after a research schooner at the Woods Hole Oceanographic Institution which, in turn, was christened *Caryn* after the tallest mountain peak rising from the floor of the Atlantic ocean. (A more renowned vessel of the Institution was the *Atlantis,* the largest steel ketch plying the seven seas, which spent most of her illustrious career pursuing oceanographic research.)

Dr. Robert Hockett, a biochemist and spokesperson for the Tobacco Industry Research Committee, informed the school of pharmacy by letter that the committee had approved our cigarette study grant for $5,500. The Temple University News reported on November 17, 1954 that *"the study will be conducted by a graduate student for his thesis and directed by Dr. David E. Mann, associate professor of pharmacology. The purpose of the study will be to determine whether tobacco residues and known carcinogenic agents (which produce cancer) are chemically related Three groups of mice will be used as follows: one group will be treated with a known cancer-producing agent; another group will be placed in a smoke filled area, and a third group will have tobacco residues placed on their bodies. Tobacco agencies wish to discover the nature of any carcinogenic agent in tobacco and eliminate it from their product. The punch line resides in the fact that if a known carcinogenic agent is administered to mice previously exposed to tobacco agents, and this slows their speed of contracting the disease, the tobacco and known carcinogenic agents are then related."*

The fall Federation Meeting at Yale was not devoted entirely to listening to research papers by others, for Dr. Munch and I went to a local cinema one evening to see "Ride Vaccaro," a film starring Anthony Quinn as a Mexican cowboy. After my ten-minute talk, Dr. Livingston and I attended a cocktail party where I was surprised to see him sipping a martini.

"I didn't know you imbibed alcohol," I remarked, after he noticed my odd expression.

"I don't," he replied, "unless someone offers me a drink at a cocktail party." (In retrospect, I think he needed a drink to calm his nerves after listening to my first presentation at a national meeting.)

In the January, 1954, issue (Vol. 110, No. 1) of the prestigious American-based journal of pharmacology, The Journal of Pharmacology and Experimental Therapeutics or JPET, my Yale-delivered paper appeared as a 32-line abstract along with several hundred others, whose authors were mainly young scientists who would include them in their C.V.'s as stepping-stones to promotion and/or salary increases. Some would add unpublished ancillary research to current work and thereby embellish papers that were coherent and important enough to warrant publication in JPET. Having published the initial study in Proc. Soc. Exper. Biol. & Med. 73:657, 1950, showing that the oral administration of potassium iodate prevented the rise in blood sugar (hyperglycemia) which followed the injection of thiourea in rats, I decided that the Yale paper required no further expansion: it was a confirmation of the original work and a realization that potassium iodide, in place of potassium iodate, elicited a similar blocking action of thiourea-induced hyperglycemia. Furthermore, having been notified that the university had received the first payment from the Tobacco Industry Research Committee, it was now time to order the essential equipment that Mr. Shanfeld and I had incorporated into our grant after spending countless hours poring over catalogues.

The most important pieces of equipment were the racks built by the Norwich Wire Works of New York, (R-D-T units), which contained 120 separate cages with wire mesh located on the front (where water bottles were held) and the bottom (which rested on steel pans filled with sawdust). The latter arrangement proved to be a very important aspect of the research for, if the mice were not permitted access to their feces, skin lesions appeared within a few days. Also, current studies involving the application of the routinely-employed carcinogenic chemical, methycholanthrene, used a special camel's hair paint brush that was dipped into a solution of methylcholanthrene (usually in acetone) and, with a single stroke, brushed gently across the shaved interscapular space on the back of the mouse. Then, the animals were placed together in large cages and allowed to rub up against one another. Because of these situations, we decided that two changes must be made to ensure success in our study:

(1) a paint brush was all right for a house, but not for a mouse. We would use a micro-pipette to deliver the carcinogen and,

(2) the animals would be placed in individual cages. Thus, the whole body contamination with carcinogen from aggregate mice crawling over one another is avoided. Previous studies by others using mice in individual cages were unsuccessful because the animals died from nutritional deficiencies in several months. Allowing the animals access to their feces provided vitamin B12 and other nutritional necessities that kept them healthy.

While Mr. Shanfeld began constructing a complicated apparatus with glass-ware, rubber tubing, and a solenoid-timing device to simulate a *cigarette smoker collecting tobacco residue in his or her lungs, a major problem had to be solved before mice could be ordered of a certain age, sexual mix, and strain: the methylcholanthrene, applied by micro-pipette to the shaved interscapular space of each mouse, had to be dissolved in highly volatile and flammable acetone and available in sealed ampoules, the tips of which would be snapped immediately before use and the micro-pipette then introduced to suck up the liquid. If larger volumes of the solution were prepared, the concentration of the carcinogen would change due to rapid evaporation of the acetone. But sealing the ampoules with a flame invited fire and a possible explosion if the temperature of the acetone were not lowered. The problem was solved by drilling holes in a block of dry ice, with a 3/4 inch bit, inserting the 10 ml. half-filled ampoules with only the tips exposed, and sealing them with a gas-oxygen torch. The temperature of the acetone within the ampoule is thereby reduced to approximately -80 degrees Centigrade at which its volatility is lost. Each ampoule was sealed in an atmosphere of nitrogen to prevent oxidation of the methylcholanthrene. This was accomplished by inserting a glass tube from a tank of nitrogen into each ampoule and counting to 15, whereupon the tip was sealed with the gas-oxygen torch. Ampoules were carefully examined for leaks before storing in a refrigerator at 0 to 5 degrees Centigrade prior to use.

Each R-D-T unit held 120 CF-1 mice (60 males and 60 females), ten weeks of age, whose basic diet consisted of Purina Laboratory Chow and tap water and was supplemented once a week with lettuce and carrots (for 2 consecutive weeks), followed by another two-week period in which the supplement consisted of bread soaked in milk pre-mixed with vitamins. Room temperature was thermostatically controlled at 25 degrees Centigrade and the cages were rotated before a window daily. This was accomplished because it is known that light can influence tumor growth.

All handling of animals and apparatus required the protection of heavy Neoprene gloves, which were washed frequently to prevent contamination of equipment. The mice were always grasped by the tail with long forceps, transferred to the left hand, and held by the thumb and index finger at the base of the tail. In this position, they were placed under the micro-pipette for application of the carcinogen. When clipping the nape of the neck each week prior to the application of the carcinogen, cloth masks were worn to prevent inhalation of toxic material.

---

* Mr. Shanfeld, using a stop watch, examined and recorded the smoking habits of literally dozens of people in order to arrive at the average time involved in placing a cigarette to the lips, inhaling, and exhaling.

As the tobacco study progressed beyond the planning stage, two gleaming, aluminum-painted R-D-T racks, each containing 120-steel cages, were delivered and placed in a large room of the cement-block animal facility, located on the roof of the dental / pharmacy building, where most of the work would be accomplished. Within a week, the methylcholanthrene, ordered from Eastman Kodak arrived in a small, tightly - stoppered, glass container, its yellow-powdered contents largely obscured by several cautionary labels indicating its tumorigenic capability.

While awaiting the arrival of the rodent participants, Mr. Shanfeld assembled his glass/rubber tubing/solenoid device, which he had created as a robot programmed with the smoking habits of the average human butt-fiend, and purchased a year's supply of cartons of the most popular, unfiltered cigarettes, i.e., Camel, Chesterfield, Old Gold, Lucky Strike, to be inserted in an aperture of the robot's glass-plate "lips" in a regular sequence as a source of tobacco residue. When 300 Carworth Farms mice (the excess number ordered to account for fatalities that might occur during transit) finally appeared in stacked cardboard boxes on the rear freight elevator, (emitting the odor of rancid butter on stale movie popcorn), they were weighed on an Ohaus spring balance, the nape of the neck carefully trimmed with scissors, and placed in individual cages with the first 60 animals selected as males and the remainder, females.

Squamous cell carcinomas appeared as small mole-like bumps in the neck region within several weeks after the application of the carcinogen alone and in the presence of the residue. Several animals succumbed to nicotine poisoning when subjected to smoke inhalation, so the frequency of exposure was reduced by half, whereupon no additional deaths were recorded. Tumors in the group receiving the methylcholanthrene application alone grew to such an enormous size that the animals were more likely to die from physical suffocation because of stricture of the throat caused by the shifting weight of the tumor than from the intrinsic physiological effects of the cancer. In contrast, application of the carcinogen in the presence of tobacco residue showed smaller growths suggesting that the residue or tar contained a substance or substances that inhibited tumor formation because of a chemical similarity to methylcholanthrene. Pathological examination of mice from all groups revealed that none had lung cancer, including those inhaling tobacco smoke. The experiment was repeated during the second year with comparable results. Mr. Shanfeld received a Master of Science degree with a major in pharmacology at the June, 1956, commencement, having fulfilled his course requirements and successfully defended his thesis entitled: "The Effect of Tobacco Smoke and Tobacco Residues on Methylcholanthrene-Induced Skin Carcinogenesis in Mice" before the graduate committee of the school of pharmacy. Because our department did not offer advanced degrees beyond the Master's, Mr. Shanfeld continued his graduate studies at another university. The Tobacco Industry Research Committee received a copy of the thesis and the study was condensed for journal publication. The editor of a prestigious journal recommended that the tumor sizes from each mouse be duplicated in modeling clay and the weights totaled for each group to accentuate treatment differences. Considering this suggestion to be extremely time-consuming and unrealistic, we decided against submitting a revised paper to this or any other journal.

In 1954, I received an official-looking envelope containing a letter personally signed by the university president, Robert Livingston Johnson, announcing my promotion from assistant professor of physiology and pharmacology to associate professor of pharmacology with tenure. In addition, the letter informed me of my association with the school of dentistry as its department chairman, a position I now held in the school of pharmacy following a lengthy discussion with dean Sprowls, who had reviewed some of the problems caused by the looming mandatory retirement of Dr. Livingston at age 65. Dr. Munch, the head of the dental school's pharmacology department, although younger than his pharmacy school counterpart, had chosen to devote his energies full time to industry where he had originally begun his brilliant career. By contrast, Dr. Livingston joined the department of pharmacology at St. John's university in New York where he spent five twilight years, occasionally visiting Temple to experience nostalgia and retrieve some of the equipment he had designed and now needed at St. John's that was hidden in our stockroom beneath coils of rubber tubing and various kinds of steel clamps.

When Dr. Munch departed, he gave me a signed copy of his landmark book on bioassays in return for lugging a huge, amphora-like, glass bottle half-filled with digitalis tincture down three flights of stairs and into the trunk of his black Lincoln sedan.

"You have been privileged to carry the most frequently assayed digitalis tincture in the United States, if not in the world," he commented, while dumping a handful of old textbooks on top of blankets that cushioned the liquid antique.

I thought about the hundreds of pigeons that would be required to assay the several gallons remaining in the bottle…more than enough to wipe out the entire bird population of Philadelphia and Camden. It was Dr. Munch who had devised the formerly used frog assay to determine the potency of digitalis prior to its replacement by the common pigeon. Each assay, however, required an endpoint of cardiac arrest, a therapeutic irony as the potency of the unknown preparation of digitalis tincture was evaluated in relation to a known standard. The frog (amphibian) bioassay involved the handling of small, slippery creatures, whose availability was seasonal, whereas the pigeon was ubiquitous and easily handled. After anesthetization with a tiny wisp of ether-impregnated cotton, a syringe needle attached to tubing and a glass pipettte was inserted into an alar (wing) vein and the known (standardized) preparation was administered until the heart stopped. In a second bird, the unknown was given until cardiac arrest also occurred. Then, a simple mathematical computation of the two measured volumes required to arrest the heart revealed the potency of the unknown (X) in relation to the standard preparation. No other biological assay of a therapeutic substance depended upon a non-therapeutic response, namely death, as an endpoint. Outwardly, it was a gross contradiction to the tenets of the Hippocratic oath. Inwardly, failure to use the bioassay might cause the demise of the human recipient, for the dose was inaccurate and deadly.

CHAPTER TWENTY-TWO

see you later, alligator!

Temple University's School of Dentistry, originally named the Philadelphia Dental College, began life as an independent institution in November, 1863, under the deanship of a young, distinguished-looking gentleman, John H. McQuillen. The following year, the number of students who graduated, all home-grown, could be counted on the fingers of two hands. Within five years, enrollment had increased to fifty, and the college's growing reputation had drawn students from Europe as well as from distant parts of the United States. In 1897, the college moved from its original location at 108-110 N. 10th Street to 18th and Buttonwood Streets, where it soon shared its physical facilities in 1907 with the newly founded School of Pharmacy (1901), which had separated from the School of Medicine and moved to a location within sight (and sound) of the Baldwin locomotive works.

In 1946, the University acquired, as war surplus (for a hundred thousand dollars), the former Packard (motor car) building at Broad Street and Allegheny Avenue, presently owned by the Bendix Corporation, and relocated both schools while spending several million dollars to alter its architecture, install equipment, and upgrade its heating system.

Dr. Gerald T. Timmons, appointed dental dean in 1942, was a native of Valparaiso, Indiana, where his father served as registrar of the University. A graduate of the pharmacy school of Valparaiso University in 1917, he received a D.D.S. degree from Indiana University, joined its faculty, and became acting dean until he left to assume the deanship of Temple University's dental school in 1942. A key figure in the planning, physical relocation, and revitalized educational goals of the school, Dr. Timmons provided exceptional skills toward these accomplishments, one of which endeared him to "his boys" (women were as rare in the classroom as hen's teeth): he never forgot a student's name and even remembered those of members of their families and where they lived! He also took great pride in being the dean of the *second* oldest dental school in the United States. (The oldest is in Baltimore.)

This was my boss in the dental school with the loquacious manner, Eisenhower resemblance, and engaging sense of humor. I recall telling him at a faculty meeting about the time I was scheduled to give a talk before a dental group in a small Pennsylvania town. The dentist who had arranged the affair gave me the wrong address and, as a result, I sat in my car for several hours waiting for someone to unlock a vacant building.

"That's nothing," he retorted. "I was scheduled to give an address at an upstate dental meeting that dragged on and on and on, way past my bedtime. Finally, the host announced that I was going to give my address. Slowly, I stood up at the podium, recited my home address, and then sat down with a feeling of extreme gratification!"

My teaching assignment in the dental school involved two-hour lectures every Friday afternoon from 2:00 to 4:00 p.m. in the sophomore year, which discussed the content of pharmacology a dentist should acquire to pass successfully the state and national board examinations given two years later in the senior year. Also, the dental accreditation committee mandated that the course provide adequate basic knowledge to enable the dentist to practice his or her profession without endangering the patient's life due to ignorance of therapeutic principles, drug interactions, dosage ranges, concepts of prescription writing and such procedures as the correct administration of a local anesthetic into subcutaneous tissues to achieve regional anesthesia prior to tooth extractions. Failure to aspirate and check the syringe for blood, indicating entry into a blood vessel, can kill the patient as the local anesthetic is swept into the circulation where it alters the electrical conductivity of the heart and arrests the beat. Incidentally, the first synthetic local anesthetic known as *Novocaine (procaine HCl), produced in a chemical laboratory by Einhorn at the turn of the century, has become the generic term for all such drugs given to a dental patient. It is rarely used today in dental therapeutics, having been replaced by safer and more potent synthetic agents with fewer side effects and short, intermediate, and long durations of action. Unlike its botanical relative, cocaine, possessing adrenaline-like actions on the cardiovascular system and nicotine-like actions on the brain (initial stimulation followed by depression), the synthetics lack these pharmacological attributes and also a potential for cocaine-like abuse that starts with habituation and eventually leads to true drug addiction. (It should be noted that workers in the tin mines of Bolivia, who routinely chewed the leaves of *Erythroxylin coca* to allay hunger and provide adequate energy while they lugged sacks of ore over mountain trails, did not become addicted, for the amount absorbed from chewing coca leaves was infinitesimal. When they were drafted into the armed services during WWII and provided with three meals a day, none craved cocaine. Furthermore, having acquired technical skills in the services that provided for a decent postwar life, few became cocaine addicts.) The first recorded case of cocaine addiction occurred when Freud gave the drug to an opium addict as a possible cure and the craving for opioids disappeared to be replaced by an intense craving for cocaine. Another factor responsible for the onset of addiction is dosage: small doses require longer periods of time for addiction to occur than large amounts. Interestingly, large doses given *slowly* experimentally by the intravenous route in mammals are not fatal, whereas small doses given *rapidly* under similar conditions are. In human beings, the fatal dose of cocaine (hydrochloride) varies from several milligrams (20 mg.) to over a gram (1.2 gm.).

The dental lecture schedule on Fridays at 2:00 p.m. coincided almost to the minute when word came over the radio on November 22, 1963 that president Kennedy had been shot in Dallas. As I discussed the pharmacology of curare and other muscle relaxants before a class of anxious, restless students, the severity of the unfolding tragedy soon became evident from the excited tone of the announcer's voice issuing from several pocket radios that had been brought into the classroom. At the end of the first hour, I mercifully canceled the second and tearfully departed for home

---

* Novocaine is the trade or proprietary name for the generic drug procaine hydrochloride whose chemical name is diethylaminoethyl p-aminobenzoate hydrochloride.

After the June commencement in 1956, traditionally held within mammoth Convention Hall in West Philadelphia, the pharmacy degree candidates and their families gallantly struggled through several miles of traffic-cluttered roads to a restaurant on Spring

Garden street where dean Sprowls, a minister, faculty representatives, and student class officers were eventually seated at a floral-bedecked head table surrounded by circular tables of newly arriving students, their friends, an occasional bored faculty member, and jocular parents, ecstatic that their educational cash flow had ended. After consuming the restaurant's famous turtle soup and herbal-enhanced fried chicken, the dean tapped a glass tumbler with a spoon and arose, laughing nervously, to introduce a middle-aged theologian, who politely requested that everyone stand with heads bowed for the short invocation, Fifteen minutes later, amid a plethora of wry necks, Dr. Sprowls introduced those seated at the head table and asked that the dedicated faculty, sporadically seated throughout the room, either stand or wave as their names were called so parents could corner them after the affair to find out why Jonathon or Kimberly hadn't received higher grade point averages. Then, settling into an oft-used oratorical mode, the dean discussed the current and future goals of pharmacy, academic and sports accomplishments of the graduating class, and finally ended with a few glowing remarks about the retirees and how they would be missed. Lastly, before a seat-shifting, yawning audience, the dean awarded three recipients of master's degrees, one of whom was Mr. Shanfeld, with a comment for the latter:

"Mr. Shanfeld has spent two years under the guidance of his mentor, Dr. Mann, showing that mice who smoke cigarettes will get cancer."

Spontaneous applause ensued, whereupon the candidates for the Bachelor of Science in Pharmacy degree were introduced alphabetically and acknowledged individually with polite applause. At the conclusion of the festivities, the theologian again arose and offered a benediction that lasted only a minute, limited by the discomfort of a full bladder.

I rushed to congratulate Mr. Shanfeld and examine his diploma, a large magnificent-looking document embossed with Temple's parthenon-like seal and signed in glistening India ink by president Robert L. Johnson, vice-president Tomlinson, provost Gladfelter, treasurer Atkinson, and the graduate school dean, George Huganir.

"I have something for you," said Joe, handing me a neatly-wrapped package.

I opened it and was amazed to see four reels of 16-mm. undeveloped colored film.

"Now you'll have a record of your trip to Scotland in July!" exclaimed Mr. Shanfeld while shaking my hand.

It was a delightful surprise! I regretted that my first graduate student had to continue his graduate studies elsewhere for lack of an in-house doctoral program at the school of pharmacy.

Mary was pleased when I showed her the thoughtful gift, but she had a worried look.

"I hope we'll be able to use them in Great Britain," she said wistfully. "I haven't heard from the immigration people about my citizenship application. Without my naturalization certificate, there's no passport and no trip."

On June 12th, the following telegram arrived:

MRS MARY E MANN=1956 JUN 12 AM 10 59
49 DEWEY ROAD CHELTENHAM PENN=

: PLEASED TO ADVISE YOU COMMISSIONER OF IMMIGRATION HAS JUST INFORMED ME FINAL HEARING RE YOUR CITIZENSHIP APPLICATION SCHEDULED FOR JUNE NINETEENTH. HOPE THIS INFORMATION HELPFUL. IF THERE IS ANYTHING FURTHER WHICH I MAY PROPERLY DO IN YOUR INTEREST LET ME KNOW
=LEVERETT SALTONSTALL UNITED STATES SENATOR

Dad had telephoned the senator at his farm in Dover to inform him of our naturalization and passport problems. He had not only responded at once (Mary received her certificate at a June 19th ceremony in the Federal Building on Chestnut Street) but, even more remarkably, a carrier rang the front doorbell and handed Dad our passports early on the morning of our intended departure for Montreal!

A week prior to our departure for Canada, I was invited by Al Rhuby, the secretary of the Needham Rotary Club, to give a talk before its members on the dangers of cigarettes. Mr. Rhuby, owner of a Moxie bottling plant on the outskirts of town on Route 128, was a successful entrepreneur who made a hobby out of rescuing faltering companies overshadowed by the competition. A millionaire several times over, Al Rhuby was one of Dad's more affluent patients who had planned to retire and travel around the world until his wife discovered the drone of airplane propellers caused painful episodes of migraine. Furthermore, ocean travel was out of the question after she had suffered through the movie "Titanic," and later "A Night To Remember," and consequently regarded any ship as a potential invitation to a watery grave. To occupy his time between automobile trips to the fishing streams of Maine, New Hampshire, and Vermont, Al had purchased the local Moxie bottling plant, whose bitter-sweet drink was once New England's favorite soft drink before the Cola wars among Coca, Pepsi, and Royal Crown caused these emerging giants to expend huge sums to tout their product not only in the United States, but throughout the rest of the world, stifling the little guys.

Apparently, Mr. Rhuby had overheard Dad talking about my tobacco study at an earlier meeting and, realizing the difficulty of obtaining speakers during the summer months and also aware that my family was coming to Needham before sailing from Quebec to England, he had asked if I'd be available in late June.

It was my first talk before a group of individuals who were pillars of the community with diverse occupations representing blue- and white-collar professions; farmers, mechanics, plumbers, bankers, physicians, dentists, builders, architects, teachers, etc. The vast majority of them had in common tobacco use in all its commercial forms: the Sergeant-at-Arms, Doc Fanning was a corncob pipe smoker; Oscar Starkweather, the builder was a cigarette butt-fiend along with Dad and at least half of the other 70 members present; the entrepreneurs such as Al Rhuby puffed on Cuban cigars. No one, however, inhaled tobacco as snuff, the only legitimate drug available in the mahogany desks of congressmen and senators by tradition.

When I began the talk, the air was a purple hue as the members lit up after a satisfying meat and potatoes meal capped with a glob of ice cream and washed down with a swig of coffee that made the taste buds on the tongue stand at attention. They were amazed when I slipped several colored slides into the projector that showed overhead views of the volcanic-like tumors on the necks of mice. After the 30-

minute discussion, there was polite applause followed by a brief question-answer period. Secretary Rhuby closed the session with a few remarks about "how proud the Needham community was to have a young man contributing something to science," and "that he expected the air to be much cleaner at the next meeting as a result of the talk."

Best of all, Dad was given an opportunity to see his son "perform" for the first and, sadly, for the last time.

The sound of the doorbell at 6:00 a.m., on the day of our departure for Canada, awakened everyone except Mom, who was sleeping on her good ear and oblivious to the piercing noise. Dad grabbed a bathrobe and ran down the stairs where he was greeted by a uniformed postal carrier, who handed him a small package that required a signature.

"It's the passports!" he cried aloud. "Saltonstall kept his word!"

Dad had met the lanky senator in the thirties when, as governor of Massachusetts, he had awarded the prestigious Saltonstall cup to the Champion Race Car Driver of the Year, Eddie Casterline. It was presented at the final Northeast Racing Association event held at the Weymouth State Fair Grounds in September. After the brief ceremony, Steward Phil Mays had introduced Dad to the governor saying: "I'd like you to meet our family physician, Dr. Mann, one of our judges and a loyal racing buff!"

To Dad's surprise, the Lincolnesque governor replied, shaking Dad's hand, "It's a pleasure to meet you, Dr. Mann. I've heard of your medical competence from a neighbor in Dover. Whenever I run into your patient, Bancroft Davis, he invariably mentions that you keep him healthy!"

Thoughts of this meeting must have entered Dad's head as he handed me the tightly-wrapped package at breakfast. An hour later, the six of us were seated in the blue and white Buick roadmaster, its huge trunk filled with suitcases, satchels, and a duffel bag. Our destination was Les and Ruth Cook's summer cottage, on the south shore of Lake Champlain in Vermont, where we'd spend the night before departing early in the afternoon for Montreal to board the boat train that evening for Quebec and eventually our ship, the Homeric.

David E. Mann, Jr.

Entering the pine needle-strewn driveway in the late afternoon, we saw Les and Ruth walking slowly heads down around the rear of their property, obviously looking for something.

"What did you lose?" inquired Dad, after greeting our old friends.

"It's nothing serious," replied Les. "Here we are next to the largest lake in New England and our faucet is down to a drip. We were looking for soggy soil that would indicate a leaky pipe, as opposed to blockage. Make yourselves at home. We'll pick up some bottled water at the market this evening. And if anyone would like a bath, there's the lake. The water is quite refreshing."

During our 21-hour stay, I went swimming off the Cook's flimsy wooden pier in water which, a month earlier, had been snow melting off the slopes of the Green Mountains of Vermont, while Dad, delicately holding a wishbone-shaped branch cut from a nearby apple tree, plodded in circles around the backyard with a crude divining rod until he finally located a spot where a professional well-driller could drive a pipe into the ground and hit an artesian well.

"I'm sure you'll find water. It worked at Mirror Lake where the conditions are similar," said Dad.

"I don't understand how the darn thing works," pondered Les. "Water should be under us everywhere, considering the proximity of the lake."

"That's the mystery behind dowsing. It can somehow differentiate between flowing and still water, and it works only when the divining rods are made of certain kinds of material. I use apple tree branches, while others use ash or birch. Some even use metal rods. Les, let me know how it turns out."

We departed for Montreal early in the afternoon of the following day, after a hurried lunch of egg sandwiches and ginger ale that left a peculiar taste in my mouth. The gustatory discomfort was still evident when we reached Felix and Yvonne's spacious granite townhouse in a posh section of Montreal, known as the Cote de Neige, and were immediately treated to cocktails. Much later, I discovered the cause was Lake Champlain algae that had entered my mouth while swimming. Within two weeks, the vile taste was gone.

Suddenly, while we were being seated at a lavishly prepared dinner table, a strange noise was heard in the basement.

"What's that?" inquired Dad. Then, turning to his host, he asked: "Felix, are you raising fur-bearing animals in your cellar?"

"No, that's our German shepherd. We keep him in the cellar because he's a little frisky and not used to people. When you're in the kitchen, don't open the door opposite the sink. He's not only frisky; he's downright crazy! What time do you have to get the children to the station? It's a ten minute drive from here."

"Eight o'clock," answered Dad.

We arrived at the station with half an hour to spare.

"Remember, we'll meet you at Quebec City when your ship docks in August. We'll be staying with Felix and Yvonne at the Chateau Frontenac. Have a wonderful trip!"

The distance between Montreal and Quebec is approximately 150 miles and should have taken a little more than 2 hours, assuming the train traveled at 65 mph. Just after 11 p.m., three hours after our departure, it stopped at a brightly-lit pier where a white-hulled ship was docked. While I presented the tickets to a uniformed official, two nurses, on the lookout for passengers with youngsters, checked the stamped tickets for our stateroom location, and promptly whisked our two children aboard. After placing our suitcases in a designated area, we walked up the Home Line's gangplank and proceeded to our stateroom where David, Caryn, and one of the nurses awaited our arrival.

It was an inside cabin, which meant it lacked portholes and was bathed in total darkness when the lights were out. On the positive side, when the ship was underway and bedtime had arrived several hours after a delicious meal, the tomb-like darkness, coupled with a gentle rocking motion, yielded an uninterrupted and lengthy sleep. Speaking of food, before settling in for the night, we attended a buffet temporarily set up on the promenade deck where we enjoyed snacks of sandwiches, pretzels, nuts, olives and celery, while observing the dockside personnel, who were now readying thick hawsers, soon to be lifted from mushroom-like steel bits and tossed into the water at the signal of a whistle rung by an officer standing on the starboard side of the bridge. Another less shrill whistle echoed from a portside tugboat, whose captain had just been notified that the ship was now at the mercy of the strong river current, having cast off its hawsers and not yet under its own power, and it was now in his hands to guide it into the deep channel. How fascinating it was to observe these maneuvers that enabled mere man to control this steel-hulled behemoth! When we returned to our dark room, the gentle vibration emanating from the engine room lulled us into a deep sleep. My final thought pertained to the reason why it took three hours to travel from Montreal to Quebec; it suddenly dawned on me that the tracks bordering the St. Lawrence River ran through farmland where cattle and other domestic animals often wandered onto the rails. The Canadian railway owners had obviously incurred the wrath of farmers whose livestock had literally been railroaded.

When we awakened the next morning, the engine room vibration had become more intense. Before entering the dining-room, we went up on deck expecting to see the restless waves of the Atlantic ocean. Instead there was land on either side, which meant we were still in the St. Lawrence river. It was three hundred miles from Quebec to the Atlantic; we had traveled less than two hundred miles since leaving the pier 8 hours earlier! Our present speed, recorded on a bulletin board at the entrance to the dining-room for those who bet on the daily mileage pool, was 20 knots. Not exactly a match for the 1940's Normandie or even the 1930's Mauretania.

The topography of the northeastern part of North America determines the number of days required for an ocean liner to reach the British Isles (Southampton, England) and western Europe (LeHavre, France). Departing from New York city, a ship, traveling at 20-25 knots requires at least six days to reach England or France; from Boston, the same ship at a comparable speed requires five days; leaving from Quebec, however, requires only four. Therein lies the great advantage of embarking at the Canadian city instead of those to the south: only four days are required to cross the Atlantic ocean whose surface rarely mimics a mill pond, but can summon 30-60 foot waves in a matter of hours. Seasonal changes, of course, rule out the possibility of sailing in months other than those during the summer and fall. The greatest disadvantage of a summer departure became evident the second day out when I looked beyond the bow, during a stroll on the promenade deck, and saw a corridor bordered by huge icebergs on either side. I recalled a statement I'd read by a Cunard Line officer who was being interviewed about his experiences on the North Atlantic.

"In all the years I've crossed the Atlantic back and forth from New York to Southampton, I've never seen an iceberg."

This was the greatest advantage of taking the southern route from New York. In the ship's library, I found a copy of Bowditch, the seaman's Bible, and was fascinated with the information available about icebergs. It was actually possible to determine the location of a *mother* iceberg from fragments called *calf* ice that break off from the main mass and were carried ahead by the ocean current. The *calf* ice was crescent-shaped like an archer's bow. To find the location of the *mother,* mentally place a hypothetical arrow in the correct position in the bow and it will point to the icy mass from which the calf originated.

Our ship, the Home Line's Homeric, formerly the Matson Line's Mariposa, was one of a fleet of magnificent, white-hulled, two-funnel vessels that sailed from San Francisco to Honolulu in the halcyon days before WWII. After the war, the vessel was purchased by a Greek shipping family, registered under the Panamanian flag, and mainly staffed with an Italian crew, few of whom spoke or understood English. This linguistic deficiency became strikingly evident during the life-boat drill, a routine and Coast Guard-mandated event that each ship, departing from North America, must comply with or face a hefty fine or confiscation. The ship's whistle sounded, early the morning of the first day, and soon passengers were running helter-skelter on the top deck carrying brightly-colored life-jackets. The atmosphere was jocular, snide references to the Titanic were uttered, and the drill ended abruptly before the proper lifeboat stations could be identified. The Italian sailors had been unable to interpret the verbal and sign language of the English and French-speaking passengers, who quickly dismissed the exercise as a time-consuming joke.

Not so on the trip back to Canada! On July 25, 1956, the Italian Line's Andrea Doria collided with the Swedish-American Line's Stockholm and sank off Nantucket. The life-boat drill on the return trip of the Homeric in August was the embodiment of seamanship perfection!

The Kungsholm of the Swedish-American Line was the cleanest ship I'd ever sailed on. On the other hand, the Matson Line's Mariposa, sold and refitted as the Home Line's Homeric, though never mentioned as such in their National Geographic advertisements, was an outstanding fun ship. It became evident as soon as passengers embarked when an atmosphere of gaiety permeated the vessel. There were three main reasons for this conviviality: (1) most of the passengers were young French-Canadians from Quebec province and a few older citizens from France, the former, public and private school teachers, the latter, college professors and scientists who, accompanied by wives and children, were either leaving the aggravations and stresses of their academic existence behind for a few months of scenic and sociologic diversion (the younger), or returning to their homeland to continue their educational pursuits with refreshed minds (the older); (2) at every meal (including breakfast) beakers, resembling Erlenmeyer flasks and brimming with Italian red wine, caused reticence to vanish; and lastly, (3) there were facilities for the entertainment of children of all ages, where parents could leave their offspring for several hours in the care of two no-nonsense nurses. Unfortunately, the toy-filled playroom, positioned in the stern section of the vessel directly above two propellers, induced severe nausea and vomiting in the novice, who required at least a day to become accustomed to the unpleasant vibrations. My unfortunate experience with incipient *mal de mer* during the 1930's, when the needle-shaped Mauretania was fighting the turbulent waters off Cape Hatteras on its way to Havana in the waning days of its remarkable and unmatched career, was responsible for my first practical pharmacologic discovery…that munching on celery and radishes prevents or relieves seasickness. Consequently, I kept a small supply of these vegetables in the stateroom for prophylactic purposes and confirmed that they successfully averted nausea in our children prior to a visit to the playroom. Of course, dramamine orally and scopolamine hydrobromide, worn as a patch behind the ear, are now available for the relief of motion sickness.

Mary and I enjoyed the fine entertainment that was offered each evening in one of the spacious lounges, featuring professional female French singers, who sang "Paris in the Spring," "I Love Paris," and other traditional Parisienne songs, accompanied by accordion and piano, whose sounds conjured up images of Maurice Chevalier movies, a struggling young Pablo Picasso, and a World War I correspondent named Ernest Hemingway, who was recuperating from war wounds incurred while driving an ambulance. In the early evening, the children, after having spent several hours in the toy-filled playland in the stern, entertained themselves mimicking an elevator operator by opening and closing the bathroom doors of our stateroom.

The four fun-filled days passed too quickly. For the first time I used Mr. Shanfeld's film to record the docking procedure at our first port of call, LeHavre, France, while the Homeric was being gently nudged alongside a concrete dock by tugs that were larger than those I'd seen working in the St. Lawrence River and in New York harbor. As the movie camera reel quietly whirred, I thought of Paul Gauguin and Claude Monet, who had walked in this vicinity many decades ago, and of Paris and the Eiffel Tower, only sixty miles upstream by rowboat on the River Seine.

Having disembarked the majority of its passengers on French soil, the ship now glided across the English Channel with a handful of Anglophiles on a slightly less than 200-mile cruise to Southampton, the famous British home port of the great Cunarders of the past, the Lusitania and Mauretania and, more recently, the Queens Mary and Elizabeth, and the latter's frequently refurbished successor, the QE-2. In the late afternoon, we sailed past the Isle of Wight, where I captured on 16-mm. film a WWII British Hawker Hurricane performing dogfight acrobatics to the delight of the shipboard crowd and, a few moments later, a sleek, wooden ketch that was bobbing in the wind-swept waves off Cowes, where aristocracy and lesser souls, once inspired by the daring exploits of Sir Francis Drake and Viscount Horatio Nelson in expanding the British Empire, were now occupied with honing seamanship skills in order to capture the America's Cup. Eventually we arrived at Southampton where a mustached Customs official, who resembled Sherlock Holmes, checked our passports and health certificates.

"Why isn't there confirmation of a recent small pox vaccination?" he asked, after glancing at the yellow certificate on which Dad had boldly written *Not Necessary* across the request for vaccination.

David E. Mann, Jr.

"My father, a physician, felt that another small pox vaccination might cause anaphylactic shock since I now have psoriasis and didn't have any skin problems as a child."

"Makes sense," replied the officer. "Enjoy your stay."

We boarded the boat train for the 75-mile trip to London, enjoyed an ice-free drink of warm ginger ale *("they don't have ice for that!" exclaimed Mary)* when we arrived at the station, and then took a taxi to the Continental Hotel off Tavistock Square, where we planned to stay for several days before heading for Scotland.

"Maybe tomorrow, we'll browse in the antique shops and visit Harrods in the morning," suggested Mary, "and then, in the afternoon, the children might like to see Madame Tussaud's wax museum, although the chamber of horrors could give them nightmares." Because leave-happy sailors from the U.S. Navy were ensconced in many of the Continental's rooms, we decided to dine at the more restrained Hotel Tavistock across the street, whose hotel brochure, available in the lobby, advertised a menu of American cuisine, which included Mary's specialty—southern fried chicken, and mine—seafood platter.

The next morning, walking past the Cunard office on the way to the antique shops, I gasped when I saw, behind a huge plate-glass window, a meticulously crafted, forty-foot model of the liner Queen Elizabeth. Such is the status of British craftmanship that, if I had poked a finger through a window of the ship's bridge and touched the helm, the rudder would have responded.

We entered only one antique shop on a street where at least a dozen stores, specializing in treasures of the past, were enticing to connoisseurs of ancient artifacts. The one we decided to explore displayed, through a dirt-streaked window, an eclectic collection of historic prints, scenic oil paintings, commemorative plates bearing the likenesses of royalty, and smaller items, one of which caught my eye—a tiny model of the Mauretania stuck in a sea of blue clay and enclosed in a glass chamber. Its price, converted from pounds to dollars, could have bought a 12-foot rowboat back in the States! Because the children were anxious to see Harrods toy department, and not wishing to waste precious time and restricted funds on over-priced gizmos, we walked several blocks to the world-famous department store.

It required only a few minutes to realize that this building warehoused treasures accumulated over centuries by British gentry, who had to liquidate some of their possessions to pay exorbitant inheritance taxes. Wall-size oil paintings of Scottish and English landscapes, bordered by thick, intricately-carved, gilded frames, portrayed sheep and long-haired cattle peacefully grazing on green hillsides while, in the distance, amid stone dwellings with thatched roofs, shabbily-clad peasants labored at all kinds of menial tasks. These depictions of Middle-Age serfdom once graced the high-ceilinged, luxurious rooms of British aristocracy as grim reminders of an ugly past comparable to what existed in America prior to the Civil War, except for the ethnic heterogeneity of the individuals involved in the latter circumstance.

The top floor was where the splendid toy department displayed its wares (as was also true for Philadelphia's old John Wanamaker store). The radio-controlled plastic yachts, black British Austins (powered by pedals and seating one child), British train sets, and European musical automatons enthralled the children. They could have remained in the enchanted room for hours, but it was now past noon and time to see Madame Tussaud's museum after a quick lunch.

Standing next to the ticket booth in the lobby, a uniformed guard was stationed with an outstretched hand, presumably to grasp the ticket that cost a shilling. However, closer examination revealed why the guard was reluctant to seize the ticket—he was made of wax. An illustrated brochure picked up in the lobby told the story of the talented Madame Tussaud, who had learned her sculpting trade by collecting severed heads of the nobility that had been subjected to the infamous guillotine of the French Revolution. Constructing molds in which hot wax was poured, she was able to preserve the features of those who were guilty of depriving French citizenry of rights to liberty, equality, and brotherhood. Mastering this technique within a short time and running out of heads when the Revolution ended, she turned her attention to molding heads of executed criminals, and eventually to preserving the likenesses of living celebrities.

So exquisite was the wax rendition of Sir Winston Churchill by the Madame's artistic successors that the veins of his head and hands seemed to pulsate. In one elaborately decorated corner of the museum, the royal family (King George VI, Queen Elizabeth, and the

Princesses Elizabeth and Margaret) appeared as though they were standing on a balcony of Buckingham Palace, gazing at an admiring crowd below.

In a special sports section, life-like mannequins of outstanding living American, British, Canadian, and Australian athletes, from the arenas of boxing, track, tennis, motor car racing, cricket, soccer, swimming, and yachting, were faithfully reproduced dressed in their respective uniforms. Black athletes were noticeably absent from the honored activities, except for boxing and track, which were dominated in the 1930's by America's Joe Louis and Jesse Owens, respectively, two of the world's greatest athletes in these domains and certainly worthy of waxen recognition. After wandering among the images of the living and dead, who had left their marks of greatness on society, it was now time to see the infamous people, whose deeds were dramatically recorded in gruesome tableaus in a subterranean room labeled the Chamber of Horrors.

As we descended the stairs into a dungeon-like room, overhead and to our left was a reproduction of a cell from Newgate, Britain's hell-hole prison, in which a bearded inmate peered sadly through iron bars at the populace. According to the sign beneath the bars, he held the unenviable record for continuous confinement—75 years! There was no mention of his prison diet, but it must have been nutritious enough to keep him alive long into his ninth decade. (Many years later, when I dined at Fisher's restaurant on North Broad Street with my professorial colleagues at Temple, I was reminded of the Newgate prisoner when I walked into the foyer and saw above me and to my left a similar aperture in the same location, but with an air-conditioning screen in place of bars. Each time I looked up I expected to see a bearded old codger peering out at me with those sad eyes!)

The gruesome sights intensified as we confronted Britain's most hideous criminals in various acts of violence. The features of Jack the Ripper, clad in a black hood and slashing the throat of an unsuspecting prostitute with a butcher's knife, were modeled after those of a known criminal, whose anonymity remained intact (according to the sign describing the incident), but whose crimes and murder weapon were similar. Who the real Ripper was remained a mystery, but several names were listed as possible suspects.

The Algerian punishment for thievery was especially gruesome: the hapless victim was suspended from the ceiling of his cell with a large meathook stuck in the abdomen. From a physiological standpoint, the tremendous blood loss would mercifully reduce pain and cause death within a few hours.

An ancient method of execution in India was graphically illustrated with a mature elephant which, at the mahout's signal, stepped on the skull of a kneeling criminal whose head was resting on a tree stump. This tableau was the first we had noticed with the movement of two key figures: the mahout raising his hand to signal the elephant and the descent of the foot touching but not crushing the unfortunate victim's skull. To have seen the head flatten on contact would have been too gross.

Emotionally drained by the dungeon-like atmosphere of the room and the many gruesome images of man's *"inhumanity to man,"* I stared at the final exhibit, a raggedly-clad skeleton slumped in an iron cage, that was hanging from a branch of an artificial tree. Its placement near the exit was obviously intended to deliver a last-minute shock. A bronze plaque affixed to the tree beneath the cage proclaimed:

Traditional punishment meted out to wrongdoers in the Middle Ages. This poor soul was apprehended for stealing a loaf of bread for his starving family. Confined in this cage, he was deprived of food and water until death, whereupon his bones, picked clean by vermin, served as a warning to the populace that thievery would not be tolerated.

"My God," I whispered, hurrying Mary and the children out the door, "I didn't realize how cruel the English were!"

"The times were especially cruel in those days," replied Mary, "so don't blame the English for all of these terrible misdeeds. Look what happened in Africa and Asia during the same period. Their legacies of cruelty were just as horrible!"

In the taxi, on the return trip to the Hotel Continental, I recalled one of Dad's predictable sayings, usually uttered when an outstanding achievement by a contemporary English scientist, author, musician, artist, or inventor was reported in a magazine or newspaper: "The English are a superior race!"

He was extremely proud of his British heritage endowed by a Scottish-born mother and an English-born father, but the scales of pride tipped slightly more in favor of the accomplishments of people from his father's homeland than those from Scotland. The next morning, we left London on the famed *Flying Scot* that traced a steel path through the gently undulating English countryside to the enticing seaside border town of Berwick-on-Tweed. Later, it pulled up to the station in the history-steeped city of Edinburgh, where Mary's parents, John and Sophia Jarvis, were awaiting our arrival in a state of nervous anticipation. After a tearful and overwhelming welcome, prompted by ohhs and aahs at the sight of their grandchildren, we took a local train that crossed the spiderweb Firth of Forth bridge, past Rosyth, where shipbreakers were reducing an old Cunarder to a waterline hull (as they had done to the Mauretania years earlier to provide ammunition casings for WW II), and eventually reached our final train destination, the town of Lochgelly. We boarded a red (do they come in any other color?), double-decker bus for Mary's home on a gentle slope of land at the foot of Bernady Hill in the quaint country town of Ballingry. That evening, after a satisfying dinner with the sun still shining brightly, Mary put our exhausted kids to bed and then joined her parents and siblings to catch up on the events that had taken place on each side of the Atlantic since her departure in 1950.

I had been forewarned by Mary and her sister, Sophie, who had married my cousin Willard and lived in Stoneham, Massachusetts, that I should expect inclement weather in Great Britain, even though we'd be in Scotland during the month of July, which happened to be the driest and hottest month of the year in Pennsylvania. From my high school days, I learned that the British Isles owed their unusual climate to the warm waters of the Gulf Stream that flowed from the Sargasso Sea to within a hundred miles of the North Carolina coast, and then veered to the northeast across the Atlantic Ocean to flow past the western coast of Great Britain. Icebergs, originating from glaciers in Newfoundland, quickly dissolved when their frigid masses encountered the bright blue waters of the Gulf Stream, without whose warm, soothing current the British Isles would resemble the Antarctic continent. The love of horticulture in Great Britain was partly due to the Gulf Stream, which brought not only a friendly climate to the land, but also adequate amounts of rain to sustain a wide variety of plants. During our four-week stay, it rained only once.

The day after our arrival in Ballingry, I climbed Benardy hill with Mary's brothers, who walked at a rapid pace past cattle grazing on the hillside. I paused to admire the cows when I noticed that one of them was a bull. In the States, it is unlawful to have a bull wandering about in a pasture without posting a notice of its presence on a substantial fence.

"Don't dally or you'll attract his attention," cautioned one of the boys.

By the time I arrived at the grassy, treeless summit, I was out of breath and fearful of having a heart attack. A few minutes lying in the grass gasping for air like a hooked bluefish, my breathing gradually returned to normal. Far below us was a lake with a tiny island on which the roofless ruins of a stone castle stood. The other brother identified the loch which is Scottish for lake:

"That's the home of Scotland's national fish—the Loch Leven salmon. And the castle, or what's left of it, is where Mary Queen of Scots was imprisoned. If you look closely, you'll see several wooden boats tied up on the shore that can be rented by those who wish to row out and investigate the ruins. There's not much to see nowadays."

On the return trip, the cattle had moved to a different location with the bull clearly in sight, eyeing us and occasionally pawing the soil with a foreleg.

"That's a bad sign. We'd better walk close to the fence in case he makes a sudden move."

We arrived home exhausted after a lovely afternoon when I was exposed to a smidgen of Scottish history.

Once weekend, our children accompanied us on a short bus trip to Scotland's Burntisland, a resort on the North Sea that is classier and less flamboyant than its English counterpart on the Irish Sea—Blackpool. In common with America's Coney Island on Jamaica Bay, these sand and gravel strips of British real estate come alive in the summer months when hundreds of winter-weary souls arrive to absorb the sun's rays and wade knee deep in the icy ocean currents. In the summer of 1956, there were no gaudy amusement rides at Burntisland—no roller coasters, ferris wheels, or boat rides in a tunnel of love to enthrall Scottish children. Instead, ponies with western-style saddles plodded in the sand near the ocean's edge, while a few yards inland, an authentic miniature copy of the famous Flying Scot circled an oval track as the only amusement rides available for youngsters.

Sitting atop the sturdy tender behind the sleek green locomotive, the youthful engineer, clad in the British version of Casey Jonese's outfit, languidly tossed lumps of coal into the firebox with one hand and manipulated the throttle or brake with the other. The children loved the train, but our three-year-old daughter was uncomfortable during the pony ride, becoming seasick due to the proximity of the ocean. The train reminded me of the narrow-gauge railroad I'd seen during a vacation trip with my parents to Byron, Maine, in the summer of 1948. Dad, rock hound and amateur geologist, had always wanted to visit the place where gold was first discovered in the United States. The town was minuscule in population with a frontier -style, two-story hotel and a gem shop that, prior to its recent conversion, had been a cracker-barrel grocery store with a wood-burning stove. Its glass-topped counters contained glistening specimens of aquamarine, tourmaline, and garnet that were mined from local outcroppings and were displayed alongside prehistoric fossils from Montana and meteorites from Arizona. Tin miner's plates for panning gold in the nearby streams were available for 5 dollars. Dad and I spent the afternoon searching for semi-precious stones on a hillside designated a gem-hunting area on the outskirts of town before returning to the hotel with a few choice mineral specimens. One of the hotel guests suggested that we take a walk to the station and see the railroad equipment that was for sale. We expected to see posters, lamps, photographs, and other railroad paraphernalia, not an entire narrow-gauge railroad consisting of several locomotives, half a dozen red-painted freight cars, as many passenger cars, and fifty miles of track! The whole kit-and-kaboodle was valued at 12,000 dollars! Months later, the Boston newspapers reported that it was sold to a junk dealer who eventually sold it to a Massachusetts entrepreneur. In the 1950's, my wife, children, and I stepped aboard the reincarnated Byron narrow-gauge train, now called the Edaville Railroad, and chugged around a cranberry bog in Carver, Cape Cod. Now our children were riding an even smaller train in Burntisland, Scotland, and experiencing what might become for them, in the distant future, a similar happy memory.

Burntisland was the birthplace of my father-in-law, John Jarvis, whose father was an intrepid fisherman on the North Sea. Several miles northeast of Burntisland was the coastal village of Kirkcaldy, famous for the manufacture of linoleum, and the birthplace of my paternal grandmother, Marion Logan.

Because Mary wanted to spend a day shopping in Edinburgh, we left the children in the care of their grandparents and boarded a double-decker for the Lochgelly train station, accompanied by her sister Margaret. We passed through the quaint town of Cowdenbeath where Scotland's most beloved soccer goalie was buried, the victim of a reckless kick to the head that cracked his skull like an eggshell. Near the grave was a professional soccer field where he once played as a youth. It was surrounded by high concrete walls in which broken bottles were embedded to discourage those who wished to see the game for free. Soccer was a game all Americans understood, unlike the game of cricket, which seemed to those in the States to be as weird as its name. My favorite comedian, Fred Allen, if asked to comment about the game, would probably have quipped that *"cricket was croquet played by tutti-frutti salesmen."* The British seemed to be fascinated with insects. What other people would assemble a rock band within two decades and call it the Beatles?

At the Lochgelly station we boarded a commuter train for Edinburgh and sat inside a beautiful wood-lined compartment on comfortable settees that looked like those on the Orient Express. As the train got up speed, another train sped past in the same direction, a creepy phenomenon I hadn't experienced back home because in the States there were two tracks near the cities for commuter trains and only one in remote areas of the West for trains like Santa Fe's Chief and Superchief, whose track was single through parts of New Mexico and Arizona. In England and Scotland, the ties held three and sometimes four tracks. It was little wonder that the British system of rail transportation was superior to those found elsewhere. Like television, the telephone, the parliamentary system of government, the macadam road, the tank of WWI, the steam engine, the jet passenger plane, the first antibiotic penicillin, etc., they either invented or discovered it!!

Suddenly, the train slowed several miles from the approach to the Firth of Forth bridge.

"It has to decrease speed because the coal beneath us is still being mined," explained Mary. "When it's gone, the tunnels will be filled in with slag and the train will then be allowed to travel at normal speed. You were too excited to notice when we first came by a week ago. The train slowed while crossing the Firth bridge and didn't resume speed until we were about to arrive at Lochgelly. And now we're about to cross it again."

It was the most unusual bridge I'd ever seen. Built in 1890 as a cantilever structure 5,330 feet long, its huge tubular steel plates wove an intricate spiderweb-type pattern that supported railroad tracks. The design proved to be serendipitous during WW II when Hitler's Heinkel bombers, attempting to destroy it, repeatedly dropped bombs and failed ignominiously to hit a single span. The Nazi's lack of success in destroying the Firth bridge was matched by the failure of their submariners to send the giant Queens to Davey Jones' locker.

Arriving in Edinburgh a few minutes later, Margaret decided to window shop on Princes street and meet us at Romanes & Paterson, the famous clothing store, in two hours, while Mary and I went to the medical school for an unscheduled meeting with the renowned pharmacologist, J. H. Gaddum. As we parted company, Mary pointed out the magnificent flower clock that lay in the ravine of the Princes street gardens and the huge rock mass surmounted upon which was an ancient edifice, Edinburgh castle.

I asked several people for directions to the University of Edinburgh medical school and received negative nods. The third person we encountered was an inebriated, ill-clad man who, though slightly tipsy, pointed out a stone-faced building of 18th-century origin where the medical facilities were housed. Thanking him for his courtesy, we walked up stone steps and entered the lobby where a student, upon learning of our desire to meet the pharmacologist, graciously accompanied us to the second floor and opened the door to Dr. Gaddum's laboratory.

"He's one of my favorite professors, but he's also a hard taskmaster," the student whispered, upon seeing that the laboratory was vacant. "He's probably in his office next door. I'll inform him that he has visitors from America."

The student disappeared before we could thank him and within a minute or so, a tall, bald-headed man in his forties appeared.

"Can I help you?" he asked. "I've been told that you came all the way from Philadelphia to see me," he added. "I'm deeply flattered."

"This is my wife, Mary, whose parents we're visiting in Ballingry. We came over to show them their American grandchildren. But I also wanted to meet you, having read your work about the mechanism of action of ephedrine. Also, Dr. Carl Schmidt, who is an acquaintance of mine, sends his regards."

"I'm no less flattered by your explanation. You're referring to my landmark paper with Kwiatkowski that presumes to explain the mechanism of action of ephedrine in acting like adrenaline (epinephrine), partly through its ability to inhibit the enzyme, monoamine oxidase, which destroys adrenaline, and partly by acting directly on adrenergic receptors in smooth muscle, cardiac muscle, and glandular tissue in the manner of adrenaline."

"Precisely. Where is Kwiatkowski now?" I inquired.

"Back home in Poland."

Our conversation continued for half an hour during which time I discussed my tobacco research and plans to continue Dr. Welsh's Harvard clam heart research with ephedrine and adrenaline, as soon as the American version of Kwiatkowski arrived for graduate study at Temple. I thanked him for his time and headed for Princes street.

Before joining Mary's sister Margaret at Edinburgh's most famous clothing store, we walked past the impressive 200 ft. Gothic spire of the monument to Sir Walter Scott, whose statue stands beneath its masonry canopy in the East Princes street gardens, where I wished to stare again in wonderment at the size and beauty of the floral clock and admire the first real castle I'd ever seen. In the States, there are many castles once owned by successful businessmen, who built huge granite domiciles to glorify their British or European heritage, and also to quell similar architectural aspirations of neighbors with neither the means nor the proper familial credentials to do likewise. The American heirs, like their British counterparts, were either forced to sell their medieval mansions to pay the gargantuan inheritance taxes incurred by the horrendous debts of WW II, or to convert them into hotels and inns where tourists could pretend to be subjects of King Arthur. The strange fascination that Americans have for castles can be attributed to the reading of fairy tales in their childhood which transformed innocent psyches into the inveigling minds of princes and princesses. It's neither a coincidence nor a theme park designer's laziness that resulted in Cinderella's castle serving as the centerpiece structure of California's Disneyland, Florida's Walt Disney World, Japan's Tokyo Disneyland, and France's Disneyland Paris. It seems that fairy tales with castles are as integral a part of childhood as the adventures of Uncle Wiggily, Winnie the Pooh, and The Cat in the Hat.

With only a few minutes remaining before meeting Margaret, I took another look at Edinburgh castle before heading for the store. "I've got to see it up close, Mary. The view of the city must be spectacular from its ramparts."

"We have a few weeks left," she answered. "Besides, the children would love to see the castle, too. I also want them to see the Edinburgh zoo. If we get there early enough, they'll see the traffic stopped to allow the penguins to cross the road with their keeper on the way to the outside pool."

We met Margaret at the store, purchased some tartan scarfs, and walked to the railroad station. The return trip was uneventful except for the stop at Lockgelly where Margaret, yanking vigorously on a leather strap to open the outside door, failed to make the latch budge. "I can't get out," she yelled as the train started to leave. However, another yank released the latch and we stepped onto the platform with a sigh of relief.

That evening my final thoughts were of admiration for Dr. Gaddum, whose spartan 19th century laboratory, lacking sophisticated gadgetry, had produced world-class research.

In the several weeks remaining of our Scottish vacation, we toured the Edinburgh zoo, where I filmed David being squashed by a chubby girl as he sat with half a dozen other kids atop an Indian elephant. Another day we went by bus to Glasgow, Scotland's other great city, stopping on the way to explore Stirling castle, whose defenders rebuffed attacks by the English in the embryonic days of Scottish separatist history. In Glasgow I bought a pair of Moroccan brown shoes that turned orange back home in the States. In a clothing store we asked the clerk if he had a Harris tweed jacket that would fit David. "We don't sell them to children that small," he muttered, "and besides he'd look like he was on a crucifix!" At the quiet, cold waters of Loch Lomond, we admired the thriftiness of local residents who were living aboard small yachts, exchanging the monetary burden of high estate taxes for the frugality associated with more modest docking fees. On an embankment overlooking the River Clyde we saw the great complex of cranes that defined John Brown's shipyard where the Cunard Queens were built and, one afternoon, I stepped into a steel cage and descended 600 feet in what was once Scotland's deepest coal mine at 1500 feet before a cave-in buried two men. In one narrow tunnel, deep underground, I saw a bricklayer busily sealing a wall behind which was a fire that had been burning for years. During our final week, I attended a soccer game in Edinburgh at the invitation of my father-in-law, who wanted me to see Scotland's best players in action against another outstanding team from Malaysia. During the opening ceremonies, Malaysian bagpipers, marching in perfect unison, entered the stadium and thrilled the crowd with their astonishing musical talents, artistic skills that obviously spilled over into athletic prowess, for the players soundly trounced the home team. Dad and I stopped at a local Pub for a glass of Guinness stout to allay our disappointment.

We wound up our vacation with a visit to Aberdeen, the Granite City on the North Sea, where we stayed overnight. After surviving the shock of seeing a newspaper photograph of the Andrea Doria's stern about to disappear in the ocean off Nantucket after colliding with the Stockholm (and realizing it wasn't our ship the Homeric), we had dinner at a local restaurant where I made a gourmet's mistake of ordering a mushroom omelette in a foreign country. That evening I had bouts of violent nausea and vomiting that kept everyone awake for hours.

"Is Daddy all right?" asked three-year-old Caryn.

"No he isn't," answered five year-old David. "He ate something bad."

Back in Ballingry, we prepared to depart for London, having spent an exciting month with Mary's parents and siblings in a beautiful country with a deeper historic past and slower pace of life than we had in America. I kissed mother Sophie and the girls and bade goodbye to Mary's brothers.

"See you later alligator," I shouted to my father-in-law, as we boarded the bus for the Lochgelly train station. "We'll be waiting for your arrival in Philadelphia within the next decade."

CHAPTER TWENTY-THREE

Father's Day

Back at London's Continental Hotel, the frivolous sailors on leave had been replaced by boisterous college students from Europe and the States. Rather than remain in our room exposed to loud noises and innane conversations from next-door strangers, we decided to unload superfluous British currency by purchasing gifts to supplement the tartan goods from Romanes & Paterson. Our boat train was scheduled to leave for Southampton at 4:00 p.m. the following day, so we had sufficient time to pack suitcases that morning and entertain the children at a movie in the early afternoon. Thus, outwardly complacent that tomorrow's departure was well organized, and inwardly happy to be returning to America, we went by taxi to Harrod's for a final shopping spree. I especially wanted to buy something more masculine than a scarf or handkerchief for Dad that was small enough to take aboard ship and also serve as an appropriate gift for a person who loved nature.

In the fine arts section of the store we found what I was looking for: porcelain statues of a bull African elephant frozen in mortal combat with a brilliantly-striped Bengal tiger. The unfortunate feline, its sinewy body firmly held by the trunk, struggled to avoid penetration by a tusk that pressed against its abdomen. Ink lettering on the sole of the pachyderm's left front foot identified the piece as: Ronzan Made in Italy 408. Its price was 34 pounds, a terrific buy compared to that of the bronze or marble statues of Grecian and Roman gods. I wondered if Dad would discover the blatant zoological mistake of the statuary: to be accurate, a small-eared Indian elephant, not a large-eared African beast would be attacked by another resident of India, the Bengal tiger. After I acquired what we later called the *nature in the raw* sculpture, the clerk gift-wrapped it and probably uttered a sigh of relief that the tasteless Yanks had removed such grotesque Italian schlock from the loftier company of bronze and marble sculptures of mythical gods. Five days later, it arrived undamaged in Canada, having survived engine-room vibrations that shook the floor of our stateroom. Adding insult to injury, the statuary was subjected to further vibration when the Homeric collided in a fog-bound harbor with Columbia, a ship of far lesser tonnage that was tied up to a pier near the Chateau Frontenac. The incident, detected as a barely perceptible jerk, occurred while we were having breakfast. Later, walking down the pier alongside the ship, we were surprised to see a large gash in the bow that required a month of repair work in Boston. According to a Canadian newspaper report, the accident cut a lifeboat in half and caused a sailor, shaving with a straight razor, to slit his throat.

In front of the Chateau Frontenac, where Mom and Dad had been vacationing with Felix and Yvonne for several days, we had a happy, but brief reunion before saying goodbye as they departed for Montreal in their Oldsmobile 98. We took off in Dad's blue and white Buick Roadmaster, stopped at a station to fill the tank with Imperial gallons, and headed for New England and an overnight layover in Maine. There, after a seafood meal, I handed him the box containing the elephant-tiger statues. His reaction was not exactly what I had anticipated. He seemed to like it, but his response was less than enthusiastic. In a few months, I would realize why.

At Temple, the pharmacology graduate program at the master of science level was starting to attract students from schools and colleges of pharmacy in New York and New England. Harvey Pless, a young man from St. John's University, studying the phenomenon of tachyphylaxis (i.e., diminished response to successive equal doses of the same drug over a short period of time), examined the effects of l-arterenol (related to adrenaline/epinephrine) in the isolated *Venus mercenaria* clam heart and received an M.S. degree along with Mr. Shanfeld, at the June, 1956, commencement. Unlike the latter student, who pursued doctoral studies elsewhere, Mr. Pless chose instead to work at the pharmaceutical house of Wyeth in Radnor, Pa., as a pharmacologist.

Another youth, Mr. Tsuneo Fujita, continued Mr. Pless's research and held the distinction of being my first master's recipient in pharmacology to publish results. While Mr. Pless had observed a tachyphylactic response on the clam heart with successive equal doses of l-arterenol that was prevented by pretreatment with ephedrine, Mr. Fujita performed additional studies that confirmed the earlier work and offered a probable explanation for the phenomenon. He received his master's degree at the June (1957) commencement and joined the pharmacology section at Smith, Kline & French Laboratories in Philadelphia.

On Father's Day of that month, the telephone rang in the early afternoon and, expecting to hear a parent's voice, I answered with an upbeat "Hi!"

Instead, a strange voice inquired solemnly: "Hello, is this David?"

When I responded affirmatively, the voice continued: "David, this is Dr. White. Do you remember me? The radiologist at the Glover Hospital."

When I answered "yes," he continued in the same formal tone: "I have sad news to report. Your father has inoperable lung cancer."

I gasped in disbelief, then asked: "How long does he have?"

"Four months at the most. I think you should come to Needham as soon as possible. Your mother is under the impression he has Hodgkin's disease and is unaware how serious his condition is."

I thanked Dr. White for being so forthright and put down the receiver as Mary entered the living-room. "It's Dad. He's got lung cancer."

She burst into tears.

Prior to hearing the devastating news, our plans were to drive to New England at the end of June and spend July and August with my parents in Needham, as we had done each summer since 1952, with the exception of 1956. Now that the children were old enough, we'd take day trips to Falmouth on Cape Cod, where they could romp on the warm sands of Old Silver Beach, or visit the Aquarium at Woods Hole. On the pier, from which ferries departed daily for the Vineyard and Nantucket, they'd watch the unloading of swordfish harpooned off Cuttyhunk and deep sea scallops dredged off the ocean bottom of Nantucket Sound to be sold at Sam Cahoon's.

Before packing to leave for Massachusetts, I had several academic obligations to fulfill: supplies had to be ordered for fall laboratory sessions; new research projects required lengthy discussions with graduate students and careful scrutiny for clarity before filing grant applications; and lastly, there were dental and pharmacy faculty meetings to attend to discuss which laggards among the borderline students could be salvaged and which should be given the boot. The latter meetings were always the most interesting of the year and the best attended because of the extenuating circumstances that frequently preserved a student's dental or pharmacy career. Of course, some students, judged hopeless as future practitioners in the health sciences, achieved success in the arts, real estate, teaching, law and politics.

At these meetings, the faculty became familiar with a wide variety of student shenanigans, ranging from petty thievery of dental instruments among undergraduates to alumni digressions of married couples, i.e., wife supported husband for four years of dental school whereupon, having met a more attractive and younger sweetie, he filed for divorce. Even worse, parents, who paid for a son's education for 8 years of college and dental school, were now living in his apartment and paying him rent. Get behind in your payments, folks, and you'll be out on the street!

After discussing these human frailties, the subject usually changed to the innovative ways in which students came up with tuition fees. Many of the young taxi drivers in Philadelphia in the 1950's were students at Temple, Drexel, LaSalle and Penn. I recall once entering a taxi at 30th Street Station for a ride to the pharmacy school at 3223 North Broad Street and going 15 mph when the traffic was sparse. The driver was a student of the cello at the prestigious Curtis Institute. "I like to drive slowly," he apologized, when I gave him a tip. "And that's how I like to play the cello. Boy, how I hate playing *"The Flight of the Bumble Bee!"*

The most unusual method of paying tuition was devised by a young dental student from Arizona. Each evening he swept up the dust in the gold casting room laboratory and meticulously extracted minute particles of the precious metal with mercury. He not only paid for tuition and board, but also had enough money left to buy a brand new Harley-Davidson motorcycle.

The summer of 1957 was a drastic departure from those of the past as Mary and I went along with Mom's assessment of Dad's condition, Hodgkin's disease, surreptitiously suggested by Dr. White to minimize stress for her and Dad. Those familiar with this condition, characterized by a progressive enlargement and inflammation of the lymphoid tissues, are aware that it runs a fatal course, the intensity and duration of which are influenced by an individual's age, i.e., youths succumb more quickly than the elderly because aging reduces the metabolic rate, an important determinant of the severity and length of the disease. To suffer from Hodgkin's disease instead of pulmonary carcinoma in a man in his mid-sixties meant that his death sentence could be postponed for several years, at which time successful treatment might be available. Conversely, the fatal outlook for lung cancer patients was almost as immediate as stepping in front of a fast-moving 18-wheeler.

Observing Dad that summer, I was appalled his addiction to the weed, despite his illness, was so gripping that he still smoked cigarettes! His favorite brand, CHESTERFIELD, however, had been replaced by one I was unfamiliar with—OASIS. Mom, on the other hand, smoked RALEIGHS, one of the first brands with filter tips and, as an added buying incentive, came with a coupon affixed to the side of each pack that was redeemable in appropriate quantities for hundreds of valuable gifts. Although Mom smoked almost as many cigarettes each day (2 packs) compared to Dad's 3, the RALEIGH'S filters were unquestionably a major factor in extending her life beyond his by three decades.

Twice a week, during July and August, Fire Chief Dick Salamone took Dad to a Boston hospital where he sat in a revolving chair for 15 minutes of x-irradiation, the target area of his bare chest delineated in India ink. This therapy had little influence on the course of the disease, but caused gastro-intestinal disturbances, hair loss, and extreme thirst. At this stage, the disease was now causing a slow but progressive loss of weight, gray discoloration of the skin, and muscle weakness (myasthenia). Yet, he continued puffing on OASIS cigarettes. These, I learned from Mom, were a replacement for the mentholated KOOL cigarettes, which he found to be more soothing to the lungs than the CHESTERFIELDS he'd smoked for decades. Yet, the aromatic ingredients of KOOLS became too harsh and had to be replaced with a cigarette he could tolerate—OASIS.

Between trips to the Boston hospital, we went to Norfolk where he enjoyed walking through the woods and visiting the pond near Noon Hill where his father had once owned a sawmill. At Mirror Lake, we drove past our bungalow, which now belonged to a Needham dentist, and parked for a few minutes to watch swimmers dive off the raft that floated over the deepest part of the lake. Some afternoons we watched the Red Sox or Braves on black-and-white television. Ted Williams was in the twilight of his illustrious career as was Warren Spahn, star pitcher of the Braves. I was reminded that the first image I ever saw on commercial television was that of Vernon Bickford pitching for the Boston Braves in 1950.

Before returning to Philadelphia in September, at the end of a summer spent watching Dad's health rapidly deteriorate, I had promised Mom I'd devote each weekend to them. I discussed the plan with Mary, who was agreeable to the arrangement, even though it meant she'd have to devote a little more than eight days per month to caring for herself and two young children. Thus, after the two-hour lectures to the sophomore dental students ended on Fridays at 3:50 p.m., I walked several blocks down Broad street to the North Philadelphia station where I purchased round-trip tickets to route 128, and waited for the train. Seven hours later, with aching muscles and an empty stomach, I stepped onto a steel and concrete platform that functioned both as a roofless station and as a display for a number of colorful billboards, one of which advertised the play *Romanoff and Juliet,* starring Peter Ustinov, which recently debuted in Boston. Mother, standing in the same well-lit area on the opposite platform each weekend, routinely waved to indicate her presence and, during our short ride to Needham in her blue and white Buick, reviewed Dad's current health status.

"The Hodgkin's disease has weakened his leg muscles. I've set up his bed in the waiting-room so he won't have to climb the stairs."

"Is he still smoking?"

"Not in bed. He does smoke a cigarette or two as he sits on the porch at the back door when the weather is warm."

Upon our arrival, Dad was sitting in the kitchen.

"How's my little Prell and the children?" he asked, as I gave him a gentle hug. Dad had a way with names: mother was frequently called *Bellura* or *Plu;* Mary was affectionately known as *little Prell;* I was simply *David,* long since having outgrown Tarzan's offspring appellation—*Boysie.* (I'd often wondered if they'd had a girl—would she have been called *Girlsie?).*

Sunday mornings, Mom would drive me to Route 128 for the seven and a half hour trip to Philadelphia. Throughout September and October I said goodbye to Mary and the children on Friday mornings and greeted them again late Sunday afternoons. On the train, I discovered that TRUE, the Man's magazine, was not only far more interesting to read than a newspaper, but also less likely to leave printer's ink on the fingers. In one article, welding problems associated with the construction of nuclear submarines were described. Cracks mysteriously appeared in the welds of those vessels which were being assembled outdoors. Engineers were at a loss to explain the reason. Not being an engineer, I nevertheless ventured a guess as to why the welds had failed: the cold weather had something to do with it! Sure enough, in a subsequent article, they discovered why the welds had failed; it was due to the extreme cold of a Connecticut winter.

By late October, Mary and I began to worry that Dad might end his earthly existence on our son's birthday, November 1st. This supposition almost approached reality when an urgent telephone call came from Dad's attending physician and close friend, Dr. Henry Gilbert, a gaunt, distinguished-looking gentleman with a pencil-thin moustache, whose physiognomy reminded me of the equestrian director of the Ringling Brothers and Barnum & Bailey circus—Fred Bradna.

"It's only a matter of days," he sighed in the same barely audible monotone he'd used when he and his wife visited us at Mattapoisett, right after hurricane Carol had devastated Cape Cod, to announce that his seaside dwelling had washed out to sea.

Assistant professor John Lynch, faculty member at Temple's school of pharmacy, devoted husband of Elizabeth Helm and longtime friend of the family, drove me to the Philadelphia International Airport on a rainy afternoon. There, in a cocktail lounge filled with nervous travelers, I had a tranquilizing beverage, thanked him profusely for being such a great friend, and boarded the plane for Logan airport in Boston. An hour later, the plane was circling Boston harbor like a seagull while I grasped the armrests with hands capped with white knuckles, wondering whether the pilot had correctly ascertained at what point the ocean ended and the tarmac began. The 12-mile trip by taxi to 863 Great Plain Avenue required another hour in the midst of sporadic bursts of heavy rain as the driver carefully maneuvered his vehicle along Commonwealth Avenue to route 128, and then down Highland Avenue, turning a sharp left at the Town Hall on Needham Square. Mom and Dr. Gilbert were sitting in the living-room when the taxi arrived in the darkness. I paid the driver, who expressed his sympathy, and climbed up the massive granite steps at the front of the house on which, years before, I had slipped on the way to the barbershop and cut my upper lip that required two stitches without anesthesia. Mom, looking drawn and piqued, informed me that Dad had entered the Glover Memorial Hospital on November 3rd and that nutritional supplements and blood transfusions were stopped because his condition was now irreversible. Mary and the children arrived by train on the 4th, the day before he expired.

The funeral was held at Needham's Evangelical Congregational Church on a beautiful November 7th morning. It was presided over by three ministers: Uncle Harold, the Episcopalian minister, the Reverend Harry Woods Kimball, founder of the Comrades of the Way, and a retired pastor of the church, and the Reverend Edward M. Condit, the present minister. The church was SRO as the magnificent service unfolded, commencing soon after a contingent of nurses, wearing dazzling white uniforms and Navy blue jackets, marched in military precision to their cordoned pews. This spectacle caused mother, somewhat composed before their appearance, to break into tears. After the service, as the limousine pulled away from the curb, I spotted Dr. William *R.* Fisher, my high school music teacher and friend, hurriedly running toward the church. I regretted I was unable to thank him for making an effort to attend the service. At the gravesite, presided over by octogenarian Dr. Kimball, I recalled what Dad had once proudly mentioned at dinner:

"Dr. Kimball told me during a housecall that I'm in his will. He's leaving me a hundred dollars!"

Before Mary, the children and I returned to Pennsylvania, Dr. Kenneth Christophe, chairman of the Board of Health dropped by to give mother two copies of a Resolution, a third to be spread upon the Records of the Town of Needham.

A RESOLUTION IN MEMORY OF DAVID E. MANN, M.D.

Whereas:

David E. Mann has served as a member and as Chairman of the Needham Board of Health from the year 1927 to 1957 and

Whereas:

During these years his constant support of a modern public health program in the Town of Needham resulted in the following new programs:

An experienced and qualified Health Officer
A Public Health Nurse
A trained Sanitarian
A Clerk-Technician
A Nutritionist
A Well-Child Conference
A Dental Clinic
An Anti-rabie Dog Clinic
A Mosquito Control Program - and

Whereas: For many years he served as a School Physician and on the Staff of the Glover Memorial Hospital and

Whereas: He was also a most competent, respected and much beloved practicing physician in the Town of Needham and likewise active in many community affairs.

Now therefore be it resolved:
That the members of the Town Meeting Assembled express a great sense of loss at his untimely death on November 5th, 1957, and our grateful appreciation for the outstanding service he gave to the Community, as we believe that the health and welfare of all the citizens of the Town have been immeasurably increased by his devotion to his many duties—and be it further

Resolved: That this Resolution be spread upon the Records of the Town of Needham, and copies sent to his family:
Signed: Kenneth Christophe, M.D. Leslie B. Cutler John D. Fernald, M.D.

CHAPTER TWENTY-FOUR

Epilogue

Near the time of Dad's demise, three events were occurring, two of which he was aware of: Russia had begun the space race by sending a basketball-size satellite called Sputnik 1 into orbit on October 4th, and the Ford Motor Company had announced it was going to introduce publicly its latest car, the Edsel, now being delivered under tarpaulin on trucks to dealerships throughout the New England area. Dad was perturbed that the United States had been scientifically usurped by the Soviets (his high school commencement address had dealt with the threat of Russia's growing dominance in science and commerce). On the other hand, he was pleased that Ford had produced an innovative car that might replace the highly touted, but commercially unsuccessful Tucker 48. The third happening took place while I was preparing for the funeral. A twenty-four-year-old pharmacy graduate of the University of Rhode Island named Ronald F. Gautieri, who was continuing Mr. Shanfeld's research area, presented a paper on cigarette smoking and cancer before the Scientific Section of the 4th Pan-American Congress of Pharmacy and Biochemistry in Washington, D.C.

Shortly after arriving at Temple's school of pharmacy to investigate its graduate program in pharmacology, he came to my office with Dean Sprowls, who made the introduction.

"Dr. Mann, I'd like you to meet another New Englander, Mr. Ronald Gautieri from Rhode Island." After a brief pause, to collect appropriate words, he continued: "Dr. Mann is from Boston. He's an associate professor of pharmacology, the department head, and our inhouse humorist."

After a lengthy discussion of departmental interests and course requirements, I handed him a copy of Mr. Shanfeld's thesis to peruse that evening at his hotel. Dr. Sprowls called the following morning to inform me of his decision to enroll in the M.S. degree program, with a major in pharmacology. At the end of our conversation, he said rather smugly, "what probably tipped the scales in our favor was mentioning that, by the time he'd earned his Master's, the doctoral program in pharmacology would be established. I told him that I'd look forward to awarding him that historic degree a few years from now. All going well, of course."

At the Pan-American Congress in November, 1957, Mr. Gautieri presented our paper introducing the concept of the MCD50, or minimal carcinogenic dose 50 of methylcholanthrene. By applying various concentrations of the cancer-producing agent, methylcholanthrene (dissolved in acetone), to the shaved necks of mice, equally divided as to sex, the study revealed that the MCD 50, a procedure that consistently produced cancers in 50% of the mice, was *504* micrograms applied through 21 biweekly applications of 0.02 ml. of a 0.12 % solution of the carcinogen. For the first time, agents that increased or decreased the onset of cancer experimentally could be evaluated precisely.

The first person to employ the MCD 50 as a standard dose/response procedure to determine factors that enhance or reduce cancer development in mice was Ronald Gautieri, who examined the effects of gonadectomy and estrogen (estradiol benzoate) therapy in his doctoral thesis. At the June 1960 commencement, when I was promoted to the rank of full professor, he received a doctorate from Dean Sprowls (who had prophesied the historic occasion), and joined our faculty with the title, assistant professor of pharmacology. The second recipient of a doctorate in pharmacology at that commencement was also my graduate student (and Gautieri's roommate), Mr. Arthur H. McCreesh, who had earned a baccalaureate degree from St. John's and chose to study at Temple. If the first letter of his last name had been alphabetically one from A through F, his doctorate presentation by Dean Sprowls would have superseded that of Dr. Gautieri as the historic first of its kind for the school of pharmacy. Several decades after Dr. McCreesh became a government toxicologist at the Army's Aberdeen Proving Grounds in Maryland, an FBI agent came to the laboratory to question me about Dr. McCreesh's loyalty. I quickly learned that his visit was prompted by President Nixon, who had selected him from an imposing list of candidates to be his drug czar. Sadly, the tragic death of a young daughter prevented him from accepting the honored post. However, Dr. Gautieri and I were proud of the scientific expertise and integrity he'd demonstrated at Aberdeen that had attracted the attention of the

President. A few years later, we were deeply saddened when Dr. McCreesh succumbed to a massive heart attack while still a young man. As a graduate student and Jack Paar look-alike, his wonderful sense of humor will also be remembered by those who knew him as a colleague and friend.

Before the doctoral program in pharmacology at the school of pharmacy came to fruition in 1960, a young man named Marvin M. Goldenberg, who was studying cyanide antidotes in mice as the original investigative component of his M.S. degree requirements, unknowingly altered the research goals of the department. After reading my Master's thesis on the subject, Mr. Goldenberg expressed an interest in searching for more effective antidotes than those in current clinical use. It required only several months for him to discover that an inorganic cobalt compound, sodium cobaltinitrite, was more effective at a lower dosage than sodium nitrite in protecting mice from lethal doses of sodium cyanide. Remarkably, the combination of sodium cobaltinitrite and sodium thiosulfate proved to be more effective in protecting mice against lethal doses of sodium cyanide than the administration of sodium nitrite and sodium thiosulfate, a combination originally tested in dogs by Eli Lilly's research director, Dr. K.K.Chen, which has since saved countless human lives to this day.

At the conclusion of this study and in the absence of an established doctoral program, Mr. Goldenberg left Temple to pursue an advanced degree offering elsewhere. He also left the department a kilogram of sodium cobaltinitrite, which had no other valid pharmacologic use than to be evaluated as a potential anti-cancer agent.

During the next twenty-five years, our graduate students performed many experiments with different inorganic cobalt compounds to observe their influence on cancer incidence elicited by the MCD 50 of methylcholanthrene in mice. Mr. Raymond F. Orzechowski, a graduate pharmacist from the Philadelphia College of Pharmacy & Science, injected sodium cobaltinitrite in increasing doses in mice and observed dose related reductions of tumor incidence.(Mr. Orzechowski's study was supported by grant CA-06239 from the National Cancer Institute, National Institutes of Health, U.S. Public Health Service, Bethesda, Md.) Mr. Ralph T.

Mancini, a graduate of Union University's Albany College of Pharmacy, noted that sequential injections of the hormone cortisone (as the acetate) in mice caused only a slight decrease in methylcholanthrene-induced tumors. As postulated, weight loss to cortisone therapy was quickly reversed after hormonal withdrawal but, surprisingly, the anti-inflammatory action anticipated with cortisone at the tumor site, during the early stages, was mimicked in mice receiving the cobalt compound in Mr. Orzechowski's study. Robert S. Thompson, a graduate of Temple's school of pharmacy, administered sodium cobaltrinitrite and sodium nitrite orally for prolonged periods to mice and found that the former agent was more effective as an antitumorigenic agent than the latter. (Mr. Thompson's research was supported by fund-grant 768 from the Damon Runyon Foundation.)

Early in his teaching career at Temple, Dr. Gautieri became intrigued with the effects of drugs on blood vessels of the human placenta. This remarkable structure, responsible for providing nourishment to the fetus, possesses vasculature that lacks innervation. Thus, drugs that alter the caliber of placental blood vessels act solely on smooth muscle and not through nerve pathways. Within the next decade, Dr. Gautieri and his graduate students published numerous papers emphasizing potential hazards to the developing fetus of commonly used drugs that alter placental blood flow. The scientific community soon recognized him as a renowned authority on placental pharmacology.

In the early sixties, when Dr. Kelsey of the FDA fortuitously delayed the approval of thalidomide for sedation in the United States, an outbreak of phocomelia occurred in the newborn of European and Canadian women who had taken the drug during pregnancy. (The condition of phocomelus is defined as a human offspring whose hands and feet appear to be attached directly to the shoulders and hips in the absence of arms and legs.) Thanks to Dr. Gautieri's expertise in the field of placental pharmacology, he and I decided to investigate the exciting field of drug-induced birth defects, a branch of pharmacology known as teratology (from the Greek word, teras, terata meaning monster). The stimulus for inducing a birth defect is known as a teratogen, which is defined *as any insult (trauma, sound, radiation, chemical, nutritional excess or deficiency)* which can disrupt the mitotic activity of neonatal cells in such a way that the newborn organism becomes incapacitated and cannot function normally within the constraints of society. Within two years, we had organized enough research and lecture material to present a graduate course entitled, Teratology and Toxicology (Pharmacology 571).

A landmark investigation, (supported by grant H D 01975-01 from the National Institutes of Health, U.S. Public Health Service), provided striking new information for our teratology course. It was a simple cleft palate study the underlying premise of which was based upon three characteristic observations that arose during our methylcholanthrene cancer induction experiments with cobaltous chloride and sodium cobaltinitrite in mice: that these inorganic cobalt compounds possessed cortisone-like characteristics, elicited anti-inflammatory actions and, lastly, manifested antitumorigenic activity. Because it is well known that cortisone causes cleft palate defects in mice, a study was designed utilizing cortisone acetate and cobaltous chloride alone and together to determine possible teratogenic and antiteratogenic effects.

The experiment is summarized as follows:

Pregnant mice receiving injections of cortisone acetate or cobaltous chloride had incidences of fetal cleft palate of 75.6% and 12.9%, respectively. When cobaltous chloride was injected prior to cortisone, the incidence of fetal cleft palate decreased from 75.6% to 12.6%. Apparently, cortisone and cobalt compounds act at similar biological sites by what is called competitive inhibition, for each agent alone caused cleft palate while together antagonism of this defect occurred. (Mr. Gilbert Kasirsky, a pharmacy graduate of Fordham, performed this work as a research requirement for an advanced degree. Upon receiving a doctorate at Temple, he studied at the College of Osteopathy in Kirksville, Missouri, and eventually returned to Pennsylvania as a physician).

The culmination of these carcinogenic and teratogenic studies appeared in the prestigious British medical journal, The Lancet (Vol. I, 699: March 1968) as a letter entitled: Ionic Hormonal Precursor Hypothesis, authored by Mann, Gautieri, and Kasirsky. The hypothesis attempted to explain the evolutionary origin of hormones from "specific cations and anions obtained by primitive organisms from their aquatic environments. As their cellular specialization increased and/or when transition from aquatic to terrestrial surroundings occurred, availability of essential ions for metabolic purposes was restricted. To make up for these discrepancies, the organism was required to include the ion in a more complicated substance (i.e., iodine in the hormone thyroxine) or to produce an entirely new compound (cortisone or posterior pituitary), whose endocrinologic actions can be increased by the presence of the original ion (cobalt and magnesium, respectively)." We concluded our letter with "to confirm the hypothesis, a more dramatic ionic response in fetal than adult tissues is required—and such has been observed in the interaction of cobalt compounds and cortisone in mice."

DAD, I DIDN'T BECOME AN M.D. OF YOUR DIMENSIONS, BUT I HAD ONE HELL OF A TIME AS A PH.D.

+++++++++++++THE END+++++++++++++

David E. Mann, Jr.

ABOUT THE AUTHOR

The only child of New England parents, the author was born in a Tennessee TB sanatorium where his mother served as a nurse and his father as director. In his youth, he sailed annually to the Caribbean on the great ships of the day. A biology major at Harvard set the stage for medical school, which he found disillusioning. Following a stint in the Navy, he studied at Purdue and joined the faculty of Temple University in 1950 where he taught physiology and pharmacology for 37 years. Also in 1950, he married Mary Jarvis, a Scottish lassie, on their third date. The marriage lasted 51 years and ended with her death from *Mycobacterium avium*, a bacterial disease, similar to that caused by the tubercular bacillus, which also destroys the lung tissue. Sadly, it was acquired while she was indulging in her favorite hobby—gardening. She left a lonely husband, three children, 7 grandchildren, and 1 great-grandchild.

www.ingramcontent.com/pod-product-compliance
Lightning Source LLC
Chambersburg PA
CBHW030002190526
45157CB00014B/94

9 7 8 1 4 0 3 3 2 0 1 3 1